Artificial Intelligence in Practice:

Driving Innovation Across Sectors

Poshan Yu
https://orcid.org/0000-0003-1069-3675
Soochow University, China

James Mulli
European Business Institute, Luxembourg

Zain Ali Shah Syed
Soochow University, China

Laila Umme
https://orcid.org/0000-0001-8050-5081
Institute of Business Management, Pakistan

IGI Global
Scientific Publishing
Publishing Tomorrow's Research Today

Vice President of Editorial — Melissa Wagner
Director of Acquisitions — Mikaela Felty
Director of Book Development — Jocelynn Hessler
Production Manager — Mike Brehm
Cover Design — Jose Rosado

Published in the United States of America by
IGI Global Scientific Publishing
701 East Chocolate Avenue
Hershey, PA, 17033, USA
Tel: 717-533-8845 | Fax: 717-533-7115
Website: https://www.igi-global.com E-mail: cust@igi-global.com

Library of Congress Cataloging-in-Publication Data

LCCN: 2026028906 (CIP Data Pending)
ISBN13: 9798260023785
Isbn13Softcover: 9798260023792
EISBN13: 9798260023808
British Cataloguing in Publication Data
A Cataloguing in Publication record for this book is available from the British Library.

Table of Contents

Detailed Table of Contents

Every day, individuals face countless decisions with the potential to shape their future. Some decisions may have relatively few consequences, such as the choice of a movie one may choose to watch or the song that they would like to listen to. On the other hand, the decision to buy a house, a car, or another significant purchase has the potential to fundamentally transform a life course. Rather than making decisions consciously, choices are often shaped by the external environment in ways in which most individuals might not be aware. For example, the majority of individuals may not realize that the structure of a shopping Centre is usually specifically designed to encourage purchases. In a similar vein, people may also not be aware of the additional effort put forth towards scent management at an overwhelmingly good-smelling bakery.

This chapter explores how Artificial Intelligence (AI) is reconfiguring strategic leadership and entrepreneurial dynamics in the food sector. As AI shifts from an operational tool to a core strategic asset, food firms face both innovation opportunities and organisational tensions. Using thematic analysis of 100 curated news articles (2020–2025), the chapter identifies seven key patterns: strategic innovation, sector disruption, operational efficiency, quality assurance, leadership and strategy, ethics and ESG, and consumer-AI interaction. Findings reveal uneven adoption across firm sizes, with larger enterprises exploiting AI at scale, while SMEs encounter infrastructural and capability constraints. Nonetheless, agile SMEs are innovating in niche areas. The chapter applies dynamic capabilities and strategic alignment theory to interpret these trends. It concludes with four actionable recommendations, including promoting AI literacy, ethical deployment, and scalable strategies. The study offers a timely lens into AI's transformative role in shaping future-ready food systems.

This paper explores the transformative potential of artificial intelligence (AI) in enhancing the livelihoods of rural smallholder farmers in Africa. While AI applications like precision farming, automated crop monitoring, and market intelligence offer significant benefits, their adoption is hindered by challenges such as digital illiteracy, poor infrastructure, and financial constraints. Drawing on educational theories, particularly Piaget/Vygotsky's and Knowles' frameworks, the study proposes the Baby Steps Theory of Development (BSTD) to facilitate incremental, hands-on learning for AI adoption. By addressing contextual learning needs and fostering gradual skill-building, this approach aims to empower farmers to effectively integrate AI into sustainable agricultural practices.

This chapter explores a conceptual study on integrating Feng Shui principles into an AI-based application for Vietnam's real estate. Feng Shui is crucial in property selection and design. A 2018 survey by the Vietnamese Royal Institute of Feng Shui found that 98% of buyers consider Feng Shui, yet few receive satisfactory guidance. This gap presents an opportunity to modernize traditional Feng Shui using AI technologies like machine learning and data analytics. The study examines how an AI-based Feng Shui application can enhance buyers' preferences by aligning with cultural values, improving interaction, increasing satisfaction, and providing tailored recommendations. Grounded in socio-technical and cultural capital theories, this qualitative research uses interviews and surveys with buyers and sellers. Thematic analysis will identify patterns in buyer perceptions and evaluate effectiveness. By merging traditional Feng Shui with AI, this study offers insights into innovation, market differentiation, and improved decision-making, contributing to buyer trust and sustainable growth.

Historically Black Colleges and Universities (HBCUs) have been vital in producing many successful graduates (Logan, 2024). HBCUs have historically faced numerous challenges, particularly funding and access to certain resources. AI resources are being advanced increasingly, and some HBCUS may not benefit from these resources because of a lack of funding. The objectives of this chapter are as follows: 1) outline historical and contemporary research and its impact on the HBCU experience (research to realization); 2) explore the impact of AI innovations on HBCUs, focusing on how these technologies can improve educational outcomes in the classroom to adopt a more inclusive and supportive learning environment; 3) provide strategic recommendations for HBCUs to enhance their AI resources, drawing on best practices; and 4) stress the importance of including AI in the classroom and community to support technological advancements at HBCUs, outlining ways to build partnerships and secure funding to sustain these initiatives.

Preface

INTRODUCTION

Artificial Intelligence (AI) has transitioned from a primarily theoretical discipline into a transformative force shaping industries, institutions, and societies worldwide. Across healthcare, marketing, agriculture, education, cultural preservation, tourism, and community development, AI technologies are redefining how organizations innovate, make decisions, and engage with people. Artificial Intelligence in Practice: Driving Innovation Across Sectors was developed to explore this dynamic evolution by bringing together interdisciplinary perspectives that examine how AI is being implemented in practical, real-world contexts. This volume reflects the growing recognition that AI's greatest impact lies not only in technical advancement, but also in its capacity to address societal challenges and create meaningful transformation across sectors.

The chapters in this edited reference book collectively examine the intersection of intelligent technologies with human systems, strategic management, and cultural realities. The contributors investigate both the opportunities and complexities associated with AI adoption, including operational efficiency, personalization, ethical governance, accessibility, sustainability, and digital inclusion. By combining conceptual discussions, empirical analyses, case studies, and exploratory frameworks, the volume provides readers with a broad understanding of how AI is reshaping organizational practices and human experiences in diverse environments.

What distinguishes this collection is its interdisciplinary and international scope. The contributors represent institutions and perspectives from Africa, Asia, Europe, North America, and the Middle East, demonstrating the global relevance of AI innovation. At the same time, the chapters reveal that successful AI integration requires contextual awareness, cultural sensitivity, ethical responsibility, and collaborative leadership. Through this collection, we aim to encourage dialogue among researchers, practitioners, educators, policymakers, and students about the future of AI-driven transformation and its implications for sustainable and inclusive development.

CHAPTER OVERVIEW

Chapter 1: Revolutionizing Healthcare Through Strategic Use of Artificial Intelligence to Optimize Patient Outcomes and Operational Efficiency examines the transformative role of AI in healthcare systems. The chapter explores applications including predictive analytics, robotic-assisted surgery, medical imaging, mental health support, and precision medicine, while also discussing operational optimization, ethical concerns, and the balance between machine intelligence and human expertise.

Chapter 2: Artificial Intelligence in Marketing: A Roadmap from Theory to Implementation provides a comprehensive examination of AI's growing influence on modern marketing practices. The chapter analyzes AI-driven approaches to customer engagement, marketing research, consumer behavior analysis, omnichannel ecosystems, advertising, and value creation across digital and traditional platforms.

Chapter 3: Enhancing Consumer Engagement Through Human-AI Collaboration: Harnessing Synergies in Marketing explores the psychological and behavioral dimensions of AI-enhanced consumer experiences. Focusing on decision-making processes and environmental influences, the chapter investigates how human-AI collaboration shapes purchasing behavior, personalization strategies, and customer engagement in modern marketing ecosystems.

Chapter 4: A Strategic Analysis of AI-Driven Transformation in the Commercial Food Sector examines how AI is reshaping innovation, leadership, and operational strategy within the food industry. Through thematic analysis of recent industry developments, the chapter identifies patterns related to efficiency, ethics, ESG considerations, quality assurance, and consumer interaction while emphasizing the differing capacities of large enterprises and SMEs to adopt AI technologies.

Chapter 5: Theory Development for AI-Driven Transformation of Rural Small Farmers in Africa explores the role of AI in supporting sustainable agricultural development among smallholder farmers. The chapter introduces the Baby Steps Theory of Development (BSTD), emphasizing incremental learning and contextual skill-building to address barriers such as digital illiteracy, infrastructural limitations, and financial constraints.

Chapter 6: A Study on AI-based Feng Shui Application for Real Estate in Vietnam: The Conceptual Idea presents an innovative conceptual framework for integrating AI technologies with traditional Feng Shui principles in Vietnam's real estate market. Grounded in socio-technical and cultural capital theories, the chapter explores how AI-driven systems can enhance buyer satisfaction, cultural alignment, and decision-making processes.

Chapter 7: Unleashing the Potential of Artificial Intelligence for the Chinese Cultural Heritage of the Forbidden City in China investigates how AI technologies are contributing to the preservation, communication, and transformation of cultural heritage. Using the Forbidden City as a case study, the chapter analyzes AI's role in collaborative cultural production, digital communication, and participatory identity formation.

Chapter 8: Digital Art Museums in the Digital Transformation of Ice and Snow Cultural Tourism examines the evolving role of digital art museums in cultural tourism and digital innovation. Through the case study of the Harbin Ice and Snow World Digital Art Museum, the chapter demonstrates how AIGC and immersive technologies enhance cultural communication, tourism engagement, and regional digital transformation.

Chapter 9: From Research to Reality: AI, the HBCU Classroom, and Community Engagement explores the integration of AI technologies within Historically Black Colleges and Universities (HBCUs). The chapter discusses challenges related to funding and technological access while presenting strategies for inclusive AI adoption, classroom innovation, community partnerships, and sustainable educational advancement.

Taken together, the chapters in this volume illustrate the remarkable breadth of AI applications across contemporary society. They reveal that AI is no longer confined to laboratories or specialized technical domains; rather, it has become embedded within healthcare systems, educational institutions, cultural organizations, agricultural communities, tourism industries, and business ecosystems. The studies presented here demonstrate how AI can support innovation, improve efficiency, strengthen decision-making, and enhance human experiences when implemented thoughtfully and responsibly.

CONCLUSION

At the same time, this collection underscores the importance of addressing ethical, social, and structural challenges that accompany AI adoption. Questions surrounding bias, privacy, accessibility, transparency, sustainability, and digital inequality remain central to the future development of intelligent systems. The contributors consistently emphasize that technological advancement must be guided by human values, interdisciplinary collaboration, and inclusive frameworks that ensure the benefits of AI are distributed equitably across communities and sectors.

As editors, we extend our sincere appreciation to all contributing authors, reviewers, and supporting institutions whose dedication and scholarly efforts made this volume possible. We hope that Artificial Intelligence in Practice: Driving Innovation Across Sectors will serve as a valuable resource for researchers, educators, students, policymakers, and practitioners seeking to better understand the evolving role of AI in contemporary society. More importantly, we hope this book inspires continued research, collaboration, and innovation aimed at harnessing artificial intelligence in ways that advance human progress, cultural understanding, and sustainable global development.

Chapter 1

Revolutionizing Healthcare Through Strategic Use of Artificial Intelligence to Optimize Patient Outcomes and Operational Efficiency

Fransisco Chindongo Ntjamba
University of Namibia, Namibia

Daniel Opotamutale Ashipala
http://orcid.org/0000-0002-8913-056X
Univerisyt of Namibia, Namibia

ABSTRACT

Artificial Intelligence (AI) is swiftly transforming healthcare by improving diagnostic accuracy, enhancing operational performance, and facilitating personalized patient care. AI applications, ranging from medical imaging and predictive analytics to robotic-assisted surgery and mental health support are fundamentally altering the methods by which diseases are diagnosed, treated, and managed. In the realm of precision medicine, AI customizes treatments according to individual genetic, clinical, and behavioral profiles, while in administration, AI optimizes workflows, lowers costs, and elevates service delivery. Notwithstanding these advancements, challenges persist regarding data privacy, algorithmic bias, and ethical implementation. The successful integration of AI necessitates a balanced approach that merges

DOI: 10.4018/979-8-2600-2378-5.ch001

human expertise with machine intelligence. This chapter explores the transformative capabilities of AI in healthcare, analyzing its impact on enhancing patient care and streamlining operational processes.

INTRODUCTION

Imagine a world where illnesses are identified prior to the onset of symptoms, personalized treatment plans are designed for each patient with remarkable accuracy, and hospitals function with impressive efficiency (Shankar, 2024). This is not a distant future; it is the reality that artificial intelligence (AI) is introducing to healthcare today. AI is transforming the manner in which medical professionals diagnose conditions, oversee patient care, and enhance hospital operations, making healthcare more accurate, proactive, and effective (Gupta, 2024).

Globally, healthcare systems are experiencing mounting pressure due to aging demographics, increasing chronic illnesses, and limited resources (Jaswal et al., 2024). AI-driven technologies are emerging to fill these gaps, providing solutions such as predictive analytics for early detection of diseases, machine learning algorithms for customized treatments, and automation tools that lessen administrative tasks. These advancements not only improve patient outcomes but can booster the overall efficiency of healthcare delivery (Gautam et al., 2025).

The healthcare industry has realized that its systems urgently require innovation and flexibility following the COVID-19 pandemic. During this global crisis, AI was essential for real-time disease monitoring, contact tracing assistance, and outbreak prediction using large-scale data analytics (Halabhavi, 2024). AI's importance in crisis preparedness was further highlighted by machine learning-powered tools that predicted hospital bed availability, managed ICU capacity, and optimized supply chains. The widespread use of digital technologies among healthcare professionals and policymakers has been spurred by these pragmatic uses of AI during a health crisis.

Additionally, AI is playing a major role in population health management by detecting risk factors and trends across communities. AI systems can identify at-risk groups and recommend targeted interventions by combining data from social determinants of health, wearable technology, and electronic health records (EHRs) (Carrasco Ramírez, 2024). By potentially lowering long-term costs and enhancing public health results, this comprehensive perspective allows healthcare systems to transition from reactive treatment to preventive care (Kosaraju, 2024).

The democratization of healthcare services is another significant change brought about by AI. In underserved areas, especially where there is a lack of human resources, AI-powered telemedicine platforms and mobile health applications are increasing access to vital health services (Joshi et al., 2025). These technologies help close gaps in care and lessen geographical disparities in health delivery by providing remote diagnoses, triage assistance, and medication advice.

Nonetheless, the swift incorporation of AI in healthcare brings forth significant questions regarding data privacy, ethical implications, and regulatory oversight. As AI-based decision-making becomes increasingly widespread, ensuring transparency, fairness, and dependability in these technologies is crucial. Achieving a balance between innovation and ethical accountability will shape AI's lasting influence on the healthcare sector (Mark & Bizz, 2025).

Clinicians trying to trust and use AI tools are faced with the challenge of the "black box" quality of many AI models, in which the reasoning behind a choice is not readily understandable (Schauer & Schnurr, 2025). As a result, the explainable AI (XAI) field has emerged with the goal of improving the interpretability of algorithmic outputs and promoting more effective human-AI interaction. This advancement is crucial for preserving clinician accountability and sustaining patient confidence.

Another important problem is bias in AI models. The findings of algorithms can reinforce disparities rather than mitigate them when they are trained on datasets that mirror current healthcare inequities or lack diversity (Fnu Zartashea, 2023). Inclusive data practices, thorough auditing of artificial intelligence technologies, and participatory design that incorporates feedback from patient groups, ethicists, and clinicians are all necessary to combat these biases (Thevißen et al., 2023).

In order to ensure ethical use of healthcare,regulatory frameworks must adapt to technological advancements. To safeguard sensitive patient data,it is necessary to adhere to regulations like the General Data Protection Regulation (GDPR) and the Health Insurance Portability and Accountabilty Act (HIPAA) (Zahra Ouariach & Ahmed, In addition, governance frameworks must be adaptable, permitting the ongoing assessment of AI systems after deployment to confirm that they are still safe, efficient, and equitable for use in the real world (Ouariach & Raghav, 2025).

A key component in the successful integration of AI is education and training. To prevent overreliance on or misinterpretation of AI outputs, it is crucial to provide healthcare workers the skills necessary to interpret and critically evaluate these outputs (Ntjamba & Ashipala, 2025). To equip the next generation of doctors for a digital healthcare environment, medical schools are increasingly integrating AI modules into their curricula. By taking this proactive stance, it is guaranteed that human oversight stays at the heart of care delivery, strengthening the idea that medical expertise should be supplemented rather than substituted by AI.

Through practical examples and emerging trends, the chapter will reveal how AI is reshaping the future of medicine and the challenges that need to be addressed to responsibly unlock its full potential.

Background

Artificial Intelligence (AI) represents a swiftly advancing domain within computer science that seeks to develop machines capable of carrying out tasks that ordinarily necessitate human intelligence. AI encompasses a variety of methods including machine learning (ML), deep learning (DL), and natural language processing (NLP). Large Language Models (LLMs) constitute a specific category of AI algorithms that leverage deep learning methods along with extremely large data sets to comprehend, summarize, generate, and predict new text-based content (Munro, 2022). LLMs have been designed to create text-based material and have extensive applicability across numerous NLP tasks, such as text generation, translation, content summarization, rewriting, classification, categorization, and sentiment analysis. NLP is an AI subfield that centers on the interactions between computers and humans using natural language, encompassing the understanding, interpretation, and generation of human language (Kulkarni et al., 2021). NLP incorporates various techniques including text mining, sentiment analysis, speech recognition, and machine translation (Olagbende, 2025).Throughout the years, AI has experienced notable changes, evolving from the early days of rule-based systems to the present age of ML and deep learning algorithms.

Figure 1.

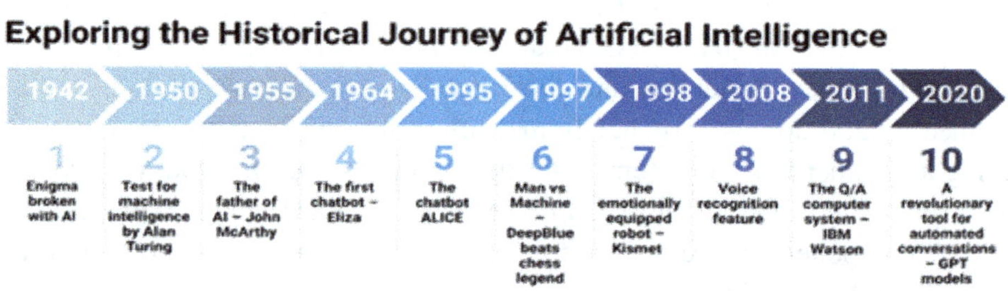

This figure is adapted from Alowais et al., 2023

Figure 2.

Understanding the Relationship Between AI, ML, DL, and NLP

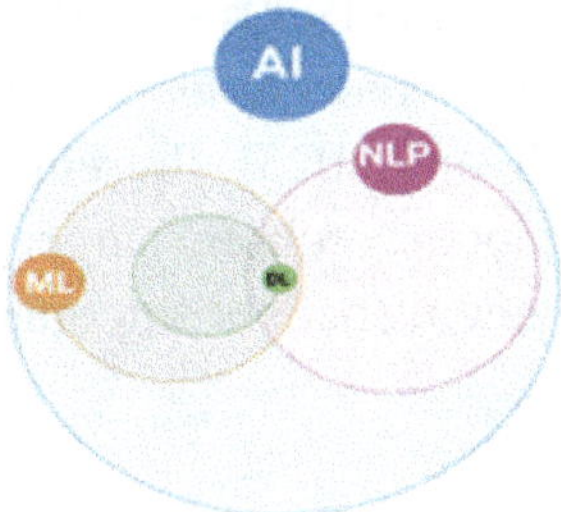

- AI is a broad field that includes anything related to making machines smart.
- NLP is the branch of AI focused on teaching machines to understand, interpret, and generate human language.
- ML is a subset of AI that involves systems that can learn by themselves.
- DL is a subset of ML that uses models built on deep neural networks to detect patterns with minimal human involvement.

This figure is adapted from Alowais et al., 2023

Tracing the Development of AI With a Deeper Insight Into the Connection Between AI, ML, DL, and NLP

Currently, AI is transforming sectors like healthcare, finance, and transportation, among others, and its influence is poised to increase (Ahmad et al., 2025). In academia, AI has been utilized to create intelligent tutoring systems, which are software programs capable of adjusting to the requirements of individual learners. These systems have enhanced student performance across various disciplines, such as math and science. In research, AI has been employed to evaluate extensive datasets and uncover trends that may be challenging for humans to notice; this has resulted in advancements in areas like genomics and drug development (Nandish, 2023). AI has been implemented in healthcare environments to create diagnostic tools and tailor treatment strategies. As AI continues to progress, it is vital to guarantee that it is developed ethically and for the advantage of everyone (Bouderbane, 2024). The swift advancement of AI technology offers a chance for its use in clinical settings, potentially transforming healthcare delivery. It is essential to record and share information about AI's function in clinical practice to prepare healthcare professionals with the knowledge and resources needed for successful application in patient care (Kambala, 2024.

Artificial Intelligence (AI) has swiftly appeared as a groundbreaking force in healthcare, changing how medical practitioners identify illnesses, create treatment strategies, and organize healthcare operations. While AI in healthcare may appear as a new advancement, its origins can be traced back to the mid-20th century when initial computer-based algorithms were crafted to aid in medical decision-making (Kumar & Joshi, 2022). Over the years, AI has progressed from basic rule-based systems to intricate machine learning and deep learning frameworks that can evaluate enormous amounts of medical data with extraordinary precision (Hatture & Kadakol, 2021).

The emergence of machine learning and neural networks in the 1990s and 2000s set the stage for more advanced AI applications, enabling computers to learn from extensive datasets without direct programming (Jain & Nigam, 2025). Presently, AI is a fundamental component of medical imaging, diagnostics, robotic surgery, and even administrative tasks in hospitals, considerably enhancing patient outcomes and operational effectiveness (Jain & Nigam, 2025).

AI-Driven Innovations in Patient Care

AI's most noticeable influence in healthcare is within patient care, where it improves diagnostics, treatment planning, and disease management. Machine learning algorithms and deep learning models are especially beneficial in medical imaging, enabling radiologists to identify conditions like cancer, fractures, and neurological disorders with high precision. AI-powered imaging tools examine patterns in radiographs, CT scans, and MRIs, frequently spotting abnormalities that human radiologists may miss. This accelerates diagnosis and enhances accuracy, reducing the chance of misdiagnosis (Ntjamba & Ashipala, 2024).

In addition to imaging, AI is revolutionizing disease prediction and early intervention. Predictive analytics models utilize extensive datasets, including patient histories, genetic data, and lifestyle elements, to pinpoint individuals at elevated risk for diseases like diabetes, heart disease, and cancer. These AI-generated insights empower healthcare providers to adopt preventive actions before diseases advance, shifting the focus from reactive to proactive care. For instance, AI models can evaluate electronic health records (EHRs) to forecast the probability of complications in diabetic patients, facilitating early intervention strategies that decrease hospitalizations and enhance quality of life (Ntjamba & Ashipala, 2025).

AI is also redefining personalized medicine by customizing treatments for individual patients. Conventional treatment methods often adhere to a one-size-fits-all approach, but AI-enhanced algorithms assess genetic, molecular, and clinical data to ascertain the most effective treatment for each patient. This is especially significant in oncology, where AI-guided precision medicine assists oncologists in selecting targeted therapies based on a patient's genetic profile. By fine-tuning treatment plans, AI reduces trial-and-error methods, lowers adverse effects, and increases survival rates (Ntjamba & Ashipala, 2025).

Enhancing Operational Efficiency With AI

Beyond enhancing patient care, AI is refining healthcare operations, making hospitals and clinics more efficient and cost-effective. One of the key applications of AI in hospital management is workflow automation. Administrative tasks, such as scheduling appointments, managing patient records, and processing insurance claims, are typically time-consuming and susceptible to errors. AI-driven chatbots and virtual assistants now perform these tasks more rapidly and accurately, alleviating administrative responsibilities for healthcare staff and letting them concentrate on patient care (Ntjamba & Ashipala, 2025).

Predictive analytics is also improving hospital resource management. AI models assess historical patient admission data, disease outbreaks, and seasonal patterns to forecast hospital occupancy rates and allocate resources accordingly. This is particularly advantageous in emergency departments, where AI-powered tools can predict patient influx and optimize staffing levels, guaranteeing timely and efficient care. During the COVID-19 pandemic, hospitals utilized AI to anticipate patient surges, manage ventilator distribution, and maximize ICU bed availability, showcasing AI's potential in crisis management (Halabhavi, 2024).

Moreover, AI-driven robotics are playing an essential role in surgical procedures and patient rehabilitation. Robotic-assisted surgery, facilitated by AI-enabled systems like the da Vinci Surgical System, enhances precision, reduces invasiveness, and shortens recovery times (Nissar & M, 2025). These robots aid surgeons in executing intricate procedures with improved dexterity and accuracy, resulting in better surgical outcomes. In rehabilitation, AI-guided exoskeletons and prosthetics are assisting patients in reclaiming mobility and independence, significantly boosting their quality of life (Ntjamba & Ashipala, 2025; Su, et al., 2018).

AI Mental Health Support

AI has the capability to transform mental health support by delivering individualized and accessible care to people. Numerous studies have demonstrated the effectiveness and accessibility of utilizing Web-based or Internet-based cognitive-behavioral therapy (CBT) as a psychotherapeutic approach. Although psychiatric professionals depend on face-to-face interaction and behavioral observation of the patient in clinical settings compared to other practitioners, AI-powered tools can enhance their efforts in various ways. AI-driven mental health applications can aid in the early identification and diagnosis of mental health issues, in addition to providing customized treatment and assistance (Punia et al., 2025). These applications are also capable of offering 24/7 support, minimizing the necessity for in-person visits and reducing waiting times. Additionally, these digital resources can be utilized to track patient progress and medication adherence, offering valuable insights into the efficacy of treatment (Ntjamba & Ashipala, 2025). The existing published research concerning the use of AI in mental health has reached the conclusion that depression is the most frequently examined mental disorder (Canady, 2019).

The Role of AI in Drug Discovery and Development

The pharmaceutical industry is also experiencing a transformation due to AI-powered drug discovery and development. Conventional drug development is a protracted and expensive endeavor, frequently requiring more than ten years and billions of dollars to introduce a new drug to the market. AI is expediting this process by examining extensive datasets to identify potential drug candidates, forecast their effectiveness, and enhance clinical trials (Singh, 2025).

AI-driven drug discovery platforms employ deep learning algorithms to investigate molecular structures and anticipate their interactions with biological targets (Chandra Kaushik & Raj, 2020). This accelerates the discovery of promising drug compounds, shortening the duration needed for preclinical testing. For example, AI models have successfully identified new antibiotic agents and potential COVID-19 treatments in unprecedented timeframes, showcasing the power of AI in addressing global health challenges (Das, 2025).

Ethical and Regulatory Challenges in AI Integration

Despite offering many advantages, AI in healthcare poses substantial ethical and regulatory challenges that must be resolved to facilitate responsible integration. A primary concern revolves around data privacy and security (Zahra Ouariach & Ahmed, 2025). AI depends on large volumes of patient information, raising worries about unauthorized access, data breaches, and the misuse of sensitive medical data (Ntjamba & Ashipala, 2025). Strong data protection strategies, including encryption, anonymization, and stringent access controls, are crucial to protect patient privacy and adhere to regulations like the General Data Protection Regulation (GDPR) and the Health Insurance Portability and Accountability Act (HIPAA) (Ouariach & Raghav, 2025)

Another important issue is the bias and fairness inherent in algorithms. AI models are often trained on historical datasets, which can encompass biases that mirror existing disparities in healthcare (Fnu Zartashea, 2023). For instance, if an AI model is trained on data that does not adequately represent certain demographic groups, it may yield inaccurate or biased forecasts for those populations. Tackling algorithmic bias demands diverse and representative training datasets, thorough testing, and continuous model assessment to guarantee fairness and equity in AI-driven healthcare solutions (Fnu Zartashea, 2023). Ensuring that AI systems adhere to safety, efficacy, and ethical standards is critical for their widespread acceptance and the trust of patients.

The Future of AI in Healthcare

The role of AI in healthcare is anticipated to grow as it develops. The future of AI in medicine is shaped by new technologies like explainable AI (XAI), which seeks to increase the transparency and interpretability of AI decision-making for healthcare professionals. The "black box" issue, in which it is hard to comprehend how AI-generated predictions are made, is solved by explainable AI, which offers explicit insights into the decision-making process. By improving confidence in AI-driven diagnoses and treatment recommendations, this facilitates the integration of AI insights into clinical practice (Schauer & Schnurr, 2025).

The integration of AI with wearable technology and remote patient monitoring is another encouraging advancement. Smartwatches, biosensors, and mobile health apps are using AI more and more to monitor real-time health information, identify early indicators of illness, and offer individualized health advice. By enabling people to take proactive steps for disease prevention and treatment, this change toward on-going health monitoring lowers the strain on healthcare institutions (Adirim, 2025).

Furthermore, AI is anticipated to be essential in bridging global health inequalities by increasing access to high-quality healthcare in limited resourced settings. By delivering medical knowledge to distant and low-resource communities, AI-powered telemedicine platforms, automated diagnostic technologies, and mobile health applications can address disparities in healthcare access. With the aid of artificial intelligence, healthcare is becoming more accessible, which may lead to better health outcomes around the world, especially in areas where the healthcare system is lacking (Joshi et al. 2025).

There is no doubt that artificial intelligence is revolutionizing healthcare by speeding up medical research, improving patient outcomes, and optimizing hospital operations (Whig et al., 2024). The incorporation of AI in healthcare is revolutionizing the industry in unprecedented ways, ranging from AI-driven diagnostics and individualized treatments to robotic-assisted operations and predictive analytics (Batra Deep Manishkumar, 2024). However, realizing the full potential of AI necessitates overcoming ethical, regulatory, and technical hurdles to guarantee its fair and responsible implementation. As AI develops, its impact on the future of medicine will be crucial in establishing a healthcare system that is more efficient, accessible, and patient-centered (Wang & Li, 2023).

By optimizing patient outcomes and operational performance through a variety of strategic applications, AI is revolutionizing healthcare. This entails enhancing patient safety and experience, streamlining administrative procedures, improving diagnostic accuracy, and facilitating individualized treatment. Healthcare practitioners may use AI to provide care that is more efficient, effective, and focused on the patient, which will ultimately result in better health outcomes and higher operational efficiency (Carrasco Ramírez, 2024).

The ways in which AI is revolutionizing healthcare:

Improved Diagnostic Accuracy (Kumar et al., 2025).

- Using medical images (such as X-rays and MRIs) and patient data, AI algorithms can analyze them to detect illnesses and abnormalities with higher precision and efficiency.
- Earlier illness detection may result from this, which would allow for more timely and successful treatments.
- By analyzing massive amounts of data and anticipating potential drug interactions, AI can aid in the diagnosis of complicated medical illnesses, facilitating the development of individualized treatment strategies.

Personalized Medical Care (Ntjamba & Ashipala, 2024)

- AI algorithms are capable of analyzing individual patient data, such as genetic information, in order to create treatment plans that are customized to each patient's specific needs.
- This facilitates more focused and successful therapies, which result in improved patient outcomes and lower expenses.
- Additionally, AI can be used to forecast how patients may react to various therapies, which helps doctors make better-informed choices.

Simplified Administrative Procedures (Asif, & Qureshi, 2024).

- Basic administrative duties like data input, invoicing, and appointment booking can be automated by AI.
- Because healthcare practitioners are able to concentrate on patient care, this results in increased efficiency and lower expenditures.
- AI can aid in the administration of massive amounts of patient data, making it simpler for healthcare professionals to access and interpret the information.

Improved Safety and Patient Experience:

- By delivering more customized and proactive treatment, AI can enhance patient interaction and happiness.

- Additionally, AI may be employed to remotely monitor patients, enabling prompt intervention and early identification of possible issues.
- AI can also help to pinpoint and lessen treatment mistakes, which results in better patient safety.

Better Operational Performance (Whig, et al., 2024).

- AI has the potential to enhance the overall efficiency of healthcare operations by optimizing resource use and lowering expenses.
- AI can also be used to forecast the demand for healthcare services, which enables providers to be better prepared for busy times.
- AI can liberate resources and increase the healthcare system's overall efficiency by automating processes and optimizing workflows.

Ethical Factors:

Although AI has many promise in healthcare, it is imperative that ethical issues pertaining to data privacy, algorithmic bias, and the necessity for strong regulatory frameworks be addressed. It is crucial to make sure that AI is used ethically and responsibly in healthcare and that its advantages are distributed fairly as it becomes more and more integrated (Ntjamba & Ashipala, 2025)

Bridging the AI Gap: Combining Human Knowledge and Technology

Human competence is still essential for making healthcare decisions, regardless of how far AI has come. The most efficient healthcare systems driven by AI operate under a human-in-the-loop paradigm, where AI supports medical professionals rather than taking their place. AI-generated insights should be used by doctors, nurses, and healthcare managers as decision-support tools rather than as conclusive solutions (Mark, & Bizz, 2025).

For instance, even if AI-enabled diagnostic tools can find abnormalities in medical images, a radiologist's knowledge is still necessary to validate the diagnosis and comprehend the clinical setting. In the same way, although AI-powered predictive models may identify patients at high risk of heart failure in cardiology, the doctor's clinical judgment is still necessary to develop a comprehensive treatment strategy. The synergy between human experience and AI guarantees that AI improves healthcare without compromising expert judgment (Wang, 2024).

Future Trends: Where AI is Heading in Healthcare

AI's role in healthcare is expected to expand into new frontiers that promise to revolutionize medicine further:

Figure 3.

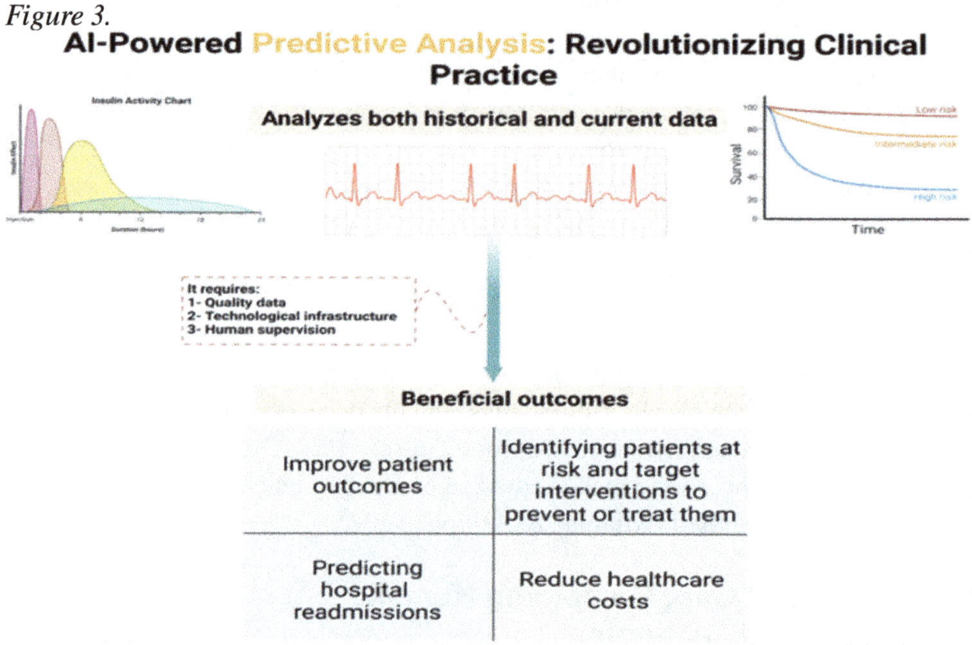

This figure is adapted from Alowais et al., 2023

Robotics and AI in Surgery: Robotic-assisted surgery will continue to advance, allowing for more accurate minimally invasive operations. Robotic systems driven by artificial intelligence will improve real-time surgical decision-making, lowering human error and raising patient recovery rates (Ntjamba & Ashipala et al., 2024).

AI-Driven Mental Health Interventions: AI is becoming an increasingly useful resource in mental health treatment, with virtual therapists, chatbots, and sentiment analysis technologies helping to identify the first indications of depression, anxiety, and PTSD. In underserved areas, in particular, AI-driven treatments can help close the gap in mental healthcare (Punia et al., 2025).

Global Health AI: Mobile health apps and telemedicine platforms powered by AI will improve access to healthcare in underserved and rural communities. By offering automated diagnoses for diseases like malaria and tuberculosis, AI can greatly enhance healthcare delivery in low-income countries (Selwyn et al., 2025).

RESEARCH SIGNIFICANCE

This chapter offers a thorough and current overview of the existing status of AI in clinical practice, including its potential uses in disease diagnosis, treatment suggestions, and patient interaction. It also addresses the related challenges, including ethical and legal issues and the necessity for human expertise. In doing so, it enhances comprehension of AI's importance in healthcare and aids healthcare organizations in the effective adoption of AI technologies.

REVIEW OF RELATED LITERATURE

The incorporation of Artificial Intelligence (AI) into healthcare systems has surfaced as a significant area of academic interest, motivated by its capacity to revolutionize almost every facet of health service delivery. The existing literature indicates that AI is shaping clinical decision-making, enhancing diagnostic accuracy, personalizing treatment, optimizing resource allocation, and even streamlining healthcare administration. As health data becomes increasingly digital and machine learning algorithms advance, AI tools are being designed and implemented to not only boost clinical efficacy but also enhance operational productivity.

Wang etal. (2024) highlight the expanding role of AI in medical imaging, one of the earliest and most extensively implemented applications within clinical environments. Their research demonstrates how deep learning models, especially convolutional neural networks (CNNs), surpass traditional imaging techniques in identifying diseases like lung cancer, diabetic retinopathy, and cardiovascular irregularities. AI-driven image interpretation tools, such as those utilized for CT scans, MRIs, and X-rays, minimize human error and facilitate early diagnoses, which is crucial for diseases like cancer where timely interventions can greatly enhance patient survival.

In addition, (Esteva & Topol, 2019), show that AI models can match or even outperform dermatologists in identifying skin cancer from photographic imagery. Their results highlight the significance of AI in making specialized diagnostics more accessible, especially in settings with limited resources or remote areas where trained professionals may be in short supply. This perspective is supported by Bairwa & Jangid, 2024), who found out an AI algorithm which is capable of detecting pneumonia from chest X-rays with accuracy levels comparable to those of radiologists.

Krishna Pasupuleti, (2024), broadens the conversation to the field of precision medicine, claiming that AI can revolutionize treatment planning by harnessing genomics and electronic health record (EHR) data. As stated by Topol, AI allows clinicians to depart from a one-size-fits-all methodology to care by taking into account individual differences in genes, environment, and lifestyle. AI applications can sift through extensive datasets to suggest tailored therapies, particularly for complex conditions like cancer, where treatment results vary significantly among patients.

In the context of chronic disease management, Rashid & Sharma (2025), reveal that AI-driven predictive models can anticipate disease progression in individuals with diabetes or heart failure. These systems assist healthcare providers in predicting complications and modifying treatment plans accordingly. For example, AI can alert clinicians when a diabetic patient's likelihood of developing retinopathy escalates, prompting immediate referrals for eye assessments.

Beyond direct patient care, the literature also acknowledges the importance of AI in enhancing healthcare operations. Mhlanga (2025), concentrate on the application of predictive analytics in managing hospital resources. Their study illustrates how AI models forecast patient admissions, discharges, and lengths of stay, enabling administrators to plan staffing and bed availability more effectively. This not only optimizes patient flow but also diminishes bottlenecks and waiting times in emergency departments and inpatient wards.

Similarly, Asif, & Qureshi (2024), examine the influence of AI on administrative processes. Their research discovers that intelligent automation alleviates the demands of repetitive administrative tasks, including patient scheduling, billing, and claims management. AI-driven virtual assistants can also assist with patient triage and appointment coordination, allowing human staff to focus on more involved, patient-centered endeavors. The improvements in administration lead to cost reductions, fewer mistakes, and enhanced satisfaction for both patients and healthcare providers.

Despite these developments, there is an increasing volume of literature that critically assesses the constraints and ethical considerations of AI in healthcare. McCall et al., (2022), express worries about algorithmic bias, particularly when AI systems are developed using datasets that lack diversity. Biased algorithms could produce inaccurate outcomes for marginalized populations, worsening existing health inequities. The authors advocate for inclusive data methods and transparency in the development of AI models to promote fairness.

In the same vein, Azith Teja Ganti, (2023), underscore the necessity of creating ethical frameworks that steer the responsible development and implementation of AI in medicine. Their research delves into matters involving patient consent, data ownership, and the lack of clarity in AI decision-making mechanisms. The so-called "black box" characteristic of numerous deep learning systems creates challenges for clinicians and patients to grasp how decisions are reached, raising concerns regarding trust and accountability.

Thevißen et al. (2023) advocate for enhanced regulatory oversight and international governance structures to ensure that AI applications conform to ethical standards and human rights. Their findings underscore the importance of cooperation among technologists, ethicists, clinicians, and policymakers to develop guidelines that reconcile innovation with public health safety. The authors also emphasize the need for ongoing monitoring and post-market evaluation of AI tools to identify unintended negative impacts.

In low- and middle-income countries (LMICs), Jones-Esan et al., (2024), point out how AI can enhance access to healthcare services by facilitating telemedicine, mobile health (mHealth) applications, and remote monitoring. Nonetheless, they also warn that limitations in digital infrastructure, low health literacy, and the absence of data privacy regulations may obstruct the widespread adoption of AI in these areas. The authors suggest capacity-building initiatives and investments in digital infrastructure to realize the potential of AI for global health equity.

Moreover, Ntjamba & Ashipala (2025) emphasize the significance of clinician-AI collaboration. Instead of replacing healthcare workers, AI should be considered a tool that enhances clinical decision-making. They introduce the notion of "augmented intelligence" — a framework where human expertise and machine intelligence operate together. This strategy acknowledges the essential contributions of human empathy, ethical judgment, and contextual awareness in delivering patient care.

Last but not least, Al-Medfa (2022), examines the psychological and behavioral effects of incorporating artificial intelligence into healthcare. Their research shows that, despite the fact that many doctors value the advantages of AI, others are worried about job security, depersonalization of treatment, and a loss of independence. For promoting trust and acceptance, it's essential to address these issues via transparency, education, and participatory AI design approaches.

The current literature, in sum, emphasizes the enormous potential of AI to improve diagnostic accuracy, customize therapies, simplify procedures, and broaden access to healthcare. Nevertheless, it also emphasizes numerous obstacles—ethical, regulatory, and practical—that must be overcome in order to fully realize the potential of artificial intelligence. The integration of AI into healthcare should be approached with caution, making sure that technological progress does not compromise human values, equity, and safety. In order to utilize AI's potential to enhance healthcare outcomes worldwide as research and real-world deployment progress, a well-rounded and ethical strategy will be necessary.

RECOMMENDATIONS AND SOLUTIONS

A comprehensive and strategic plan is required to ensure the successful and ethical integration of artificial intelligence (AI) into healthcare systems. The moral dilemma surrounding algorithmic bias, data protection, and transparency is one of the most urgent problems facing the adoption of AI in the healthcare industry. It is essential to create strong ethical and governance structures at both the national and institutional levels in order to solve these problems. This involves establishing ethical oversight boards inside healthcare facilities and creating clear national rules governing the use of AI. To lessen inherent biases, especially those that may result in disparities in diagnosis or treatment outcomes between different demographic groups, these policies must make sure that AI systems are trained using datasets that are representative and diverse.

It is also crucial to promote a collaborative approach in which AI enhances, rather than replaces, human clinicians. AI should be used as a complementary tool that improves clinical decision-making while preserving the function of healthcare practitioners. The "human-in-the-loop" strategy, in which AI-generated insights are evaluated by humans, is necessary for this, as it guarantees that compassion, morality, and professional judgement are still at the core of patient treatment. Through focused training initiatives that provide healthcare practitioners the skills to critically analyze and interpret AI outputs, this cooperative environment should be strengthened.

The fair use of AI is still a key issue. If left unaddressed, AI technologies could make the current disparities in healthcare access and outcomes even worse. For this reason, designers should create AI solutions that are practical, inexpensive, and adaptable to low-resource environments. To make sure that these innovations are contextually relevant and beneficial, pilot programs should be implemented in underrepresented areas. In order to back such initiatives and ensure equitable distribution of AI technologies, collaborations between the public and private sectors can be essential.

Multidisciplinary collaboration is also essential for the integration of AI in healthcare to be successful. The combined knowledge of doctors, data scientists, ethicists, engineers, and legislators is necessary to create ethical and effective AI solutions. This collaboration can be facilitated by establishing innovation hubs or research laboratories within healthcare facilities, which promotes the co-creation of AI systems that are tailored to the needs of end users. Additionally, encouraging joint research initiatives that concentrate on addressing real-world clinical issues can help guarantee that AI development is aligned with actual healthcare priorities.

FUTURE DIRECTIONS FOR RESEARCH

AI has the potential to revolutionize clinical practice, but several challenges must be addressed to realize its full potential. Among these challenges is the lack of quality medical data, which can lead to inaccurate outcomes. Data privacy, availability, and security are also potential limitations to applying AI in clinical practice. Additionally, determining relevant clinical metrics and selecting an appropriate methodology is crucial to achieving the desired outcomes. Human contribution to the design and application of AI tools is subject to bias and could be amplified by AI if not closely monitored. The AI-generated data and/or analysis could be realistic and convincing; however, hallucination could also be a major issue which is the tendency to fabricate and create false information that cannot be supported by existing evidence. This can be particularly problematic regarding sensitive areas such as patient care. Thus, the development of AI tools has implications for current health professions education, highlighting the necessity of recognizing human fallibility in areas including clinical reasoning and evidence-based medicine. Finally, human expertise and involvement are essential to ensure the appropriate and practical application of AI to meet clinical needs and the lack of this expertise could be a drawback for the practical application of AI.

Addressing these challenges and providing constructive solutions will require a multidisciplinary approach, innovative data annotation methods, and the development of more rigorous AI techniques and models. Creating practical, usable, and successfully implemented technology would be possible by ensuring appropriate cooperation between computer scientists and healthcare providers. By merging current best practices for ethical inclusivity, software development, implementation science, and human-computer interaction, the AI community will have the opportunity to create an integrated best practice framework for implementation and maintenance. Additionally, a collaboration between multiple health care settings is required to share data and ensure its quality, as well as verify analyzed outcomes which will be critical to the success of AI in clinical practice. Another suggestion is to provide appropriate training and education that starts from the undergraduate level of all the health care practitioners and proceeds to the continuous development and improvement for the practitioners working in the current practice to ensure the proper adaptation which provides the best patient care and avoid any legal or ethical issues or misinterpretation of the outcomes without verifying the results. Medical schools are encouraged to incorporate AI-related topics into their medical curricula. A study conducted among radiology residents showed that 86% of students agreed that AI would change and improve their practice, and up to 71% felt that AI should be taught at medical schools for better understanding and application. This integration ensures that future healthcare professionals receive foundational knowledge about AI and its applications from the early stages of their education.

CONCLUSION

From early diagnosis to customized therapy and real-time clinical decision-making, Artificial Intelligence (AI) is set to revolutionize the healthcare industry.

By utilizing massive medical databases, sophisticated analytics, and machine learning, AI allows for the identification of intricate patterns, enhances diagnostic precision, and speeds up clinical processes. Its increasing value is highlighted by its proven success in areas like medical imaging, where it can compete with or even outperform human specialists in identifying illnesses like cancer and cardiovascular disorders. Additionally, during the age of precision medicine, AI enables clinicians to combine clinical, genetic, and behavioral information in order to create very personalized treatment regimens. AI has enormous potential for providing more efficient, personalized, and effective healthcare worldwide as these technologies continue to advance.

REFERENCES

Adirim, T. (2025). Digital health technologies: Virtual health and remote patient monitoring. In Digital health, AI and generative AI in healthcare (pp. 11–26). DOI: 10.1007/978-3-031-83526-1_2

Ahmad, N., Fatima, I., Alam, M., Khan, I. R., & Mahmoud, A. (2025). Transforming healthcare through generative AI. In Generative intelligence in healthcare (pp. 259–279). DOI: 10.1201/9781003539483-11

Al-Medfa, M. (2022). [Title not provided]. DOI: 10.2196/preprints.38137

Almeida, D., & Barr, N. (2025). Innovations in health data protection: Ethical, legal, and technological perspectives in a global context. In Navigating privacy, innovation, and patient empowerment through ethical healthcare technology (pp. 171–196). DOI: 10.4018/979-8-3693-7630-0.ch007

Alowais, S. A., Alghamdi, S. S., Alsuhebany, N., Alqahtani, T., Alshaya, A. I., Almohareb, S. N., Aldairem, A., Alrashed, M., Bin Saleh, K., Badreldin, H. A., Al Yami, M. S., Al Harbi, S., & Albekairy, A. M. (2023). Revolutionizing healthcare: The role of artificial intelligence in clinical practice. *BMC Medical Education*, *23*(1), 239. Advance online publication. DOI: 10.1186/s12909-023-04698-z PMID: 37740191

Asif, F., & Qureshi, M. F. (2024). AI and quality and patient safety. In Advances in healthcare information systems and administration (pp. 217–232). DOI: 10.4018/979-8-3693-7051-3.ch009

Avaji, G. M., & N, D. G. (2024). IoT innovations in healthcare: Enhancing patient care and operational efficiency. *International Journal of Research Publication and Reviews*, *5*(5), 11479–11485. DOI: 10.55248/gengpi.5.0524.1414

Azith Teja Ganti, V. K. (2023). Exploring the intersection of bioethics and AI-driven clinical decision-making: Navigating the ethical challenges of deep learning applications in personalized medicine and experimental treatments. *Journal of Material Sciences & Manufacturing Research*, *4*(6), 1–10. DOI: 10.47363/JMSMR/2023(4)192

Bairwa, H., & Jangid, R. (2024). Pneumonia detection from chest X-rays using the CheXNet deep learning algorithm. *Preprints*. DOI: 10.20944/preprints202407.0104.v1

Batra Deep Manishkumar, P. (2024). Revolutionizing healthcare platforms: The impact of AI on patient engagement and treatment efficacy. *International Journal of Science and Research (Raipur, India)*, *13*(2), 273–280. DOI: 10.21275/SR24201070211

Bouderbane, A. (2024). Using AI tools to prompt knowledge appropriately and ethically in higher education. In General aspects of applying generative AI in higher education (pp. 327–345). DOI: 10.1007/978-3-031-65691-0_17

Canady, V. A. (2019). Study explores use of AI to treat patients with depression. *Mental Health Weekly*, *29*(42), 3–4. DOI: 10.1002/mhw.32116

Carrasco Ramírez, J. G. (2024). AI in healthcare: Revolutionizing patient care with predictive analytics and decision support systems. *Journal of Artificial Intelligence General Science*, *1*(1), 31–37. DOI: 10.60087/jaigs.v1i1.p37

Chandra Kaushik, A., & Raj, U. (2020). AI-driven drug discovery: A boon against COVID-19? *AI Open*, *1*, 1–4. DOI: 10.1016/j.aiopen.2020.07.001

Das, U. (2025). Generative AI for drug discovery and protein design: The next frontier in AI-driven molecular science. *Preprints*. DOI: 10.20944/preprints202504.0512.v1

Esteva, A., & Topol, E. (2019). Can skin cancer diagnosis be transformed by AI? *Lancet*, *394*(10211), 1795. DOI: 10.1016/S0140-6736(19)32726-6

Farinu, U. (2025). Fairness, accountability, and transparency in AI: Ethical challenges in data-driven decision-making. *SSRN*. DOI: 10.2139/ssrn.5128174

Gautam, R., Kaur, P., & Sharma, M. (2025). Predictive analytics for early cancer detection using machine learning and generative AI. In Generative intelligence in healthcare (pp. 150–175). DOI: 10.1201/9781003539483-7

Gupta, K. (2024). Transforming healthcare: The role of artificial intelligence in revolutionising patient care. *MMJ*, *1*(4). Advance online publication. DOI: 10.62830/MMJ1-04-6a

Halabhavi, B. M. (2024). AI-driven predictive analytics in healthcare: Improving patient outcomes and resource allocation. *African Journal of Biomedical Research*, *27*(4S), 9288–9297. DOI: 10.53555/AJBR.v27i4S.5427

Hatture, S. M., & Kadakol, N. (2021). Clinical diagnostic systems based on machine learning and deep learning. In Demystifying big data, machine learning, and deep learning for healthcare analytics (pp. 159–183). DOI: 10.1016/B978-0-12-821633-0.00011-8

Jain, S., & Nigam, H. (2025). AI-powered diagnostics and personalized treatment: Enhancing patient outcomes in modern healthcare. *Journal of Neonatal Surgery*, *14*(6S), 349–359. DOI: 10.52783/jns.v14.2241

Jaswal, S., Sharma, S., & Parihar, N. (2024). Artificial intelligence-driven predictive analytics for early detection of chronic diseases in primary healthcare. In *2024 International Conference on Innovative Computing, Intelligent Communication and Smart Electrical Systems (ICSES)* (pp. 1–6). DOI: 10.1109/ICSES63760.2024.10910314

Jones-Esan, L., Somasiri, N., & Lorne, K. (2024). Enhancing healthcare delivery through digital health interventions: A systematic review on telemedicine and mobile health applications in low and middle-income countries (LMICs). *Research Square*. DOI: 10.21203/rs.3.rs-5189203/v1

Joshi, H., Bansal, J., & Shukla, P. (2025). Revolutionizing patient access through AI in healthcare. In Advances in artificial intelligence for healthcare applications (pp. 208–227). DOI: 10.1201/9781003503385-15

Kambala, M. (2024). AI-powered healthcare: Transforming patient outcomes with machine learning. *Journal of Medical Science and Clinical Research, 12*(8), 34–47. DOI: 10.18535/jmscr/v12i08.07

Kosaraju, D. (2024). Predictive analytics in healthcare: Leveraging AI to anticipate disease outbreaks and enhance patient outcomes. *Galore International Journal of Health Sciences and Research, 8*(3), 73–79. DOI: 10.52403/gijhsr.20230312

Krishna Pasupuleti, M. (2024). Quantum AI and precision genomics: Redefining cancer care and personalized medicine. In Precision genomics and quantum AI: Revolutionizing cancer treatment and personalized health (pp. 40–55). DOI: 10.62311/nesx/46080

Kulkarni, A., Shivananda, A., & Kulkarni, A. (2021). Natural language processing and artificial intelligence overview. In Natural language processing projects (pp. 1–37). DOI: 10.1007/978-1-4842-7386-9_1

Kumar, A., & Joshi, S. (2022). Applications of AI in healthcare sector for enhancement of medical decision making and quality of service. In *2022 International Conference on Decision Aid Sciences and Applications (DASA)* (pp. 37–41). DOI: 10.1109/DASA54658.2022.9765041

Kumar, V., Iqbal, M. I., & Rathore, R. (2025). Natural language processing (NLP) in disease detection. In AI in disease detection (pp. 53–75). DOI: 10.1002/9781394278695.ch3

Mark, B., & Bizz, P. (2025). Ethical implications of AI-driven decision making in healthcare information systems. *Preprints*. DOI: 10.20944/preprints202503.0078.v1

McCall, C. J., DeCaprio, D., & Gartner, J. (2022). The measurement and mitigation of algorithmic bias and unfairness in healthcare AI models developed for the CMS AI health outcomes challenge. *medRxiv*. DOI: 10.1101/2022.09.29.22280537

Mhlanga, D. (2025). AI and predictive analytics and risk assessment in healthcare. In AI in hospital administration (pp. 89–104). DOI: 10.1201/9781003475804-6

Munro, K. (2022). Natural language processing (NLP). In The handbook of data science and AI (pp. 273–316). DOI: 10.3139/9781569908877.010

Nandish, M. (2023). How has the rapid growth of artificial intelligence impacted business management? A case study of prominent organizations that have experienced significant incline in growth due to adaptation to AI advancements. *International Journal of Research Publication and Reviews*, *4*(6), 4124–4130. DOI: 10.55248/gengpi.4.623.47437

Nissar, H., & M, M. S. (2025). AI-enhanced robotic surgery: A machine learning approach to improving surgical precision and patient safety. TechRxiv. DOI: 10.36227/techrxiv.174430796.67882981/v1

Ntjamba, F. C., & Ashipala, D. O. (2024). Impact on and ethical considerations of artificial intelligence on human healthcare. In Advances in computational intelligence and robotics (pp. 1–36). DOI: 10.4018/979-8-3693-9158-7.ch001

Ntjamba, F. C., & Ashipala, D. O. (2025). Beyond reality: Integrating AI and Metaverse technologies for enhanced scientific collaboration. In Advances in computational intelligence and robotics (pp. 99–118). DOI: 10.4018/979-8-3373-0340-6.ch005

Ntjamba, F. C., & Ashipala, D. O. (2025). Innovative pathways: AI and the Metaverse in interdisciplinary research and knowledge discovery. In Advances in computational intelligence and robotics (pp. 119–138). DOI: 10.4018/979-8-3373-0340-6.ch006

Ntjamba, F. C., & Ashipala, D. O. (2025). Ethical dimensions of AI in the Metaverse. In Advances in computational intelligence and robotics (pp. 195–212). DOI: 10.4018/979-8-3373-0340-6.ch010

Olagbende, R. (2025). Natural language processing (NLP) techniques for Afan Oromo text analysis. *Preprints*. DOI: 10.20944/preprints202503.0841.v1

Ouariach, S., & Raghav, A. (2025). Legal accountability in AI-enhanced healthcare. In Computational intelligence in healthcare law (pp. 72–85). DOI: 10.1201/9781003538462-5

Punia, A., Bhakuni, P., & Sharma, N. (2025). Development of AI- and NLP-driven chatbots and virtual assistants for mental health support. In Advances in psychology, mental health, and behavioral studies (pp. 229–262). DOI: 10.4018/979-8-3693-9341-3.ch010

Rachh, R., Allagi, S., & Shravan, B. (2021). Machine learning algorithms for prediction of heart disease. In Demystifying big data, machine learning, and deep learning for healthcare analytics (pp. 247–275). DOI: 10.1016/B978-0-12-821633-0.00010-6

Rashid, M., & Sharma, M. (2025). AI-assisted diagnosis and treatment planning. In AI in disease detection (pp. 313–336). DOI: 10.1002/9781394278695.ch14

Schauer, A., & Schnurr, D. (2025). *Fostering trustworthy human-AI collaboration through explainable AI and uncertainty quantification*. AEA Randomized Controlled Trials., DOI: 10.1257/rct.15730-1.0

Selwyn, E. J., Tiwari, A., Pandurang, G. G., Das, S. S., & Vardhan, R. (2025). Healthsyncc: An AI-powered mobile application for health management. In Modern technologies in healthcare (pp. 241–266). DOI: 10.1201/9781003481959-11

Shankar, K. (2024). Transforming healthcare: The role of artificial intelligence in revolutionising patient care. *MMJ*, *1*(4). Advance online publication. DOI: 10.62830/MMJ1-04-06a

Singh, A. (2025). LLM-assisted drug discovery: Examining the potential of LLMs in identifying new therapeutic targets, predicting drug efficacy, and streamlining the drug development process. *SSRN*. DOI: 10.2139/ssrn.5213574

Su, H., Li, G., & Fischer, G. S. (2018). Sensors, actuators, and robots for MRI-guided surgery and interventions. In The encyclopedia of medical robotics (pp. 201–231). DOI: 10.1142/9789813232303_0008

Thevißen, P., Santiago Wolf, L., & Stommel, D. (2023). *AI or human? Applicants decisions in discrimination settings*. AEA Randomized Controlled Trials., DOI: 10.1257/rct.12308

Wang, J., & Li, A. (2023). Revolutionizing healthcare with AI: Predicting future medication models and improving patient outcomes. In Machine learning & applications (pp. 25–35). DOI: 10.5121/csit.2023.131003

Wang, Z., Xue, L., & Wei, L. (2024). *Overcoming medical overuse with AI assistance: An experimental investigation*. AEA Randomized Controlled Trials., DOI: 10.1257/rct.13376-1.0

Whig, P., Dutta, P. K., Yathiraju, N., Bhattacharya, P., & Batra, I. (2024). Enhancing healthcare operations: Harnessing generative AI for efficiency and innovation. In Studies in systems, decision and control (pp. 273–287). DOI: 10.1007/978-3-031-75771-6_18

Whig, P., Dutta, P. K., Yathiraju, N., Bhattacharya, P., & Sharma, S. (2024). Transforming healthcare operations with generative AI: Optimizing resource allocation, workflow management, and predictive analytics. In Studies in systems, decision and control (pp. 289–308). DOI: 10.1007/978-3-031-75771-6_19

Zahra Ouariach, F., & Ahmed, S. (2025). Ethical and legal dimensions of AI in healthcare. In Computational intelligence in healthcare law (pp. 86–98). DOI: 10.1201/9781003538462-6

Zartashea, F. (2023). Bias and fairness in AI-driven healthcare: Addressing disparities in machine learning models. *International Journal of Science and Research Archive*, *9*(1), 835–846. DOI: 10.30574/ijsra.2023.9.1.0359

ADDITIONAL READING

Bohr, A., & Memarzadeh, K. (2020). *Artificial intelligence in healthcare*. Academic Press.

Ejiofor, V. O., Obieli, C. J., Gyasiwaa, M. Y., Shonubi, J. A., Maale, F. D., & Ilori, G. J. (2025). Leveraging artificial intelligence for predictive analytics and decision support in healthcare information systems: Enhancing patient outcomes and operational efficiency. *INTERNATIONAL JOURNAL OF ARTIFICIAL INTELLIGENCE IN LIFE SCIENCES*, *3*(1), 1–17. DOI: 10.34218/IJAILS_03_01_001

Elmassarawy, A. (2025). *The transformative power of AI applications in healthcare*. Ayman Elmassarawy.

Machine learning: The power and promise of computers that learn by example. (2017).

Manjula, D., Uma, G., Pradeep, T., & Nandhinidevi, R. (2025). Integrating artificial intelligence in healthcare for improved decision - Making, patient outcomes, and operational efficiency. [IJSR]. *International Journal of Science and Research (Raipur, India)*, *14*(2), 346–350. DOI: 10.21275/SR25201114043

Mehta, V. (2023). Artificial intelligence in medicine: Revolutionizing healthcare for improved patient outcomes. *Journal of Medical Research and Innovation*, *7*(2), e000292. DOI: 10.32892/jmri.292

Ogunmola, T., & Aloba, C. O. (2025). Revolutionizing cancer immunotherapy: Artificial intelligence accelerates personalized treatments and optimizes patient outcomes. *International Journal of Research Publication and Reviews*, *6*(1), 3353–3370. DOI: 10.55248/gengpi.6.0125.0512

Sakly, H., Guetari, R., & Kraiem, N. (2025). Scalable artificial intelligence for healthcare. DOI: 10.1201/9781003480594

Singh, J. P., Mishra, N., & Singla, B. (2025). Diagnosis to treatment. *Revolutionizing Healthcare Services*, 123-141. DOI: 10.1201/9781003531272-7

Singla, B., Shalender, K., & Stamer, K. Revolutionizing the healthcare sector with AI. (2024). *Advances in Medical Technologies and Clinical Practice*. Advance online publication. DOI: 10.4018/979-8-3693-3731-8

KEY TERMS AND DEFINITIONS

Artificial Intelligence (AI): A branch of computer science that focuses on creating systems capable of performing tasks that normally require human intelligence, such as learning, reasoning, problem-solving, perception, and language understanding. In healthcare, AI is used for diagnostics, treatment planning, and improving operational efficiency (Shankar, 2024).

Data Privacy: The right of individuals to control access to their personal health information. AI systems must adhere to legal and ethical standards to protect patient data from unauthorized access or misuse (Almeida & Barr, 2025).

Ethical AI: The design, development, and use of AI systems in a manner that aligns with ethical principles such as fairness, transparency, accountability, and respect for human rights (Farinu, 2025).

Machine Learning (ML): A subset of AI that enables computers to learn from data and improve their performance on tasks without being explicitly programmed. Machine learning algorithms are often used in predictive analytics and disease risk modeling in healthcare (Rachh et al., 2021).

Operational Efficiency: The ability to deliver healthcare services in a cost-effective and timely manner. AI contributes to operational efficiency by optimizing resource allocation, automating administrative tasks, and improving patient flow (Avaji & N, 2024).

Predictive Analytics: The use of statistical techniques and AI algorithms to analyze current and historical data to make predictions about future events, such as disease outbreaks or patient deterioration (Kosaraju, 2024).

Chapter 2
Artificial Intelligence in Marketing:
A Roadmap From Theory to Implementation

Parihar Suresh Dahake
http://orcid.org/0000-0002-2286-4120
Ramdeobaba University, Nagpur, India

ABSTRACT

In recent years, artificial intelligence (AI) has become more and more important for. AI in marketing is currently breaking new grounds and provides a broad variety of pursuing opportunities including marketing research, strategy, and action, advertising, customer acquisition, relationships, and experience as well as new trend, niche, and competitive analysis across conventional and new media channels, and the leveraging of social, environmental, and mobile marketing. Clustering in marketing of AI applications and practices is taking on characteristics of industry and market-specific omnichannels and vertical ecosystems. Also, AI may revolutionize marketing along the entire supply and demand side: customers, consumer behavior, and engagement, the marketing mix and customer value proposition.

1. INTRODUCTION

In this digital era, marketing theory and practice are being transformed by increasingly bigger and real-time data, by big data-driven companies such as Google, Facebook, and Amazon, and by their novel ways to analyze and exploit these data (Hermann, 2022). One of the primary analytic tools concerns Artificial Intelligence (AI). AI can be conceptualized as "the use of computational machinery to emulate

DOI: 10.4018/979-8-2600-2378-5.ch002

capabilities inherent in humans" and refers to "programs, algorithms, systems or machines that demonstrate intelligence." AI improves decision-making through real-time insights, pattern recognition, and personalization while addressing industry challenges like ethics and job automation(Campbell et al., 2020). Marketers' interest and investment in AI technologies have grown considerably, in particular in machine learning and natural language processing (NLP) analytics. In more methodological terms, AI can be defined as "a system's ability to correctly interpret external data, to learn from such data, and to use those learnings to achieve specific goals and tasks through flexible adaptation" (Kumar et al., 2015). AI has the potential to revolutionize industries like finance, healthcare, manufacturing, and retail, with its impact ranging from finance to healthcare to manufacturing to retail(Y. K. Dwivedi et al., 2021).In another attempt to specify the AI technology field definition, AI marketing applications (AIMA) denote an assortment of technologies and instruments allowing for data-driven insights, predictions, and decisions based on customer and market behavior analysis, preferences, attributes, emotions, and opinions related to online or offline interactions. Realizing the threats and potentials, AI in marketing, is taking new routes in research and is drawing the attention of practitioner's world- and nation-wide. AI playing a transformative role in tourism, emphasizing enhanced customer experiences through automation, personalization, efficiency, and innovative technologies(Samala et al., 2022). It is evident that the AI in marketing research, strategy, and actions is poised to redraft the scenario laid out by McCarthy's seminal 1960 '4P' marketing framework. The 4Ps (product, price, place, promotion) aim at maximizing economic functions, shaping a determinate value to be transferred from producers to consumers. To design effective marketing strategy and action plans in such complex, crowded, and hyperspeed global and domestic contexts, marketers can rely on quite sophisticated AI marketing decisions, insights, and predictions provided by a nascent profession of AI marketing consultants, expert in a range of applications and methods at the state of art.

2. THEORETICAL FRAMEWORK OF AI

This section provides information regarding some fundamental aspects of AI as a primer to the subsequent discussion on AI application in marketing. "AI" is the science and engineering of making intelligent machines, especially intelligent computer programs. AI is related to the similar task of using computers to understand human intelligence, but AI does not have to confine itself to methods that are biologically observable. IoT and AI in modern marketing, highlighting hyper-personalization, predictive analytics, improved customer engagement, and ethical data usage as transformative factors driving strategic advancements(N. S. Dahake,

Gajghate, et al., 2024).AI researchers have turned to other domains for insights. Cognitive science is the interdisciplinary study of mind and intelligence, mostly as studied by psychology, computer science, linguistics, neuroscience, and philosophy. AI researchers are interested in human learning and thinking in order to better understand how to design more effective learning algorithms and build intelligent systems. In the management field, marketing is the study of strategy planning and applications with respect to the sales, distribution, and promotion of goods and services. So far, the application basics of AI in marketing are presented (Hermann, 2022). Another part of this paper is also applied as the basis of the second part which discusses how AI, through its subparts can be applied to review, research theory, and real-world studies of marketing researches. All these would give a better understanding of how AI (marketing intelligence) benefits marketing researchers.

2.1 Definition and Scope of AI

In recent years, the field of artificial intelligence (AI) has made significant progress, so that the term is dominating arena of discussion. However, there is still a lack of clarity on the subject and often distorted misconceptions. This roadmap article is precisely identifying the scope of AI in the context of marketing applications and disclosing the necessary knowledge that marketing professionals should have to understand the possible advantages of AI in their work. First, the general basics and key concepts of AI theory are provided. After discussion, the concepts of AI's detailed applications in the marketing area are discussed, and illustrative examples are given.

The roots of AI as a theoretical field come from 1956, though the dreams of creating synthetic intelligence date back much further. AI decision-making in the era of Big Data faces challenges and offers twelve research propositions for information systems researchers to advance AI research and development(Duan et al., 2019). Since its inception, the dream has been to create intelligence that mimics every aspect of human intelligence. In the early years, a similar approach was followed, creating programs that do tasks that usually require human intelligence. For that, an initial study was carried out on how to create a program that can play chess, a strategy game and a good example of a problem whose solution requires creativity, intelligence, and manipulation of knowledge. However, the current situation encountered in many fields is different from the futuristic image that appeared in the cinema. As in the medical field, with the coming of the electronic circuits and digital world, the power of fast computations has become a significant tool used today even in domestic environments. In particular, over the past 10 years, developments in AI have made a particular acceleration, making its impact more widespread. Therefore, it has gradually entered non-advanced professional fields too. One of these is marketing.

So, starting by what is really AI is a critical part of this roadmap article. As marketing professionals gather more knowledge in this field, grasping this definitional part in order to analyze and understand gains and implications of marketing will provide an advantage. This can also be beneficial in establishing a common understanding and discussion platform between AI experts and marketers (Hermann, 2022). AI is a complex structure that combines many scientific disciplines. Therefore, the basic concept explanations are described in a categorical structure. Whether AI studies will be started from the theory of learning or looking at the day in the life of an AI model, presenting that how guiding in model improvement. Apart from this, AI can be viewed by giving a list of algorithms and models widely used by how to study knowledge of many concepts. The concepts and visual explanations are chosen to simplify the subject. On the other hand, there are examples for each concept that is understandable with simple mathematics and reasoning followed.

2.2 Key Concepts in AI

Conversational AI is a generic term introduced to refer to the machine intelligence and supplementary automated processes that enhance users' conversational experiences with a brand. Chatbots and virtual agents nowadays are programmed with conversational AI. However, there are also AI applications embedded in organizations' back-end computer systems that feed the chatbot or virtual agent with real-time answers to be delivered to convey solicitous consumer value. A machine learning framework for is very vital for customer purchase prediction in the non-contractual setting(Martínez et al., 2020). Another thrilling AI for marketing application, termed as conversational analysis, utilizes sophisticated natural language processing and machine learning algorithms to unravel, organize, and act on the voluminous amount of data flowing in and out of these platforms. . Usually, the output is a deep understanding of customer insights, channel conversation success rates, and efficiency, predicted customer actions or feelings after the interaction. As the AI involved is omnipresent and all-encompassing, building disruptive marketing intelligence frameworks for companies. Edge AI, or edge intelligence, is a promising new field that enables AI applications and services at the network edge, unleashing the potential of edge big data(Zhou et al., 2019) On this premise, the AI marketing roadmap corresponds to a joined and synergistic developing and implementing AI strategy, spanning: the core constructs of conversational and other AI in marketing, their array and operation in a novel taxonomy comprising real-world examples and the most recent industry users and trends, needed data, personnel, and technology readiness, and what lies ahead in governance, monitoring, and evaluation. Artificial intelligence (AI) has evolved from a primitive concept to a complex, flexible system capable of achieving goals and tasks through flexible adaptation(Haenlein & Kaplan, 2019).

There is a rich, growing literature in computer science, design, and Internet policies that target AI applications and its ethical, legal, and societal problems. Rigorous research studies tackle how to conduct effective and non-malicious AI development, deployment, and use to maximize positive externalities for society and minimize AI-related harm. Artificial Intelligence also emerges as a prime driver of the digital transformation of markets, with significant implications for data security and data protection legislation and policy making. In the context of AI's growing presence, the focus will be on how regulators, scholars, and businesses are dealing with the evolving regulatory framework of AI Ethics and its possible implications for marketing. Also, the novel development, 'datafication of everything', on how it enables modern AI applications and can upgrade marketing competitiveness, will be discussed. The discussion explains the potential impact on AI marketing, especially the most relevant provisions and ethical questions. Finally, a general outlook will be given on how to develop an ethical AI-fueled marketing insight.

2.3. Historical Development of AI

The term AI was first used in 1956. This term became popular in the 1970s encompassing task-oriented systems using logic, search, and heuristic algorithm as the basic tools for problem solving. Nevertheless, the approaches to developing the system were too ambitious, and the condition of computing machinery was not too advanced. The progress of AI research seemed to be slow. In the 1980s commercial applications of AI outside research labs began, such as expert systems and computer vision.

In the late 1990s, there was great progress in artificial neural networks and other machine-learning techniques. Since the turn of the millennium, there has been a significant rise in the availability of tools, platforms and data needed to develop AI-based marketing applications. Today, the combination of big data, faster & cheaper computers, and more sophisticated algorithms have made AI and advanced machine learning viable, enabling a broad set of systems in the form of marketing automation tools, predictive analysis and segmentation, among others(Keegan et al., 2024a). AI is no longer viewed as fiction but as a matter of possibility. Deep learning, a subset of machine learning models, has become feasible due to falling computational cost, rapid advances in hardware and the huge troves of data on the Internet. Artificial intelligence significantly change marketing strategies and customer behaviors, with its effectiveness being enhanced when it augments human managers rather than replacing them(Davenport et al., 2020). In fact Machine learning and deep learning are crucial for intelligent systems, automating analytical model building and solving tasks, but face challenges in human-machine interaction and AI servitization(Janiesch et al., 2021).

With the availability of the mentioned resources, the Artificial Intelligence Model Institute has been flourishing, giving rise to applications for a wide range of human activities, from computer games to fire control; from tax assessment to restaurant planning; from environment monitoring to medical diagnosis; from automatic control to teaching aids; from personal activities to large decision making. Consequently, efficiency and productivity are much improved. Recognizing the importance of this circumstance, the Govt. of India has passed the National AI Strategy outlining its vision for implementing AI and data analytics techniques across various sectors through policies, infrastructure, and supporting framework to make proper use of vast data generated in rapidly growing India. Former activities have been discerned in education and health services. AI research has also been announced in preventing encroachment on forest areas and preventing illegal mining on beaches.

3. RESEARCH QUESTION

Artificial intelligence and machine learning have received great attention in the emerging fields of marketing and have influenced a variety of industries and workplaces worldwide, especially the relatively advanced economies. Marketing is fundamental for different types of organizations as it has substantial economic value. As the transition of marketing practice from traditional to digital gains momentum, the influence of AI and ML in marketing is expected to increase further, and India is not an exception. The Indian economy with its global power and relations—especially its service industries, which play key roles in its economic growth and the development of society as a whole desperately needs to stand on the shoulders of these technological giants. Due to the complexity and the pace of changes, the proper management and employment of AI in marketing requires advanced knowledge and practical know-how. Though a number of works have been carried out in the application of AI in marketing, a consolidated framework from a theoretical and practical point of view and a detailed roadmap for India to develop and captivate the anticipated potential in these intermingled fields is missing in the current academic literature (Hermann, 2022). In response to the need mentioned above, it is the aim of this article to develop a consolidated framework for AI/ML in marketing and a roadmap by which the Indian government, industries, market researchers, and academics could be benefited. The roadmap suggests the necessary actions to understand the basics of AI, data, software/hardware, and their relationships as applied to marketing context; programs underpinning the development of theoretical knowledge and technical skills in AI and big data; and infrastructure needed to involve industries and market researchers to apply this knowledge in practice. This article fosters a multidisciplinary and practical roadmap that leverages AI/ML in marketing, with a special canonical example in view of developing a roadmap for India. It recognizes the advancement of the theoretical and conceptual foundations of AI, data, and software engineering as applied to the marketing domain, alongside the practical endeavours taken outside the academic frontier. This manuscript aspires to bridge this gap in understanding and readiness of the Indian tech and service industries to capitalize on expected AI dividend.

4. GAP IDENTIFICATION

Artificial Intelligence (AI) has the potential to be one of the most transformative and most disruptive technologies to emerge in the last decade. Businesses and marketers globally are adapting to AI quickly to take advantage of its capabilities. There is, however, a dearth of structured guides to help marketers adopt AI. This paper looks at the application of AI in marketing from theory to implementation, documenting the insights gained and the problems faced while implementing AI technology in marketing efforts in companies in India.

The approach to the study, adoption of AI in marketing, begins with a background of AI and its applications in marketing, followed by how AI is adapted to B2B scenarios. This will then pass to the research methodologies employed. The study then tells how AI technologies are introduced in marketing departments, strategies are developed, marketing processes are altered, and finally evaluations are completed. This is followed by a high-level roadmap describing how these stages come together in the development and implementation of an AI marketing strategy.

Although the tech industry has made great strides in adopting AI, B2B marketers qualify their marketing approach as "traditional", meaning it is based predominantly on direct sales, outbound marketing and traditional events, such as networking. The B2B marketing sector accounts for 50% of all B2B entities, and up to 70% of those are small to medium-sized, and marketers are not in a position to redesign their business models and processes to fit industry. As a result, the potential for viral AI marketing is not being realized. Personality or psychometric content is not being made use of due to lack of analytical skills, resources and tools. Modern digital outreach is also beyond the means of most B2B marketers. The preoccupation with event marketing is something of a legacy approach ingrained by experience. It is observed that online events, which promise the most cost-effective route to market, suffer from a relatively low return on investment. There is a cautious approach to adoption of novel practices and technologies among B2B marketers. Meanwhile, the techniques of digital consumer (B2C markets) are not easily replicable in B2B marketing given the professional nature of the transactions and interactions (James Keegan et al., 2022).

5. NEED OF THE STUDY

Artificial intelligence (AI) - commonly defined as systems that display intelligent behavior and perform tasks under complex environments without assistance from human intelligence - has become an extensive focus of science. The explainable artificial intelligence (XAI) and its applications, highlighting the need for better understanding of deep neural networks and their strategies in various applications(Samek et al., 2021). Despite the considerable research on AI and its growing roles and applications across industries, relatively little research on AI on the marketing side in the context of Indian agriculture has been identified. Up to now, much of the AI research has centred on the social and ethical consequences of AI technology in individual domains. AI methods, such as machine learning, pattern recognition, and deep learning, can efficiently solve complex structural engineering problems, save time, and increase computational efficiency(Salehi & Burgueño, 2018). On the other side, artificial intelligence is generally perceived as a potential strategic asset by companies to gain a sustainable competitive advantage. Thus, further marketing-oriented research on AI, especially the individual domain and country, will be helpful for both academia and practice. India, as one of the largest global economies, poses a suitable platform for the study of AI in agriculture, given the rapidly growing application of AI technologies in numerous domains. Also, regarding the marketing strategy, agriculture plays an important role in the rapidly transforming and modernizing Indian economy. From the marketing tactical perspective, the liberalization of the Indian economy has significantly influenced the agricultural marketing strategy in India. Now, centralized systems of procurement by governments are continuously being guided towards market-driven systems allowing the private sector and foreign companies. However, little has been emphasized on marketing strategy in agriculture, more particularly the role of artificial intelligence in Indian agriculture at macro, meso, and micro levels. That builds a persistent need to be addressed.

6. REVIEW OF LITERATURE

One of AI effects main goals is to map the evolving field of AI in marketing and to build an agenda for AI in marketing that is able to guide the structuration of relevant research and the implementation of best practices in the field. An analysis of the Most Relevant Terms includes the title of the 330 most prominent keywords derived from the abstracts and title texts included in the database in the period 1980–2020, as AI in marketing or as AI in adjacent areas with strong connections to marketing (such as advertising and consumer behaviour). Font size represents keyword importance, with the largest words being more important.

Figure 1. Countrywise Coupling Analysis Vosviewer

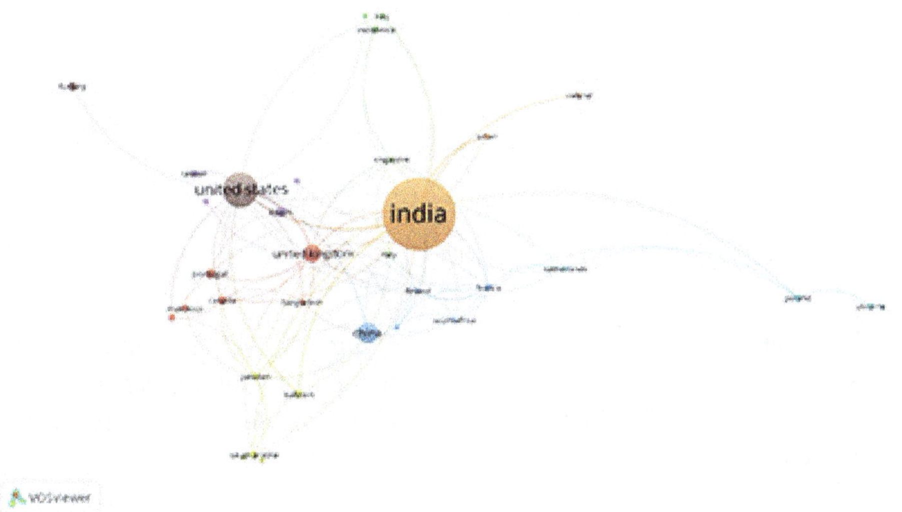

Co-occurrence networks of the most relevant terms were filtered to reduce the noise in the network and focus on the strongest connections among the keywords. Lines are weighted by strength (number of times that keywords appear in the same record) and are made up of individual keywords (rather than being connected to a group of terms).

Figure 2. All Keywords Analysis Vosviewer

Coloring is applied according to cosine similarities, which ultimately results in clusters of keywords that are conceptually similar and, therefore, they represent cohesive thematic areas. The intensity of the color indicates the strength of the thematic cluster.

Figure 3. Author Keywords Analysis Vosviewer

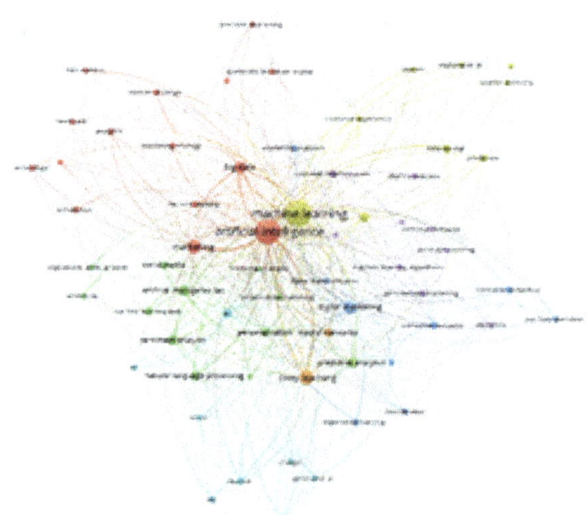

The overall map was thus based on the strength of the co-occurrence proximity in the same record of title and abstract text records. The central nodes in the map correspond to the 26 most relevant terms. They play a central role in the map because they appear in articles that are the most relevant among the records altogether.

Figure 4. Index Keywords Analysis Vosviewer

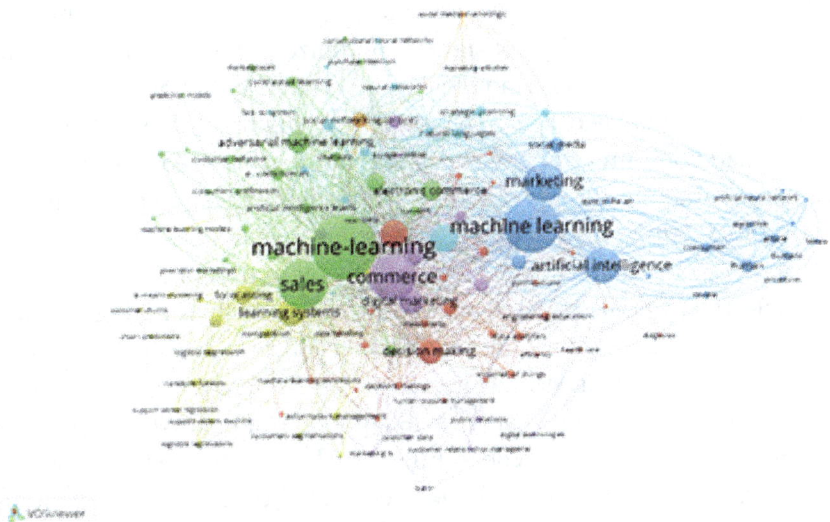

These records were used to identify the articles that are closest to them in the conceptual space of all keywords. Regarding this article, central nodes can facilitate the thematic structuration of text networks by pointing to the main conceptual areas explored (AI in business, AI in healthcare, AI in retailing, and consumer behaviour). These main fields of interest further spill over the development of the theoretical frameworks for the work by pinpointing the knowledge gaps or areas that are worth further investigation. A detailed list of the 26 central terms is provided to clarify why they play essential roles in the map. In this digital era, marketing theory and practice are being transformed by increasing complexity due to information availability, higher reach and interactions, and faster speeds of transactions. These have led to the adoption of intelligent agent technologies (IATs) by many companies. As IATs are relatively new and technologically complex, several definitions are evolving, and the theory in this area is not yet fully developed. This paper proposes a marketing-centric definition and a systematic taxonomy and framework(Kumar et al., 2016) also, (Russell & Norvig, 1995) offers a comprehensive introduction to artificial intelligence for modern applications, covering search methods, game playing, reasoning, planning, and decision-making, In the similar vein (Barredo Arrieta et al., 2020) proposes a novel definition of explainable Machine Learning and a taxonomy for explaining AI models, aiming to advance the field towards responsible AI implementation in real organizations. (Miller, 2019) Consequently Artificial intelligence should build on existing research in philosophy, psychology, and cognitive science to improve its understanding of human explanation processes. Leading to this (Huang & Rust, 2021a) emphasizes AI application in marketing research, segmentation, targeting, positioning, and actions across marketing's 4Ps/4Cs, The integration of Artificial Intelligence (AI) in marketing represents a transformative shift in how businesses engage with consumers, optimize processes, and innovate strategies. The literature on this subject reveals a diverse array of applications and implications, beginning with (Kose & Sert, 2017) who highlight AI's potential to enhance content marketing processes. Their study outlines how AI can facilitate intelligent content marketing through current web technologies, emphasizing its role in improving marketing effectiveness and efficiency analogously (Soni et al., 2019)expand on this foundation by exploring the broader impact of AI on business models. They identify key areas affected by AI integration, such as innovation and market deployment, and underscore the significance of deep learning advancements that have revitalized interest and investment in AI across various sectors, including marketing. This comprehensive review provides a contextual backdrop for understanding the transformative capabilities of AI in the business landscape. Furthermore (Braga, 2021) narrows the focus to consumer decision-making, comparing generational differences in the adoption of AI tools like voice assistants and chatbots. His findings reveal how marketers are striving to create intimacy between AI technologies and consumers, indicating a shift in marketing strategies that leverage AI to enhance user engagement and service delivery. In the same vein (Hermann, 2022) addresses the ethical dimensions of AI in marketing, conducting a systematic review of existing literature to map the ethical principles relevant to AI applications. This critical examination of AI ethics not only sheds light on the responsibilities of marketers but also emphasizes the need for ethical considerations in the deployment of AI technologies. It Should be noted that, (Keegan et al., 2024a)delve into the implications of AI for traditional B2B marketing practices, highlighting the role of AI in automating strategic decision-making processes. Their analysis points to the significant enhancements AI brings to marketing functions, while also revealing gaps in the understanding of marketing managers' perceptions and the challenges posed by AI adoption.

(Varghese et al., 2022) contribute to the discussion by analyzing AI's influence on decision-making in the context of big data. They explore how cognitive algorithms and immersive technologies shape consumer behaviors and engagement strategies, further illustrating the multifaceted impact of AI on marketing. (Nelson et al., 2023) provide a broader perspective by examining the societal implications of AI applications in manufacturing, which parallels marketing in its reliance on data-driven insights and predictive analytics. Their systematic review highlights the diverse applications of AI across sectors, reinforcing the interconnectedness of AI's role in business processes. Eventually (Haleem et al., 2022) explores the transformative impact of Artificial Intelligence (AI) on marketing, focusing on customer insights, personalization, and automation. In Fact (Mustak et al., 2021) systematically reviews AI's impact on marketing, identifying ten key themes like consumer sentiments, customer trust, and strategic marketing. AI aids in optimizing marketing practices through automation, emotion analysis, and data-driven decisions by given that (Mariani et al., 2022) highlights AI's influence on consumer behavior and decision-making, offering automation and data-driven insights to optimize marketing strategies, Furthermore, (Kumar et al., 2021) analyzes four transformative technologies IoT, AI, ML, blockchain emphasizing their adoption impact on marketing outcomes and strategies, Building on these premises (Noble & Mokwa, 1999) developed a managerial theory to understand the factors influencing the implementation of marketing strategies, and tested it in a study of mid-level marketing managers in two organizations. Also it should be noted that (Zhang & Lu, 2021) provides an overview of artificial intelligence, highlighting its background, drivers, technologies, and applications, while providing insights for practitioners and future study.as stated differently (J. Lee et al., 2018) explores the current state of AI technologies and the ecosystem needed to harness their power in Industry 4.0-based manufacturing systems. Together, these articles illustrate a comprehensive landscape of AI in marketing, highlighting its transformative potential, ethical challenges, and the necessity for a human-centered approach in its implementation. The synthesis of these insights provides a roadmap for understanding the evolution of marketing practices in the age of AI. Artificial intelligence (AI) powers many marketing applications, automating repetitive tasks and personalizing the customer experience. Despite the potential and growing application of AI in consumer marketing, there is still limited understanding in academic circles, especially in the context of the fast-growing economy of India. We address this research gap and provide a roadmap supporting academics and practitioners. This study first provides a systematic review of the extant literature, examining the current state of the art, and highlighting opportunities for future research. Next, the study reports the findings of a qualitative study with 18 marketers in India, investigating the enablers, challenges, outcomes, and ethical implications of AI in marketing. Insights distilled from these two components of research are synthesized into a roadmap offering a structured overview of the theoretically and practically most meaningful pathways from AI knowledge to its practical application.

7. ARGUMENTATION BASED ON THE KEYWORDS ANALYSIS OF VOSVIWER

INDIA Implementation of Artificial Intelligence in Marketing Research INSTITUTIONAL SETTING AI and ML have become ubiquitous in marketing due to their capacity to store information, transform it, and provide the best solution to a condition. AI-based technologies have grown be integrated into pre-existing systems seamlessly. An Amazon customer, for example, is accustomed to getting personalized suggestions based on previous purchases and searches. Predictive algorithms rapidly evaluate sales performance as new data arrives. Meanwhile, spatial sensors identify how customers navigate through a populated commercial space, allowing the analysis of where things are going right and wrong. Lattice Engines executed utilization of AI-based data-mining technologies effectively with multiple non-competing clients applied to CRM systems hosting the raw operational and demographic data across a wide base of customer segments. The AI can be tailored to learn who custom software. The assignments can then be implemented back into the CRM system workflow so that marketing can then execute campaigns based on these models. Non-competing clients increase the success of the task because training data can be gathered from diverse sources differently. New features can be incorporated, and solutions tailored based on customer sector and geography can be created. Meanwhile, competition cannot make the same use of a single integrated system. Winchester College's new tools were designed especially for the school and informed marketing and recruitment strategies as narrowly as possible, meanwhile recommending general approaches to retention improvement that were valuable to all schools. It could help businesses jump-start utilization with reduced costs and lower associated risks

8. AI TECHNOLOGIES IN MARKETING

AI applications in marketing aim at data-driven personalisation and customisation of products, services, and the marketing mix variables along the entire customer journey to engage customers and optimise experiential value (Hermann, 2022). Survey findings highlight that artificial intelligence marketing services have commenced to escalate in adoption, and developers see the significant potential and opportunities, but supporting the education market obsession, is a strategic requital plan to the pandemic, and it should be employed. Tech unparalleled players have expedited their AI investments and market entry on a global scale.AI personalizes strategies, predicts trends, optimizes budgets, and enhances customer targeting(Ziakis & Vlachopoulou, 2023)

The activities of AI have been classified for an enhanced arrangement and conversation, L1 "Introduction of AI," L2 "Selecting the AI activities of developers," L3 "AI road metaphor for developer," and a category of L4 "AI environment for developers." The smart city era is possible for intelligent marketers, and an understanding of AI is necessary for the era, and education is suggested as the target market of AI developers, which their feed-forwards must be strategic inputs "AI support for education market" obsessed. In a global scale, viewing the recent key events and enabling them with an AI-BERT big data approach, survey findings refer to a sudden trend of AI strategies. Regarding developers (providers of AI services), a "Introduction of AI" preference algorithm and §AI) usage of the big data analytics mostly used standard AI technologies and AI GPT variations. With increasing AI activity levels, a growing interest, seeing it as a panacea or the new marketing strategy, would expect a more sophisticated application, and thus, the "Selecting the AI activities of developers" could help the organisations on a gradual and well-prepared AI adoption.AI's focus on predicting customer behavior and optimizing marketing mixes in initial and tactical stages. It underscores AI's transformative potential to enhance decision-making, offering recommendations for marketing practitioners(Ljepava, 2022)

In this regard, selecting and commissioning market research and a comparative approach has been developed and recommended to enhance effectiveness (James Keegan et al., 2022). Eventually The interconnected nature of these elements within the ecosystem and their role in optimizing touchpoints, deriving insights, and supporting workflows(Malthouse & Copulsky, 2023). Future research directions include privacy concerns, stakeholder interactions, and the evolution of AI in marketing practicesCurrently, AI marketing solutions have already picked up, and notable big techs came out with a user-friendly way to implement AI in their marketing platforms. For instance, AWS and Sagemaker have been utilised for simplifying the building of AI models and its application shell architecture. In a relational approach, the domestic market rhythm and its catching up position are outlined with related AI-consumer perspectives and effects on the marketing strategies while focusing on the "host" country in this sense India.

8.1. Machine Learning

Machine Learning in Marketing includes the practice of machine learning techniques on marketing strategies. In recent years, this strategy has become increasingly popular due to its high accuracy rate and lower cost compared to other data mining or artificial intelligence (AI) techniques (Bayoude et al., 2018). The Machine Learning in Marketing market is influenced by many distinct variations. 3.1 Machine Learning Out of the fundamental marketing toolbox required for any businesses there have been many means to centralize on the classification and predictive data analytics for realizations out. Machine learning practices have been burgeoning in their rise for data discovery projects with enterprises having begun to invest in AI as well as looking to use the art of models for optimization of consumer statistics. In this literature, a print is presented to paint on how machine learning potentials are transforming the more workmanlike of the digital marketing exercises. From the decision model's consumers, prescriptive suggestion engines are rising for campaign optimization in discretionary assignments such as spending, content e-mail, electronic media copy it, in addition to other personalization suggestions. Apart from the use of clustering for sequester examination and inventory segmentation, machine learning methods are evincing beneficial in the practice of recommendation systems. Video, digital image plenipotentiaries, and the unanticipated achievement of machine learning structures underpin industrial projects as streamlined as influencer marketing, and with the success of the identification of faced gadgets, further blockchain ramifications may be seen. Machine learning on marketing tactics is not new, having actually existed for approx three decades, but recent developments on processing power, algorithmic advancements, together with the explosive growth in industries regarding big data and a wealth of fresh marketing numerical functions to analyze, have formulated a fast-evolving modification in the state of machine learning in marketing. Varied by domain, applications, likewise algorithmic sophistication, a summary markets analysis is provided. At present, a majority of particular industries do not have the technological wherewithal or the desire to commence further than the subsection application. The service outlines a guidance on entering the machine learning marketing strategy for further promotion.

8.2. Natural Language Processing

Artificial intelligence (AI) has completely transformed many sectors in the past decades – retail, healthcare, banking and office setting, are a few of the prominent ones. However, the implementation and use of artificial intelligence in the marketing sector is still lagging. This text provides a comprehensive report of not just what AI is, but a complete roadmap on how it can be implemented for the marketing sector in India. The second section covers the wide span of AI. The third section discusses the various tools of AI, and how these can be used. The fourth section covers the specifics of the use of specific AI models and machine learning in marketing in India. And this fifth section contains the survey of AI usage among consumers in India. Concise yet comprehensive, on conclusion of this text, a reader not only knows what AI is, but has a complete comprehensive guide and framework of how to use AI for the marketing sector in India, from theoretical backings to real-world implementable practical knowledge.

8.3. Computer Vision

In the world of marketing today, one thing that has become incredibly essential is having visually appealing content. Due to the plethora of devices proliferating the mass media market, such as high-resolution screens in devices ranging from mobile phones to tablets, and wide-known initiative to bring reasonably priced connectivity and devices available for people in emerging nations, the importance of engaging visual content is greater than ever before. It is for this reason that visual content has been at the forefront of a lot of company marketing strategies in recent years(Mayahi & Vidrih, 2022).

In particular, the last decade has seen an enormous push in visual marketing from many social media platforms. Especially, the rise in popularity of such platforms as Instagram, Snapchat and facebook has seen the distribution and creation of visual medium as the dominant form of information exchange on the net. As cloud infrastructure has become more easily available, it has become adaptable to store images and videos in the cloud, enabling visual interaction to continue to rise across the world. Paid social is now the go-to marketing strategy for many businesses with high quality visual content undoubtedly being the driver in viewers engaging with the advert or content and becoming shared; cloud services and the instantaneous nature of the net mean that putative high quality content often comes from processed and professional sources and the knowledge to shoot and run high quality visualised content is neither available nor within budget for smaller companies with immensely remotely stationed target markets.

8.4. Robotic Process Automation

Robotic Process Automation (RPA) is the system made for automating monotonous, high-volume, and time-consuming jobs. The system comprises many capabilities that together help in segregating and as well in some cases entirely computerizing jobs by accessing the user's interface for manipulating data and leading through programmed actions. Automation encourages performance through the utilization of robots developed by scripting and competing at an introduced application's GUI. These systems can be setup be it both front and back office of an organization. In the US, there is a projection that Robotic Process Automation (RPA) would do away with approx 16% of workforce and create a hike in national income by 0.3% by the early 2030s. Though, assessing the Robotic Process Automation case study in India, the situation is quite different due to the political and legislative issues which could make India resist for automation. jobId is provided along with automation, automation can schedule unattended robots to run the process-sets and robots do the certain tasks for entirely occupied in it. The "Open Span" setting, user has the capability to create jobs and also to form the employment arrangement appointing the hours that the Bot ought to work each day in a week, and in this as well, the Robotic Process Management (RPM) orchestration aids a program that is to be scheduled and implements the models accordingly (ANAGNOSTE, 2018). Additionally, at first while the action has been completed, the executive can be oriented, tasklist and assign the work to the relevant team and this kind of upshot assists in the prevention of backlogging. The outcome provides the descriptive analytical characteristics of the RPAM using the RPA Tool Congnizant Process Automation (CPAP) IDE solution and the extra operations that have been performed to accomodate the 02 main sets of activities. The complete automation screen-shots for the initial and approach runs are also displayed. The assimilated setting utilizes the simple setup method where every data preceding to an execution like column labels, row labels, and etc. should possess precise Phone setup entered in the rundown list and the setup file also is comprised. It emphasizes the real-time process of RPA in Customer Service business area and spotlighted on the employment arrangement of Robots through RPAM(Gajjar et al., 2022)

9. APPLICATIONS OF AI IN MARKETING

The results disclosed that a consensus among Indian marketers regarding the roadmap for implementation of AI in marketing cannot be reached as there are vast contrasts in factorial loads across the factors and unclear cut-off values. Not sequencing the factor groups seems to be an oversight, the lack of discussion and appropriateness is a limitation (James Keegan et al., 2022). The perception of technology is proposed to consist of well-being, AI video, AI audio, AI headset, and futuristic hologram projection. The adoption of technology is proposed to consist of AI implementation challenges, AI ethics, AI marketing applications, AI marketing analytics, and AI marketing strategy. Emotional Intelligence in Marketing is proposed to consist of actionable AI marketing strategies, AI marketing education, HI, and CI in Marketing, and MI-T starting strategies. The moderation of market turbulence is proposed to reflect CVI, CVT, CVT in marketing, and IPLM 400 actions.

Artificial intelligence is one of the newest and most important marketing strategies that, when analyzed and interpreted correctly, will allow companies to derive details essential for strategic planning. Several marketing activities in a marketing firm may have data implications, and this study explores and predicts the use of AI (Hermann, 2022). By using AI techniques for deep learning, activity guessing is also conducted behind the scenes. This provides a holistic view of upcoming marketing activities to help deepen interpretative insights supporting information technology development, informing managers on their selection of relevant technologies, anticipating knowledge growth, and prompting strategic preparation of technology implementation.

9.1. Customer Segmentation

Customer segmentation is a widely used approach to group customers into different categories, which has shown its importance to companies' bottom line. With the continuously increasing of mobile phones' capabilities, the services provided on mobile phones now are more numerous and concise. In order to access these services, operators also put large sum on advertising to attract people's attention. Accordingly, in this paper partnership is developed with an operator that provides on-phone services to perform basket analysis on users' actions around banner ads. The data covers three recent days' worth of user's behavior to study its purchasing activities. The task is to predict which banners user will click on and improve the conversion rates of banner ads. A new user features based on the number of clicks on basket or the inter-arrival time between basket click and the clicks on banner is introduced. These novel user features significantly improve the performance of the model being able to separate high-value user from others almost one hour in advance. On the other hand, a novel RFME model is proposed to concurrently track the static and the behavioral features of customers. The dynamic feature of RFME improves the prediction accuracy of LTV models by more than 25%. Finally, the proposed pipeline is trained and tested in an end-to-end manner (Kumar Pandey et al., 2024). Operators typically offer various discounts and package deals in an attempt to induce customers to both retain the service and increase data revenue. To this purpose, customer segmentation is vital in order to obtain suitable offer sets targeted to each customer segment. In one exploitation approach, operators apply the proposed customer segmentation to cluster the lung customers and respectively plan marketing strategies, and also make experiment to lunch discounted offers specially devised for these segments to promote user shop. The results show that with the proposed approach, the click and purchase rates of banner ads have a significant increasing trend and obtain a few more high-value users, whereas without, customers on the analysis just keep shopping as they used to be. Nowadays, marketing is one of the high cost activities for any online platforms. In order to optimize marketing campaigns, customer segmentation is playing a fundamental role.

9.2. Predictive Analytics

Predictive analytics represents a statistical technique combining historical data, predicting future events based on such information. In marketing, predictive analytics utilizes data mining models, sifting through data for answering queries regarding customer behavior, in this case purchase history. Predictive tools and emphasizes the need for realistic evaluations to ensure tangible business benefits from data-driven decision-making(Dahake et al., 2024). Sales trends enable retailers for optimizing their resources, such as aligning their inventory with higher profitability merchandise and reducing data inventory. The hierarchical nature of sales data encourages the necessity of forecasting sales at multiple aggregation levels. The synergy of AI and big data, revolutionizing industries through predictive analytics, automation, customer-centric solutions, and efficiency. It emphasizes AI's ability to handle complex business challenges like fraud detection, trading, and personalized recommendations, driving transformative growth across sectors and enabling strategic decisions for modern enterprises(Ghimire et al., 2020). Predictive Analytics forecast consumer preferences, optimize product alignment, and enhance marketing strategies for improved competitiveness(P. S. Dahake, Parchake, et al., 2024)It would not only benefit manufacturers or retailers at a high organizational level, but would also gain benefits at lower levels by assessing daily sales. A stock-keeping unit (sku) is described as a unit of product with unique attributes identified by a part number, enabling sellers to uniquely distinguish the item. The owner of the retail outlet of a franchise of Apparel Tour Ltd receives weekly sales data regarding 0030 different stock keeping units (Bayoude et al., 2018). An artificial intelligence framework was suggested, automating an end-to-end predictive analytics process. First data, then base (DMB) along with rapid-miner tools were utilized for tackling apparel sales forecasting utilizing a mobile platform, contributing comprehensive explanation of the process.

9.3. Personalization

Imagine now a space of probability distributions, one for each desired representation. Given a user's information need and past interactions, it is highly probable that a user's next interaction will be with a different representation of the system. It may be inferred that traps are regions of low probability in this space, while those other representations are regions of high probability. To comprehend the person of this spatial representation, consider a user going to a page at which point the user can choose either to link to page Pil, or a competitor page, Pil could be designed in such a fashion that if a user was to go to it, the user would become highly probable to visit a distribution that was "designed" to contain traps, and which led the user to leave the site altogether(Ramakrishnan, 2002) and personalized engagement, ethical considerations, and technology's impact on customer relationships(Wang, 2023) In this way, competitive sites or services are valorized over the users. the transformation from traditional segmentation to AI-driven methods that predict behavioral patterns, personalize strategies, and optimize campaign efficiency.

9.4. Chatbots and Virtual Assistants

The teaching of noticeable advances in Information and Communications Technology (ICT), particularly online news, has led to improved language teaching and a more diverse range of instructional methods. Nonetheless, proper use of machinery benefits learners who have the knowledge to operate them. Technology allows learners to advance at their speed with their diverse set of abilities (Varghese et al., 2022). However, knowing when and how to integrate it with core educational programmes is difficult for educational officials, administrators, and language practitioners. The introduction of AI-based language learning technology will profoundly enhance the affordability and diversity of learning. Growing research in AI and language learning and instructing suggests that language acquisition will be fostered considerably.

For example, the artificial intelligence chat programme enables learners to rapidly construct written language. This is consistent with the work of, who opposes writing like reading which has a reciprocal procedure. Language acquisition increases significantly by creating collaborative language production. Internet-based language learning applications have grown rapidly. To maximise study results, it is crucial to identify the factors that encourage the students' long-term use of these resources. This software provides a versatile method which, in contrast to a traditional classroom, enables students to pursue their instruction in a more portable and self-organised way. A study found that as underachievers enhance their language skills, AI-supported web-based language learning programmes should promote their engagement in writing behaviour.

10. CHALLENGES OF IMPLEMENTING AI IN MARKETING

India represents one of the most enticing and growing economies in the digital context. This fact has not been overseen by industrial and institutional players in the country, who are willing to invest and exploit the potentialities of Artificial Intelligence (AI) technologies in the frame of marketing and related practices. In this paper the landscape of the usage of AI in the frame of marketing is unraveled as it is developing in India. Afterwards, a roadmap is delineated showing the next steps that should be addressed in the following years to overcome the current limitations.

An Activity Theory framework is exploited to interpret the existing conditions and derive insights on the actions to undertake to overcome the current limitations in setting the basis for successful AI campaign.AI's potential to boost satisfaction, loyalty, and revenue through innovative and data-driven marketing approaches(Durai et al., 2024).

Industrial marketing activities represent a chain of operations entailing the management of various dimensions of company-client relationships in a structured and planned manner. Since the start of the millennium, there have been several attacks criticizing a specific market feature, which has been dubbed skepticism. In this sector it is assumed that actors might take part in potential marketing operations on a non-verbal basis, but there is a complete lack of structured and effortful commercial transaction. Traditional marketing operations are then restricted to structured commercial operations, with a preferential focus on buyers and suppliers. A burgeoning debate has stemmed about these criticisms, and a variety of counter-appraisals have been proposed. The bulk of this debate has been carried within the strategic marketing field. On the other hand, there is a limited body of work on the actions that marketing actors and companies can deploy to enhance the quality of such business opportunities (James Keegan et al., 2022).

10.1. Data Privacy Concerns

Artificial Intelligence (AI) applications significantly improve marketing performance and boost the substantial economic benefits for companies and countries. the role of AI-assisted social media marketing in enhancing SME performance, customer acquisition, and business management(Basri, 2020).However, the knowledge of converting marketing theory into AI implementation is limited, especially in India. Based on a broad and scientific literature review of marketing theory and AI-coordination case studies, this provides a five-stage roadmap for the successful integration of marketing theory with AI applications in Indian companies and organizations. This roadmap starts with a baseline analysis of a marketing strategy, a marketing-mix strategy, and a target customer strategy. AI applications in digital marketing, emphasizing content curation, predictive analytics, dynamic pricing, and enhanced ROI strategies(Nair & Gupta, 2021). Afterward, it roves on to AI-derived target customer profiling followed by AI-determined product attributes. At the fourth stage, the AI-determined customer-group strategy is then elaborated. The last stage is the concurrent utilization of various types of AI applications to facilitate automated marketing implementation. By Cognitive insights derived from facial expressions and neuro-signals, offering managers tools to enhance customer engagement and boost conversion rates(Srivastava & Bag, 2024), thorough the AI There are several critical open ethical and privacy questions that accompany the use of data mining and predictive modeling techniques. By enabling localized data processing, Edge AI minimizes reliance on cloud storage, reducing vulnerabilities associated with personal and confidential consumer data(Sachdev, 2020) Through effective use of data profiles derived from smart band and mobile phone usage data, companies can use analytics to improve individual consumers' shopping experiences. AI distinguishes fake news from real news by analyzing emotional appeal, revealing negativity in fake news content(Paschen, 2019). Although AI applications significantly enhance marketing performance and achieve considerable economic benefits, the implementation of AI still lags in conducting marketing practices in companies, especially in India. Despite the outstanding performance of AI, the sustainability goals have received little attention in AI marketing. In addition to these environmental goals, social goals have also been included in the sustainability objectives and social stability and social inequality have been introduced (Hermann, 2022).

10.2. Integration with Existing Systems

The following need to be addressed with AI adoption to smooth and secure deployment. A framework to steer and optimize AI investments: There are many options for AI technologies and vendors but there needs to be logic, rather than intuition, to prioritizing, selecting and rejecting certain of these. Steps to integrate AI solutions with existing IT systems: Systems have to continue to inter-operate, security and privacy concerns must be managed, as well as the implications of issues characteristic of the IT context, such as lock-in. Establishment of ethical guidelines and oversight for AI: Marketing activities, such as targeted advertising, can raise concerns about customer manipulation. This can be addressed by ethical guidelines and oversight for marketing.AI Playing a pivotal role in marketing by automating tasks, enhancing personalization, and addressing ethical challenges effectively(Rivas & Zhao, 2023)

In this digital era, marketing theory and practice are being transformed by increasing complexity due to information availability and higher interaction among stakeholders at the transaction and exchange levels. AI & ML-powered chatbots' ability to provide real-time customer support, marketing assistance, and enhanced user engagement(Kushwaha & Kar, 2020). This research underscores the framework's capability to overcome data scarcity challenges and empower businesses to adopt interactive AI-driven tools effectively AI-driven model predicting review are very helpful the Key findings include credibility's significance over ratings and sentiments of the consumer(M. Lee et al., 2021) .Overwhelming information from various sources, such as Web analytics, social media, and sales transactions, and constantly changing dynamics are affecting marketing processes (Kumar et al., 2015). Moreover, interactions are occurring at faster speeds between stakeholders, such as companies and government regulations, universities and students, and buyers and sellers. There is a growing stream of research and technical capability being developed to address these concerns. Marketing faculty members at academic institutions in the United States and Canada ranked intelligent agent technologies (IATs) as the third most important topic for research and teaching. The primary purpose of this study is to provide structure and guidance for the general and broad marketing community to comprehend and further develop research. It provides a marketing-centric definition and a systematic taxonomy and framework to understand and investigate organized applications of IATs in marketing.

11. SUCCESSFUL AI IMPLEMENTATIONS

During data collection within organisations and the research community, marketers can rely on AI approaches for scaled selection and survey of samples derived from specific demographics and deidentified sales data analysis for performance insights overlooked by internal applications. AI advancements in efficiency, computational power, and specialized applications(Talib et al., 2021) AI-led data-driven product development is then possible when computer models are trained on these diverse sources for optimisation. This might involve the prototyping of draft advertising rationale generated by models fine-tuned to predict KPIs. Given this experience, whatever use case brought by the client, an impressment of competence for attentive problem-solving goes a long way to placate stakeholders. In simpler terms; listen, advise and assuage. Not only AI But also predictive analytics in marketing, showcasing its ability to forecast consumer preferences, optimize product offerings, and enhance strategies for improved competitiveness(P. S. Dahake et al., 2023) Clients will vary greatly in their prior knowledge, interest or enthusiasm for AI technologies, so it's important to bridge that gap. More seasoned customers can tackle the nitty-gritty of their data in a way casual users won't, yet such exposure might leave them fretful about infringed IP. businesses can adopt chatbots without historical data, leveraging user-generated content for improved customer engagement(Kushwaha & Kar, 2024), Decode models' internal weights using techniques to assure understanding doesn't compromise a black-box' apophenia. This will spur curiosity since laypeople might be familiar with 'regular' ML but less so all the cool and whacky new stuff.

11.1. Emerging Technologies

Over the past 50 years, artificial intelligence (AI) has gathered support for both utopia and dystopia. Although the EU is increasing its investment in AI policy, individual countries are actively creating AI policies. For example, the UK, France,

China, and Japan are targeting different strategies to secure and lead the emerging AI industry. As a part of these ongoing efforts, India is planning to guide the industry policy of its AI and data analysis further in 2025/2026. Considering India's huge potential for AI development, the following strategic directions are proposed, heading from survival to leadership:

i) Reinforcement of intelligent software production capabilities;
ii) Reforming the industry policy toward ai- and data-intensive sectors;
iii) Transparent strengthening of the institutional structure of AI policy; and
iv) Intervention to improve the market dynamics surrounding AI technologies.

By adopting these directions, India wishes to effectively confront global competition in the emerging sector while roughly addressing its long-standing socioeconomic challenges. An AI capability in a firm leads to increased organizational creativity and performance(Mikalef & Gupta, 2021). Artificial intelligence includes numerous technology platforms that allow systems to perceive, act, learn, and reason. Advances in computing power, progress in algorithms, and the availability of digitized data have led to AI achieving impressive results in various domains. AI also offers promising marketing applications, such as smart advertising, consumer behavior prediction, automatic chatbots, recommendation systems, and custom content filtering.

11.2. Impact of AI on Consumer Behavior

Research on consumer behavior can benefit and empower AI innovations in marketing like consumer behavior in an era of algorithm-driven marketing changes, the implications for marketing decision-making, societal well-being, or consumer welfare and privacy applications (Sun et al., 2022). Marketers' decisions shape the thinking of AI models, affecting the type and amount of advertising content that consumers are exposed to and thus influence consumer behavior. Knowing the mechanisms and effects of these cycling driver-receiver relations helps to achieve a better understanding of the wide-ranging consequences of AI innovations in marketing on individual consumers and society as a whole. Furthermore, research on these topics can provide a starting point or support to the development of concrete measures, best practice guidelines, normative recommendations, self-regulation, or responsible AI frameworks. Given the current development and the rapid pace of innovative AI applications in marketing, an interdisciplinary approach is suggested—a solid collaboration and knowledge exchange between computational social science, marketing, and computer science would be particularly beneficial and have a substantial impact. The digitalization of products and services fueled the generation of copious amounts of data, which laid the foundation for the application of artificial intelligence in marketing. With the widespread use of artificial intelligence, the body of marketing decisions has been reshaped. How marketing invests in AI, and how AI models are framed, training data is selected, pre-processing is conducted, or outputs are used has been shown to have unintended negative side effects, promoting the acceleration or exacerbation of racial inequality. New AI-powered technologies significantly transform the customer experience, impacting each stage of the shopping journey and creating experiential value(Hoyer et al., 2020).Artificial intelligence technology manages the handling and studying the data of disturbance consumer trends to recognize and solve problem more efficiently and effectively, which can enforce the performance and quality of decision making abilities there by games in consumer communication behaviors.

11.3. Ethical AI Practices

Artificial intelligence has unprecedented implications for marketing and promises vast potential for improving marketers' capabilities, performance, and impact across the marketing process. Utilizing a mixed-method research approach, this study presents a quantitative study and a qualitative study designed to understand the current state of artificial intelligence in marketing and to provide a comprehensive roadmap for applying artificial intelligence in marketing practice. The study deviates from the current discourse, by not just exploring the numerous opportunities and implications, but also considering implementation challenges to assess the state of artificial intelligence in marketing. The developing country context of India is focused upon, allowing for new insights that can be considered on a wider international scale.

Naïve Bayes and decision trees are the most commonly used artificial intelligence methods in practice, while decision trees have the highest perceived impact on marketing performance. Niche is the marketing domain where artificial intelligence is considered the most successful in practice, while brand and advertising are labeled as the marketing domain with the highest potential for artificial intelligence. A significant percentage of marketers in India conduct market research, and a notable percentage develop strategy formulations and conduct customer segmentation as their main activity. The perceived effectiveness of artificial intelligence on process activities is higher than on analytical or strategic activities. There is no consensus on the future job replacement of artificial intelligence, with a portion of marketers noting a possibility, indicating the low likelihood of replacement, and holding no opinion. The vast majority of marketing companies in India are developing or acquiring artificial intelligence capabilities, while concerns about ethical risks, damaging the company brand and reputation, and scooping the market by competitors, are directly at the top of the challenge list.

11.4 Building an AI-Driven Marketing Strategy

User reviews the operational relevance of AI marketing, providing a macro view on adoption, strategies, and possible detrimental effects within a system of collaborating B2B organizations, based on the theoretical and practical interpretations from an activity system-focused view. consumer expertise as a crucial factor influencing the performance of AI-driven recommendation technologies, providing valuable insights for marketing strategies and system optimization(Chinchanachokchai et al., 2021). In the digital age, AI-powered marketing is enjoying exponential growth. AI and ML integration in customer journeys, highlighting personalization, automation, interactive tools, and enhanced brand engagement(Rana et al., 2022). Fine arts were the lens of digital marketing in the initial stage. The flourishing of digital media technologies shaped digital marketing as it is today. Such technologies involved the ability to conduct data analytics, generate insights and extract meaningful patterns for strategic decision-making. More recently, AI and machine learning (ML) became the driving force of digital marketing innovation. Applications developed by large technology firms exploit large amounts of digitized customer data, enabling them to generate valuable insights. This explains their growing part of strategic industry market estimates of US $322.6 billion and 34% growth until 2027.

Amid the unprecedented growth in data storage and processing capacities, AI is playing a progressively more strategic role in the swift-paced and complex contemporary industrial marketing industrial ecosystem. First, AI has been intensively employed for price setting and prediction as a response to increasingly perfect information about commodities, i.e. digital commoditization that significantly mitigates information opacity. Furthermore, AI responsiveness modeling has been developed alongside buyer behavior. An AI & machine learning approach to analyze social media data, identifying and prioritizing customer needs for businesses(Kühl et al., 2020). It allows firms to anticipate demand patterns in an individual dimension with high accuracy and adjust sales efforts correspondingly. Going beyond microeconomic models of the market, marketing theory often explores humankind interactions within markets viewing them as networks. Given the system-dynamical nature of such networks it is difficult to describe the systemic feedbacks at a simplified level, which results in inconclusive (and sometimes contradictory) empirical findings. AI as a technology might substitute high-level managerial activities with automated processes extending the scope of the professional area in which automation is thought. Due to the rise of such groundbreaking technologies numerous emerging markets are constantly shifting.

11.5 Measuring Success

Measuring the success or failure of any given theoretical concept such as AI could also be a potential area for a research application. AI's transformative role in shaping preferences, enhancing efficiency, and personalizing experiences(Braga, 2021) Measurement could be taken as an important driving factor in adopting newer technologies, and AI in marketing refers to introducing newer metrics and observing the fluctuations in these metrics (James Keegan et al., 2022). The present study is focused on AI in marketing and on measuring the ability of this and other innovations to change the market-related metrics in IT firms. For the purpose of the present study, AI in marketing refers to introducing AI in marketing and observing the fluctuation in the related metrics. A primary goal of the research community is the development of new algorithms and models, the provision of solutions that are better than hand-engineered solutions. An inherent contradiction exists between the ultimate goal of providing marketing metrics from an AI systems approach and the current state of machine learning practice. As indicated in many areas, the supposed gains of AI in marketing do not manifest, or never materialize even after extensive research. One potential reason for this is that certain problems of marketing have an under-defined set of probability models. It is difficult to learn a single model, with a training data informative of how to do well in the task of interest. Many researchers have made efforts to push forward AI in marketing, some of them successful.

12. THE DISTINCTION BETWEEN THE THEORETICAL DISCUSSION AND APPLIED CASE EVIDENCE

At this early stage of the adoption of AI in marketing, the theoretical development of this context is more advanced than its evidenced application. Written accounts of the issues concerning AI in marketing appear in contemporary marketing theory development texts, but practical research case evidence is scarce. Specifically, the positive impacts of AI on contemporary B2B business marketing practices appear to be unnoticed. There is a corresponding outcome of the theoretical discussion that marketing managers and practitioners view AI adoption and application as a threat rather than a vehicle. AI seems ripe for malicious activities in B2B marketing practice and contains unaddressed ethical challenges in its application. Early responsibility and accountability in the marketing of AI will favor human beings over the less ethical exploitation of AI-generated vehicles in business marketing. Caution against hyping AI appears to be warranted (Kumar et al., 2015), especially regarding the threat that AI poses to displaced marketing practice roles and positions. AI could raise investment costs, thereby rejecting some current marketing practitioners. With these concerns about AI's malignant transformation in traditional B2B marketing, academic discussions adopting Latin regarding AI's design and application are ahead of real-world cases of AI marketing practices. A corresponding outcome of this practical stage is the accountability of authors, detailing investigations and results rigorously(Keegan et al., 2024b).

The real-world marketing cases of AI appear ahead of the concern and awareness about challenges in the ethical adoption and application of AI marketing vehicles in academic research. Marketing vehicles based on AI tools are designed and launched, although investigation of the moral aspects of such applications are noticeably lacking. Issues of AI marketing and business ethics are minimized, and there is potentially an undue favor towards the application side of the marketing strategy. Coinciding with the recency of AI adoption in B2B marketing practices, it is not surprising that there is a dearth of case studies of AI-sustained marketing practice modes. Consequently, while aware of the potential reaction against AI-sustained B2B marketing practice, theoretical cautious optimism is sounded, rather than neglecting AI's vehicle sides in its development and adoption in practice

13. RESULT & DISCUSSION BASED ON THE LITERATURE

This section discusses the results and implications of an extensive literature review of intelligent agent technology adoption in marketing as well as findings from 50 in-depth interviews with managers from various industries in the City of Nagpur, India. Based on the literature, a marketing-centric definition of IATs and a taxonomy and adoption framework are proposed. The findings reveal that with the increase in online marketing, companies are gradually shifting from traditional CRM to more advanced intelligent agent technologies. Customers undergo a three-step acceptance generation process, with social influence, hedonic motivation, and anthropomorphism positively influencing performance expectancy and effort expectancy, ultimately influencing(Gursoy et al., 2019)

The goal of this study is to provide a systematic review of the research framework, strategies, and applications of IATs in marketing based on literature from the fields of marketing, information technology, and general business. AI can enhance marketing research, strategy, and actions by automating repetitive functions, processing data, and analyzing interactions and emotions(Huang & Rust, 2021b). Also A notable transition is observed from domain-specific to organizational applications of AI(R. Dwivedi et al., 2023). Further, an IAT adoption framework is proposed for understanding and predicting the adoption of IATs by companies. An empirical examination of the framework having six constructs - perceived benefits, perceived barriers, technology readiness, firm size, technology-related government regulations, and competitive intensity - is also conducted. On the basis of the research aim, a number of research gaps, propositions, and practical implications are discussed (Kumar et al., 2015). The visually standard content is easier to process, evoking positive affect and increasing engagement(Philp et al., 2022), In today's digital era, marketing theory and practice are being transformed by increasing complexity due to the vast availability of rich information, higher reach and interactions among markets and market participants across multiple platforms, and the faster speeds of transactions. VIs, driven by AI and machine learning, are reshaping consumer attitudes, behaviors, and decision-making processes on social media platforms. Key findings reveal VIs' unique appeal, enhanced consumer engagement, and potential integration into marketing strategies(Laszkiewicz & Kalinska-Kula, 2023) These have led to the adoption of intelligent agent technologies (IATs) by many companies to assist marketers in their decision-making processes and in the execution and operations of marketing strategy. As IATs are relatively new and can be technologically complex, there is a confluence of several semantic definitions, attributes, applications, and a wide spectrum of related technologies that continue to evolve at a fast pace, and thus the existing marketing-centric theory in this area is not yet fully developed.There are multiple avenues to address these critical employment, gender gaps, and inequities. First, collectively and based on existing industry frameworks and suggested public policies and collaborations, companies should conduct a gap analysis to better understand their own house as well as various societies in which they operate. This could initially include how the AI skills measure up, are individuals best suited for their assigned tasks, and how diverse and equitable is the team composition and hiring process, in addition to the bias audits for AI talent assessments, tools, models, and outputs. Next, companies could experiment with scenario builders envisioning the productive uses of various AI tools and any ethical policy gaps. Such results could be tied into the newly proposed AI policies and the national strategic initiatives reported above for a collaborative approach toward addressing gaps with less duplication.

Additional avenues for research include longitudinal studies to assess the effectiveness of various AI in marketing applications and any unintended consequences on brand, value, sentiment, and reputation. Marketing researchers want to solicit cross-disciplinary research and theories to study and inform on these key AI imbedded paradigm shifting changes in the field. For example, information system researchers are sought to study, develop, and provide better, smarter, more interpretable, efficient, worthy, and inexpensive tools with enhanced user design and experience; to study the online digital divide, consciousness bots, and metaverse application; and to experiment, expand, and report the next quantum leap in data privacy protection with AI to repel the AI-generated scathing attacks. There is a clear need for both industry and academia to better leverage the unprecedented price-performance-competitiveness-wholesale-minimizing-power computing and operating infrastructure of AI, such as the AUPT, TPU, mPPA, and silicon photonic and optical computing, as well as the ingenuity of the software and phenomenons scenarios like the GAN diffusion and NPC, for responsible evaluations, new inventions, and unprecedented discoveries.

All of these calls for lensing should factor in the research gaps in understanding how to properly and responsibly apply such tools for marketing research and strategy development. Therefore, marketing researchers are encouraged to complement their knowledge of marketing with the collaborative study of responsible AI and its applications through jointly proposed courses, books, conferences, and for a more responsible better world(Hermann, 2022).

14. CONCLUSION

Nowadays, artificial intelligence and machine learning find widespread applications in marketing. AI and machine learning techniques are revolutionizing fundamental sciences, transforming research and promoting new research trends(Xu et al., 2021).Artificial intelligence (AI) has the potential to revolutionize marketing, setting new opportunities, and posing new challenges in the field. Companies use machine learning algorithms to mine historical purchase or viewing data for information on future purchases or viewing behavior. Digital companies use then this gained knowledge to target individual business buyers and personalize online content for them. This has been demonstrated to increase website engagement growth by a factor of 6 Given this upsurge of AI in marketing, the article provides a roadmap for students and practitioners in marketing and computer science on theory, methods and techniques, as well as challenges and research directions from an Indian perspective on how to effectively use and harness AI to improve marketing. AI adoption in construction engineering and management is rapidly transforming the industry, enhancing automation, risk mitigation, efficiency, digitalization, and computer vision(Y. Pan & Zhang, 2021).

This article is based on an extensive literature review as well as some preliminary results from a consumer survey in India and hence lays out a clear future research agenda for researchers. While advanced economies are already entering the “fourth industrial revolution”, with profound technological advances in AI and machine learning that are already beginning to impact almost all industries, India still struggles with basic infrastructural challenges. Yet, the principles of AI-driven marketing can still be expected to be transferable and beneficial to Indian firms. AI & Machine learning's ability to personalize experiences, optimize assortments, analyze consumer emotions, and enhance financial strategies, offering practical insights for businesses to improve marketing efficiency and consumer satisfaction(De Mauro et al., 2022). Moreover, the ideas provided in this article form a comprehensive framework for taking the seminal steps of analyzing and entering the AI-driven marketing. AI-driven marketing is essentially about using AI to understand, predict, and influence customer behavior. This can aid firms in better locating potential buyers or segments and on this basis to more efficiently productize market offerings, create targeted packages, and optimize pricing schedules.

Notably The increasing responsibilities of AI marketers for unintended negative side effects of their AI applications on diverse societal and human welfare dimensions calls for comprehensive and ethically designed AI marketing applications to maximize benefits while minimizing risks and harms. Ethical AI marketing can help restore consumer trust in AI marketing applications as their implementation proliferates. personalized consumer insights, behavior analysis, and enhanced decision-making efficiency(Ullal et al., 2021). The AI enabled hyper-individualization of customer communication can exacerbate the 'transparency paradox' in marketing, i.e. while customers express privacy concerns with respect to non-transparent privacy practices, they still want to reap the benefits of personal engagement through individualized services and online marketing measures. Based on thsese theories digital marketing, emphasizing AI-driven strategies, integrating analytics, personalization, ethics, and customer-focused solutions(Chaffey & Smith, 2022). Understandably, identifying and modeling the causal relationships between marketing actions and consumer actions is crucial to optimize these actions. With the emergence of big data and improvements in machine learning algorithms, it is now possible to model advertising response functions or consumer choice models. This research roadmap introduces a significant amount of this evolving literature and is accompanied by basic economic definitions and theoretical models to facilitate entry into this interdisciplinary field. To try to enrich the review, some of these models will be exemplified at the end of abstract consumer survey results from India. Until now, such a comprehensive framework from a dual Indian (management and marketing) and Western (data and computer science) perspective on AI-driven marketing, consumer behavior schedules in the Indian digital commerce industry, and implications for this developing economy, have not been provided.

15. FUTURE SCOPE OF THE STUDY

The proposed study aims to develop a roadmap for the comprehensive and efficient implementation of artificial intelligence (AI) in marketing in the private sector in India and provide a reference model for practical applicability. The study adopts the theoretically grounded approach of the stage model of AI implementation in marketing to scrutinize both theoretical and practical aspects. Unifying large language models and knowledge graphs can enhance their abilities in natural language processing and artificial intelligence(S. Pan et al., 2024). In doing so, it significantly advances research on marketing intelligence regarding AI and beyond, illuminating a promising multi-dimensional, well-prepared, and sociotechnical path for the organizational and strategical conceiving and effective and efficient utilization of AI in companies' marketing activities (Hermann, 2022). On an empirical level and from a marketing perspective, this study provides a novel contribution by conducting, analyzing, and adapting an extensive in-depth study in the private sector in India. Thereby, the findings advance the current understanding of the situation on the ground as well as the opportunities, challenges, and improvements for the future leveraging of AI. Through automation and data-driven insights, AI simplifies business decisions, optimizes digital marketing strategies, and provides a competitive edge for future advancements in marketing practices(Sarath Kumar Boddu et al., 2022).

There are also good examples of the implementation of artificial intelligence. Therefore, many companies consider their use in marketing. It is a big opportunity to implement artificial intelligence applications in the marketing environment in a relatively early stage, compared with other companies. However, without specific knowledge and know-how, most companies have trouble applying them to practical applications. Also, the existing artificial intelligence model is difficult to apply to companies without modification. In this paper, a roadmap is provided to examine the artificial intelligence system, so that the company can apply it to their marketing strategy starting from understanding of artificial intelligence to the application on specialization in business. As case studies, companies successfully marketing simulated applications

The multiplication of research streams has driven scholars and practitioners to arrive at a better understanding of AI implantation from a marketing perspective. The beneficial discrimination of this study is its multidimensional and enlightening focus on marketing intelligence leveraging through AI, as well as well-thought sociotechnical groundwork for results most in line with practice. The results are based on an extensive in-depth examination in the private sector in India, scrutinizing different perspectives and sources of evidence. The study's main contribution is to offer companies a multi-dimensional stage model, a roadmap towards the implementation of AI in marketing that combines theoretical grounding, practical applicability, and sociotechnical preparedness. This roadmap provides guidance and valuable workings to marketing practitioners and their endeavors toward successful, efficient, and ethical leveraging of AI.

REFERENCES

Barredo Arrieta, A., Díaz-Rodríguez, N., Del Ser, J., Bennetot, A., Tabik, S., Barbado, A., Garcia, S., Gil-Lopez, S., Molina, D., Benjamins, R., Chatila, R., & Herrera, F. (2020). Explainable artificial intelligence (XAI): Concepts, taxonomies, opportunities and challenges toward responsible AI. *Information Fusion, 58*, 82–115. DOI: 10.1016/j.inffus.2019.12.012

Basri, W. (2020). Examining the impact of artificial intelligence (AI)-assisted social media marketing on the performance of small and medium enterprises: Toward effective business management in the Saudi Arabian context. *International Journal of Computational Intelligence Systems, 13*(1), 142. DOI: 10.2991/ijcis.d.200127.002

. Braga, F. M. I. (2021). The influence and impact of artificial intelligence in the consumer decision-making process: Comparing Generation X with Millennials.

Campbell, C., Sands, S., Ferraro, C., Tsao, H.-Y., & Mavrommatis, A. (2020). From data to action: How marketers can leverage AI. *Business Horizons, 63*(2), 227–243. DOI: 10.1016/j.bushor.2019.12.002

Chaffey, D., & Smith, P. (2022). *Digital marketing excellence: Planning, optimizing and integrating online marketing* (6th ed.). Routledge., DOI: 10.4324/9781003009498

Chinchanachokchai, S., Thontirawong, P., & Chinchanachokchai, P. (2021). A tale of two recommender systems: The moderating role of consumer expertise on artificial intelligence based product recommendations. *Journal of Retailing and Consumer Services, 61*, 102528. Advance online publication. DOI: 10.1016/j.jretconser.2021.102528

Dahake, N. S., Gajghate, A., Dahake, P. S., & Bansod, S. (2024). A transformative impact of IoT and AI by predictive analytics on modern marketing strategies: A study of its significance & accuracy. In 2024 2nd International Conference on Advances in Computation, Communication and Information Technology (ICAICCIT) (pp. 1150–1158). https://doi.org/DOI: 10.1109/ICAICCIT64383.2024.10912388

Dahake, N. S., Suresh Dahake, P., & Gajghate, A. (2024). Beyond the hype: A critical evaluation of predictive analytics accuracy in high-stakes business decisions. In 2024 2nd DMIHER International Conference on Artificial Intelligence in Healthcare, Education and Industry (IDICAIEI) (pp. 1–6). DOI: 10.1109/IDICAIEI61867.2024.10842857

Dahake, P. S., Chandak, S., Mohare, R. V., Wadhwani, K., & Bhadade, P. (2023). The crystal ball of marketing: How predictive analytics is reshaping the industry? In *2023 Second International Conference On Smart Technologies For Smart Nation (SmartTechCon)* (pp. 304–311). https://doi.org/DOI: 10.1109/SmartTechCon57526.2023.10391334

Dahake, P. S., Parchake, S., Somani, N., & Ingale, A. R. (2024). Predictive analytics in automotive consumer behaviour: An insight into personal use 4-wheeler purchase decisions. In 2024 2nd International Conference on Advances in Computation, Communication and Information Technology (ICAICCIT) (pp. 1143–1149). https://doi.org/DOI: 10.1109/ICAICCIT64383.2024.10912242

Davenport, T., Guha, A., Grewal, D., & Bressgott, T. (2020). How artificial intelligence will change the future of marketing. *Journal of the Academy of Marketing Science*, *48*(1), 24–42. DOI: 10.1007/s11747-019-00696-0

De Mauro, A., Sestino, A., & Bacconi, A. (2022). Machine learning and artificial intelligence use in marketing: A general taxonomy. *Italian Journal of Marketing*, *2022*(4), 439–457. DOI: 10.1007/s43039-022-00057-w

Duan, Y., Edwards, J. S., & Dwivedi, Y. K. (2019). Artificial intelligence for decision making in the era of Big Data – Evolution, challenges and research agenda. *International Journal of Information Management*, *48*, 63–71. DOI: 10.1016/j.ijinfomgt.2019.01.021

Durai, S., Manoharan, G., Priya, T. S., Jayanthi, R., Razak, A., & Ashtikar, S. P. (2024). Quantifying the impacts of artificial intelligence implementations in marketing. In Irani, H. R., & Nozari, H. (Eds.), (pp. 120–144). Advances in marketing, customer relationship management, and e-services. IGI Global., DOI: 10.4018/979-8-3693-1339-8.ch008

Dwivedi, R., Nerur, S., & Balijepally, V. (2023). Exploring artificial intelligence and big data scholarship in information systems: A citation, bibliographic coupling, and co-word analysis. *International Journal of Information Management Data Insights*, *3*(2), 100185. Advance online publication. DOI: 10.1016/j.jjimei.2023.100185

Dwivedi, Y. K., Hughes, L., Ismagilova, E., Aarts, G., Coombs, C., Crick, T., Duan, Y., Dwivedi, R., Edwards, J., Eirug, A., Galanos, V., Ilavarasan, P. V., Janssen, M., Jones, P., Kar, A. K., Kizgin, H., Kronemann, B., Lal, B., Lucini, B., & Williams, M. D. (2021). Artificial intelligence (AI): Multidisciplinary perspectives on emerging challenges, opportunities, and agenda for research, practice and policy. *International Journal of Information Management*, *57*, 101994. Advance online publication. DOI: 10.1016/j.ijinfomgt.2019.08.002

Gajjar, N., Rathod, K., & Jani, K. (2022). A systematic literature review on Robotic Process Automation security (Version 1). arXiv. https://doi.org//arXiv.2212.05544DOI: 10.48550

Ghimire, A., Thapa, S., Jha, A. K., Adhikari, S., & Kumar, A. (2020). Accelerating business growth with Big Data and artificial intelligence. In 2020 Fourth International Conference on I-SMAC (IoT in Social, Mobile, Analytics and Cloud) (I-SMAC) (pp. 441–448). https://doi.org/DOI: 10.1109/I-SMAC49090.2020.9243318

Gursoy, D., Chi, O. H., Lu, L., & Nunkoo, R. (2019). Consumers acceptance of artificially intelligent (AI) device use in service delivery. *International Journal of Information Management*, *49*, 157–169. DOI: 10.1016/j.ijinfomgt.2019.03.008

Haenlein, M., & Kaplan, A. (2019). A brief history of artificial intelligence: On the past, present, and future of artificial intelligence. *California Management Review*, *61*(4), 5–14. DOI: 10.1177/0008125619864925

Haleem, A., Javaid, M., Asim Qadri, M., Pratap Singh, R., & Suman, R. (2022). Artificial intelligence (AI) applications for marketing: A literature-based study. *International Journal of Intelligent Networks*, *3*, 119–132. DOI: 10.1016/j.ijin.2022.08.005

Hermann, E. (2022). Leveraging artificial intelligence in marketing for social good—An ethical perspective. *Journal of Business Ethics*, *179*(1), 43–61. DOI: 10.1007/s10551-021-04843-y PMID: 34054170

Hoyer, W. D., Kroschke, M., Schmitt, B., Kraume, K., & Shankar, V. (2020). Transforming the customer experience through new technologies. *Journal of Interactive Marketing*, *51*(1), 57–71. DOI: 10.1016/j.intmar.2020.04.001

Huang, M.-H., & Rust, R. T. (2021a). A strategic framework for artificial intelligence in marketing. *Journal of the Academy of Marketing Science*, *49*(1), 30–50. DOI: 10.1007/s11747-020-00749-9

Huang, M.-H., & Rust, R. T. (2021b). A strategic framework for artificial intelligence in marketing. *Journal of the Academy of Marketing Science*, *49*(1), 30–50. DOI: 10.1007/s11747-020-00749-9

Janiesch, C., Zschech, P., & Heinrich, K. (2021). Machine learning and deep learning. *Electronic Markets*, *31*(3), 685–695. DOI: 10.1007/s12525-021-00475-2

Keegan, B. J., Dennehy, D., & Naudé, P. (2024a). Implementing artificial intelligence in traditional B2B marketing practices: An activity theory perspective. *Information Systems Frontiers*, *26*(3), 1025–1039. DOI: 10.1007/s10796-022-10294-1 PMID: 35637917

Keegan, B. J., Dennehy, D., & Naudé, P. (2024b). Implementing artificial intelligence in traditional B2B marketing practices: An activity theory perspective. *Information Systems Frontiers*, *26*(3), 1025–1039. DOI: 10.1007/s10796-022-10294-1 PMID: 35637917

Kose, U., & Sert, S. (2017). Improving content marketing processes with the approaches by artificial intelligence. arXiv. https://doi.org//arXiv.1704.02114DOI: 10.48550

Kühl, N., Mühlthaler, M., & Goutier, M. (2020). Supporting customer-oriented marketing with artificial intelligence: Automatically quantifying customer needs from social media. *Electronic Markets*, *30*(2), 351–367. DOI: 10.1007/s12525-019-00351-0

Kumar, V., Dixit, A., Javalgi, R. G., & Dass, M. (2016). Research framework, strategies, and applications of intelligent agent technologies (IATs) in marketing. *Journal of the Academy of Marketing Science*, *44*(1), 24–45. DOI: 10.1007/s11747-015-0426-9

Kumar, V., Ramachandran, D., & Kumar, B. (2021). Influence of new-age technologies on marketing: A research agenda. *Journal of Business Research*, *125*, 864–877. DOI: 10.1016/j.jbusres.2020.01.007

Kushwaha, A. K., & Kar, A. K. (2020). Language model-driven chatbot for business to address marketing and selection of products. In Sharma, S. K., Dwivedi, Y. K., Metri, B., & Rana, N. P. (Eds.), *Re-imagining diffusion and adoption of information technology and systems: A continuing conversation* (Vol. 617, pp. 16–28). Springer., DOI: 10.1007/978-3-030-64849-7_3

Kushwaha, A. K., & Kar, A. K. (2024). MarkBot – A language model-driven chatbot for interactive marketing in post-modern world. *Information Systems Frontiers*, *26*(3), 857–874. DOI: 10.1007/s10796-021-10184-y

Laszkiewicz, A., & Kalinska-Kula, M. (2023). Virtual influencers as an emerging marketing theory: A systematic literature review. *International Journal of Consumer Studies*, *47*(6), 2479–2494. DOI: 10.1111/ijcs.12956

Lee, J., Davari, H., Singh, J., & Pandhare, V. (2018). Industrial artificial intelligence for industry 4.0-based manufacturing systems. *Manufacturing Letters*, *18*, 20–23. DOI: 10.1016/j.mfglet.2018.09.002

Lee, M., Kwon, W., & Back, K.-J. (2021). Artificial intelligence for hospitality big data analytics: Developing a prediction model of restaurant review helpfulness for customer decision-making. *International Journal of Contemporary Hospitality Management*, *33*(6), 2117–2136. DOI: 10.1108/IJCHM-06-2020-0587

Ljepava, N. (2022). AI-enabled marketing solutions in marketing decision making: AI application in different stages of marketing process. TEM Journal, 11(3), 1308–1315. https://doi.org/DOI: 10.18421/TEM113-40

Malthouse, E., & Copulsky, J. (2023). Artificial intelligence ecosystems for marketing communications. *International Journal of Advertising*, *42*(1), 128–140. DOI: 10.1080/02650487.2022.2122249

Mariani, M. M., Perez-Vega, R., & Wirtz, J. (2022). AI in marketing, consumer research and psychology: A systematic literature review and research agenda. *Psychology and Marketing*, *39*(4), 755–776. DOI: 10.1002/mar.21619

Martínez, A., Schmuck, C., Pereverzyev, S.Jr, Pirker, C., & Haltmeier, M. (2020). A machine learning framework for customer purchase prediction in the non-contractual setting. *European Journal of Operational Research*, *281*(3), 588–596. DOI: 10.1016/j.ejor.2018.04.034

Mayahi, S., & Vidrih, M. (2022). The impact of generative AI on the future of visual content marketing (Version 1). arXiv. https://doi.org//arXiv.2211.12660DOI: 10.48550

Mikalef, P., & Gupta, M. (2021). Artificial intelligence capability: Conceptualization, measurement calibration, and empirical study on its impact on organizational creativity and firm performance. *Information & Management*, *58*(3), 103434. Advance online publication. DOI: 10.1016/j.im.2021.103434

Miller, T. (2019). Explanation in artificial intelligence: Insights from the social sciences. *Artificial Intelligence*, *267*, 1–38. DOI: 10.1016/j.artint.2018.07.007

Mustak, M., Salminen, J., Plé, L., & Wirtz, J. (2021). Artificial intelligence in marketing: Topic modeling, scientometric analysis, and research agenda. *Journal of Business Research*, *124*, 389–404. DOI: 10.1016/j.jbusres.2020.10.044

Nair, K., & Gupta, R. (2021). Application of AI technology in modern digital marketing environment. *World Journal of Entrepreneurship, Management and Sustainable Development*, *17*(4), 383–396. DOI: 10.1108/WJEMSD-08-2020-0099

Nelson, J. P., Biddle, J. B., & Shapira, P. (2023). Applications and societal implications of artificial intelligence in manufacturing: A systematic review (Version 1). arXiv. https://doi.org//arXiv.2308.02025DOI: 10.48550

Noble, C. H., & Mokwa, M. P. (1999). Implementing marketing strategies: Developing and testing a managerial theory. *Journal of Marketing*, *63*(4), 57–73. DOI: 10.1177/002224299906300406

Pan, S., Luo, L., Wang, Y., Chen, C., Wang, J., & Wu, X. (2024). Unifying large language models and knowledge graphs: A roadmap. *IEEE Transactions on Knowledge and Data Engineering, 36*(7), 3580–3599. DOI: 10.1109/TKDE.2024.3352100

Pan, Y., & Zhang, L. (2021). Roles of artificial intelligence in construction engineering and management: A critical review and future trends. *Automation in Construction, 122*, 103517. Advance online publication. DOI: 10.1016/j.autcon.2020.103517

Paschen, J. (2019). Investigating the emotional appeal of fake news using artificial intelligence and human contributions. *Journal of Product and Brand Management, 29*(2), 223–233. DOI: 10.1108/JPBM-12-2018-2179

Philp, M., Jacobson, J., & Pancer, E. (2022). Predicting social media engagement with computer vision: An examination of food marketing on Instagram. *Journal of Business Research, 149*, 736–747. DOI: 10.1016/j.jbusres.2022.05.078

Ramakrishnan, N. (2002). The traits of the personable (Version 1). arXiv. https://doi.org/DOI: 10.48550/arXiv.cs/0205022

Rana, J., Gaur, L., Singh, G., Awan, U., & Rasheed, M. I. (2022). Reinforcing customer journey through artificial intelligence: A review and research agenda. *International Journal of Emerging Markets, 17*(7), 1738–1758. DOI: 10.1108/IJOEM-08-2021-1214

Rivas, P., & Zhao, L. (2023). Marketing with ChatGPT: Navigating the ethical terrain of GPT-based chatbot technology. *AI, 4*(2), 375–384. DOI: 10.3390/ai4020019

Russell, S., & Norvig, P. (1995). Artificial intelligence: A modern approach. *Choice (Chicago, Ill.), 33*(3), 33–1577. DOI: 10.5860/CHOICE.33-1577

Sachdev, R. (2020). Towards security and privacy for edge AI in IoT/IoE based digital marketing environments. In *2020 Fifth International Conference on Fog and Mobile Edge Computing (FMEC)* (pp. 341–346). https://doi.org/DOI: 10.1109/FMEC49853.2020.9144755

Salehi, H., & Burgueño, R. (2018). Emerging artificial intelligence methods in structural engineering. *Engineering Structures, 171*, 170–189. DOI: 10.1016/j.engstruct.2018.05.084

Samala, N., Katkam, B. S., Bellamkonda, R. S., & Rodriguez, R. V. (2022). Impact of AI and robotics in the tourism sector: A critical insight. *Journal of Tourism Futures, 8*(1), 73–87. DOI: 10.1108/JTF-07-2019-0065

Samek, W., Montavon, G., Lapuschkin, S., Anders, C. J., & Muller, K.-R. (2021). Explaining deep neural networks and beyond: A review of methods and applications. *Proceedings of the IEEE, 109*(3), 247–278. DOI: 10.1109/JPROC.2021.3060483

Sarath Kumar Boddu, R., Santoki, A. A., Khurana, S., Vitthal Koli, P., Rai, R., & Agrawal, A. (2022). An analysis to understand the role of machine learning, robotics and artificial intelligence in digital marketing. *Materials Today: Proceedings*, *56*, 2288–2292. DOI: 10.1016/j.matpr.2021.11.637

Soni, N., Sharma, E. K., Singh, N., & Kapoor, A. (2019). Impact of artificial intelligence on businesses: From research, innovation, market deployment to future shifts in business models (Version 1). arXiv. https://doi.org//arXiv.1905.02092DOI: 10.48550

Srivastava, G., & Bag, S. (2024). Modern-day marketing concepts based on face recognition and neuro-marketing: A review and future research directions. *Benchmarking*, *31*(2), 410–438. DOI: 10.1108/BIJ-09-2022-0588

Sun, H., Zafar, M. Z., & Hasan, N. (2022). Employing natural language processing as artificial intelligence for analyzing consumer opinion toward advertisement. *Frontiers in Psychology*, *13*, 856663. Advance online publication. DOI: 10.3389/fpsyg.2022.856663 PMID: 35769737

Talib, M. A., Majzoub, S., Nasir, Q., & Jamal, D. (2021). A systematic literature review on hardware implementation of artificial intelligence algorithms. *The Journal of Supercomputing*, *77*(2), 1897–1938. DOI: 10.1007/s11227-020-03325-8

Ullal, M. S., Hawaldar, I. T., Soni, R., & Nadeem, M. (2021). The role of machine learning in digital marketing. *SAGE Open*, *11*(4), 21582440211050394. Advance online publication. DOI: 10.1177/21582440211050394

Varghese, M., Raj, S., & Venkatesh, V. (2022). Influence of AI in human lives (Version 1). arXiv. https://doi.org//arXiv.2212.12305DOI: 10.48550

Wang, C. L. (Ed.). (2023). *The Palgrave handbook of interactive marketing*. Springer., DOI: 10.1007/978-3-031-14961-0

Xu, Y., Liu, X., Cao, X., Huang, C., Liu, E., Qian, S., Liu, X., Wu, Y., Dong, F., Qiu, C.-W., Qiu, J., Hua, K., Su, W., Wu, J., Xu, H., Han, Y., Fu, C., Yin, Z., Liu, M., & Zhang, J. (2021). Artificial intelligence: A powerful paradigm for scientific research. *Innovation (Cambridge (Mass.))*, *2*(4), 100179. Advance online publication. DOI: 10.1016/j.xinn.2021.100179 PMID: 34877560

Zhang, C., & Lu, Y. (2021). Study on artificial intelligence: The state of the art and future prospects. *Journal of Industrial Information Integration*, *23*, 100224. Advance online publication. DOI: 10.1016/j.jii.2021.100224

Zhou, Z., Chen, X., Li, E., Zeng, L., Luo, K., & Zhang, J. (2019). Edge intelligence: Paving the last mile of artificial intelligence with edge computing. *Proceedings of the IEEE*, *107*(8), 1738–1762. DOI: 10.1109/JPROC.2019.2918951

Ziakis, C., & Vlachopoulou, M. (2023). Artificial intelligence in digital marketing: Insights from a comprehensive review. *Information (Basel)*, *14*(12), 664. Advance online publication. DOI: 10.3390/info14120664

Chapter 3
Enhancing Consumer Engagement Through Human-AI Collaboration: Harnessing Synergies in Marketing

Parihar Suresh Dahake
http://orcid.org/0000-0002-2286-4120
Ramdeobaba University, Nagpur, India

Nihar Suresh Dahake
http://orcid.org/0009-0007-8971-9247
Dr. Ambedkar Institute Of Management Studies And Research, Nagpur, India

Rahul Vijay Mohare
Ramdeobaba University, Nagpur, India

ABSTRACT

Every day, individuals face countless decisions with the potential to shape their future. Some decisions may have relatively few consequences, such as the choice of a movie one may choose to watch or the song that they would like to listen to. On the other hand, the decision to buy a house, a car, or another significant purchase has the potential to fundamentally transform a life course. Rather than making decisions consciously, choices are often shaped by the external environment in ways in which most individuals might not be aware. For example, the majority of individuals may not realize that the structure of a shopping Centre is usually specifically

DOI: 10.4018/979-8-2600-2378-5.ch003

designed to encourage purchases. In a similar vein, people may also not be aware of the additional effort put forth towards scent management at an overwhelmingly good-smelling bakery.

1. INTRODUCTION

Artificial Intelligence (AI) applications have become increasingly significant for marketing over the last five years. The current discussion focuses on adaptations of AI relevant for marketing in an ethical perspective with a particular emphasis on how marketers, corporations, and policy-makers can leverage AI innovations for social goods, for example for enhancing consumers' experiential values, for marketing cities, and for supporting start-ups(Hermann, 2022) and Consumer Engagement (CE) significantly affects brand outcomes. However, CE in the context of developing countries is under-researched. Employing a qualitative approach, the study observed one of India's most affluent and urban markets, exploring customer engagement with AI advertisements and focusing on the CE aspect.

In the year 2022, Artificial Intelligence is at the forefront of business solutions and is regularly used across consumer targeting, analysis of consumer sentiment, and the production of content and the Internet offers a powerful platform for customer engagement in product innovation, enabling co-creation of value with customers at various stages of the New Product Development process(Sawhney et al., 2005). However, considering the apparent success in the domains of advertising, marketing, and customer service, no research was found that examined how consumers might engage with AI advertisements in a marketing context or collaborate with AI. This research adds to the body of knowledge on Artificial Intelligence and CE by presenting empirical findings in this realm. Findings provide preliminary consumer perspectives on types of advertisements, consumer collaboration with Artificial Intelligence, and strategies employed in response to engagement with AI advertisements. Wealth and awareness are personally derived factors that affect CE. Furthermore, the growing advertising space of online platforms affects CE. These advertising communication channels have increased competition among various brands, making it more difficult for brands to capture their consumers' attention.

The remaining parts of this research are structured according to the first presentation of the literature, which outlines the definition and development of CE. The second literature review explores Artificial Intelligence and consumer engagement, focusing on how AI has been adopted in marketing and the analysis of the CE antecedents of knowledge, attitudes, and intentions. The study is discussed in the methodology section, followed by the presentation of the results. In developing countries, research from the developed world predominates, resulting in a gap in understanding CE in the Indian context. This research focuses on customer engagement with AI advertisements in a developing country-specific context, namely the Indian capital. One approach is to introduce marketing-to-consumer (M2C) and consumer-to-marketing (C2M) AI concepts, which address three consumer-relevant ethical questions: explainability, privacy, and trustworthiness. Another approach expands contemporary Lauterborn's 4Cs to 7Cs of city marketing, adding Consumer Needs, Consumer Satisfaction, and Consumer Value to address three AI-related concerns of entropy, transmuted ethics, and reinforcing inequalities (Sun et al., 2022). A third approach presents a case study on a human-AI start-up collaboration. The current framing of AI for marketing fits into three marketing phases: one-way road, speed boost, and mutual aid. However, the AI decision-making in the era of Big Data faces challenges and offers twelve research propositions for information systems researchers to advance AI research and development (Duan et al., 2019).

Artificial Intelligence (AI) in marketing has broken new ground in marketing research, strategy, and action, particularly in terms of creating executable content, enhancing customer relationships, and improving customer experiences. One essential research stream involves AI marketing applications that distill fine-grained and data-driven personalization and customization of products, services, and various marketing mix variables throughout the entire customer journey. These applications all aim to engage consumers, bringing the experiential value of the latter to the greatest extent possible.

2. UNDERSTANDING CONSUMER ENGAGEMENT

Due to the rapid advancement of artificial intelligence, corporate houses are harnessing marketing campaigns to execute a profound collaboration between human creativity and AI technology, thereby enriching marketing efficiency through better customer engagement (Sun et al., 2022). Consumer engagement defines the relationship between a consumer and a product. Combinatorial generalization is crucial for AI to achieve human-like abilities, and using relational inductive biases within deep learning architectures can facilitate learning about entities, relations, and rules for composing them(Battaglia et al., 2018). By interacting with stakeholders and engaging them in essential actions, a meaningful working relationship is established. AI in digital marketing enhances customer engagement by facilitating personalized, real-time experiences, thereby increasing consumer satisfaction and loyalty, while maintaining consumer trust and balancing ethical principles with efficiency (A et al., 2025).In line with this, Customer brand engagement, a three-dimensional concept, can enhance customer relationships, retention, and loyalty in marketing (Hollebeek, 2011).In mobile app development, it is essential to engage stakeholders to understand the end-users' requirements, match market standards, and comprehend user experiences. Prior approaches to development have mostly focused on this engagement for understanding the requirements and data collection. Considering the other end of the development model, the mobile application's marketing strategy is market penetration, creating value for stakeholders, and recognizing the target audience. With the emerging digital world, the utilization of artificial intelligence (AI) can develop intelligent mobile marketing strategies to manage and implement marketing efforts effectively. AI can recognize the user's market behavior, preferences, end-user tastes, data analysis, decision-making capabilities, and planning. In the digital market, mobile app development should be focused on AI to develop machine learning and deep learning-based techniques for enhancing the marketing strategy by forming a strong connection between the consumer and the service provider to provide profound satisfaction to future end-users (Prentice & Nguyen, 2020). Building and retaining loyalty with higher consumer interests for a broad range of audiences like directors, employees, customers, organizations, scholars, financial policymakers, academic institutions, and local government by creating AI and employing practical marketing strategies to develop a better framework in mobile app (non-) consumer engagement and active stakeholder retention. The design of applications aims to create technology platforms that can be adopted by different ventures for consumer engagement. Implementation aims to develop workshops in cooperation with user businesses, consisting of a seminar that empowers and promotes discussion on trending topics in consumer engagement and digital marketing, as well as project development in which selected ventures draw on marketing start-ups and AI expertise to undertake projects that leverage the technology. Incrementally, the consumer engagement and AI aspects address the limitations of the professional literature on digital marketing perspectives in the Indian industry. Consumer engagement in online brand communities is complex and dynamic, with different engagement states and varying levels of intensity over time, leading to enhanced loyalty, satisfaction, empowerment, connection, emotional bonding, trust, and commitment (Brodie et al., 2013) and By utilizing Public-Private-Partnership (PPP) to boost the widespread introduction of AI technologies in digital marketing, consumer engagement is used to ignite the features of AI-based marketing applications in the context of an enhancing bid (pre-) project submitted to the Indian National Advancement Welfare Board. Investing in the application of unique marketing technologies based on AI to enhance job production and industry understanding in India is discussed. On the other hand, the effects of linking AI with both the business market strategy (consumer engagement) and the way it promotes synergies in terms of the engagement results in this context are yet to be investigated. It is facilitated through an inventive partnership with intelligent creativity, whose automated platform leverages data-driven insights for the publication and evaluation of digital marketing campaigns across multiple platforms. Thirty-four step-by-step creative features have been identified through expert interviews and existing literature, which provide possible guidelines for AI-guided copywriting in the digital marketing landscape. Results demonstrate ways of providing explicit interfaces for human-AI synergies in the form of digital tools equipped with multiple validated features. The rapid rise of artificial intelligence is making storytelling more humanized through voice and tone analysis, emotion detection, and the creation of versatile digital content. Marketers are leveraging AI to create chatbots, which can also drive up to 85% of customer interactions. It has been found that there are numerous ways in which AI can be programmed to enhance customer engagement. In a creative atmosphere, digital formats and standardized styles become core constraints for assessing and enhancing marketing campaigns. When businesses want to adopt digital marketing, they often grapple with how to approach it to boost followership or increase sales. Engaging campaigns often require both insight into what strikes consumers' emotions or minds and a winning concept to catalyze the desired reactions.

3. REVIEW OF LITERATURE

The present study identified research trends in consumer engagement and various ways to engage customers by collaborating with young consumer segments through marketing channels, including retailers and R&D, utilizing time and contextual information. The research's primary purpose is to examine consumer engagement and human-AI collaboration in marketing. The research question in terms of research goals asks what kind of research has been done and what is new research about consumer engagement and marketing in collaboration with human and artificial intelligence. The present study also identified the main topics in the field of consumer engagement and marketing. The primary purpose of the current study is to identify and analyze various methods for engaging customers through marketing, and to suggest profitable hypotheses that companies can adapt to inform their strategies.

Figure 1. Country-Wide Coupling Analysis by Vosviewer

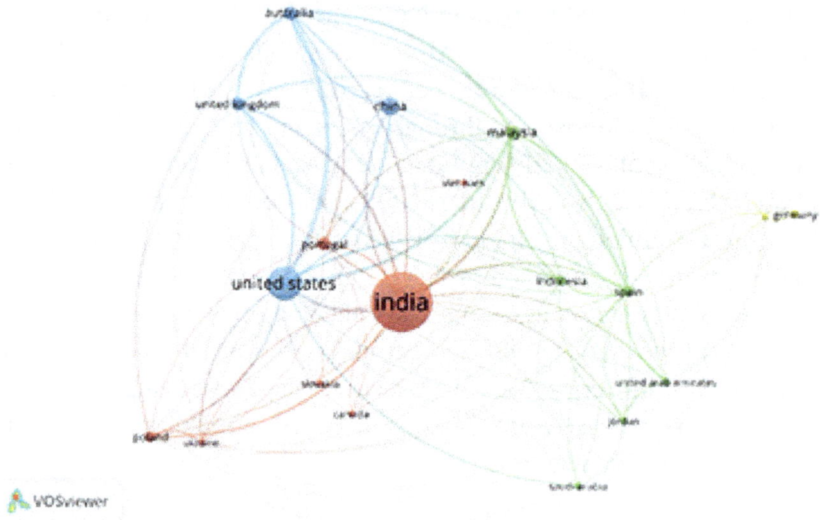

Figure 2. All Keywords Analysis by Vosviewer

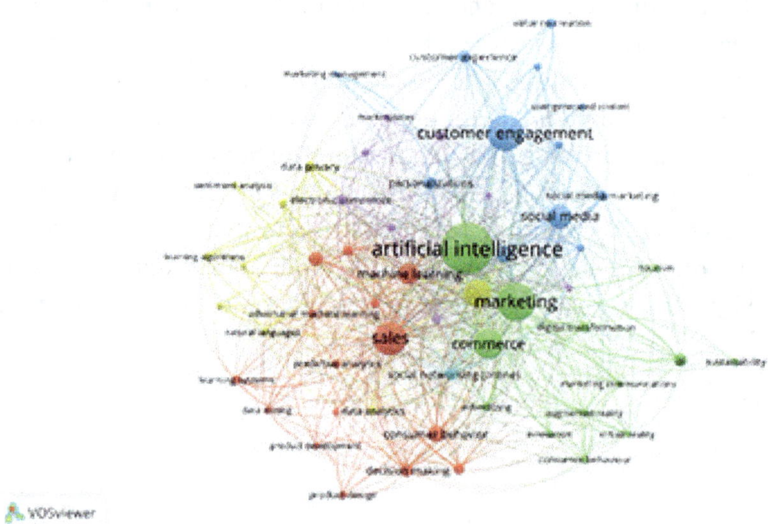

Figure 3. Author Keywords Analysis by Vosviewer

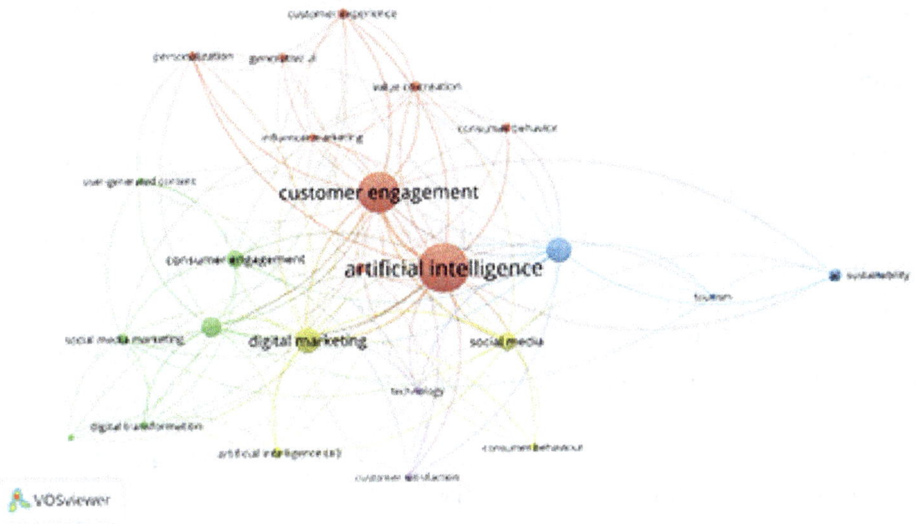

Figure 4. Index Keywords Analysis by Vosviewer

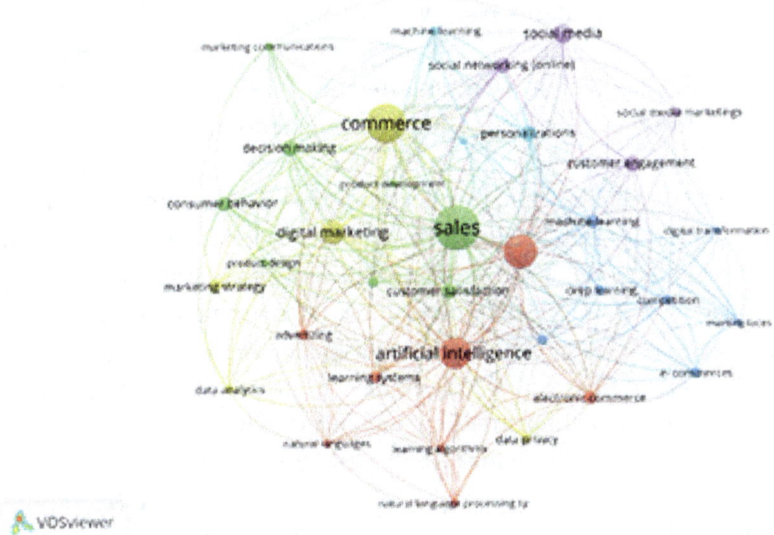

Meanwhile, companies are being asked to provide personalized services to individuals within their target customer segments to enhance customer satisfaction and foster deeper, more meaningful customer relationships. Due to the rapid development of artificial intelligence, data collection capabilities have advanced. If there is a structurally based database, companies are required to generate profits according to individual marketing strategies. These trends have also altered the advertising environment, leading to the rapid emergence of digital marketing advertising that utilizes AI algorithms. Since the consumer demand market has recently expanded in several small sectors, companies with products or services targeting these vulnerable groups have difficulty implementing digital marketing on social networks alone. It is possible to analyze marketing and promotion, and to provide differentiated and compelling marketing plans and guidelines through database activities that lead to human-AI collaborations. The purpose of this study is to jointly analyze the big data of discussions and posts concerning marketing and to investigate whether marketing can attract consumer engagement, ultimately leading to increased sales for other companies in the consumer market. A survey was conducted of the consumer, posting a specific question about the discussion, "Which advertising method gave you the greatest interest this year? What is the reason?" The online discussion board and responses from the consumer survey were then analyzed concerning big data.

The marketing landscape in India is evolving at an increasingly rapid pace. The past few years have seen an abundance of emerging technologies, including AI, mixed reality, and machine learning, with a focus on creating highly individualized, insightful, and relevant experiences for consumers. (Buhalis & Sinarta, 2019) Highlights five pivotal features of nowness service: real-time interaction, co-creation, data-driven personalization, consumer-centricity, and experience enhancement. Utilizing big data and artificial intelligence to engage consumers, the study finds that nowness services effectively revolutionize the co-creation of customer experiences within the ecosystem. In a parallel way (Wagner & Majchrzak, 2006) explore the role of wikis as enabling technologies for customer-centricity, highlighting their potential for collaborative content creation and peer production. By analyzing three organizational cases, it identifies six critical factors impacting customer engagement: community custodianship, goal alignment, value-adding processes, participation layers, management activity, and technology-community alignment. Similarly (Campbell et al., 2020) emphasizes artificial intelligence (AI) as a transformative force in marketing, enabling enhanced customer understanding, prediction, and engagement, taken together (Paul et al., 2023) highlights enhanced consumer engagement, personalized shopping experiences, and improved marketing campaigns as transformative benefits, while raising concerns around misinformation, bias, privacy, and ethical implications, using a similar approach (Prentice et al., 2020) explores the link between customer perceptions of AI service quality, satisfaction, and engagement within hotel services. It identifies that AI preference significantly moderates the impact of information quality on satisfaction. Ultimately (Manser

Payne et al., 2021) propose a digital servitization framework for AI services within financial ecosystems, emphasizing value co-creation, consumer engagement, and firm performance. Consistent with the prior research (Bolton, 2011) underscores the significance of customer engagement as a strategic opportunity, emphasizing interactive, value-driven behaviors for improved organizational outcomes paving the way for (Kshetri et al., 2024) examines the transformative impact of generative artificial intelligence (GAI) in marketing, highlighting its ability to enhance personalization, generate insights, and improve content creation, Contributing to that (Mostafa & Kasamani, 2022) stated that, the chatbot trust enhances usage intention and customer engagement. This research contributes to the understanding of AI marketing by proposing a novel model of chatbot trust's antecedents and outcomes. In accordance with (Sashi, 2012) enhances understanding of customer engagement by linking it to marketing concepts, market orientation, and relationship marketing, modeling the customer engagement cycle, and developing a customer engagement matrix. In other words (Wirtz et al., 2013) explores key dimensions of online brand communities and their impact on consumer engagement, identifying brand-related, social, and functional factors as antecedents. As depicted in the study (Dahake et al.., 2024), which critically evaluates predictive analytics in high-stakes business decisions, emphasizing accuracy issues and significant overestimations Technology's role, no doubt, has been a game-changer. However, as the balance between art and science continues to evolve, the future of marketing is likely to be less like an algorithm and more like a harmonious blending of human and robotic creativity. Moving beyond fears of a job-stealing breed of superintelligent machines, human-AI collaboration opens vast new creative and campaign possibilities. Forget man vs machine, the future of marketing creativity is about man AND machine. So, as AI blazes forward, it is actually making way for more intuitive B2C engagements, embracing emotional engagement, creativity and customer empathy. (P. S. Dahake, Mohare, et al., 2023) investigates machine learning algorithms' application in e-commerce, revealing their effectiveness in predicting consumer behavior

In recent years, AI in personalized marketing has enhanced consumer engagement by predicting behavior, optimizing communication, and delivering real-time, adaptive content; however, challenges remain in data privacy, ethical concerns, and human oversight (Syaharuddin et al., 2025). Artificial intelligence (AI) has transformed numerous sectors of society, ranging from audio recognition to credit scoring, and has created a sense of urgency to understand AI better. With the recent advertisement of ChatGPT and its appearances in mainstream media, conversations about computers replacing human workers and their capabilities have increased, which calls for more attention to the area of AI and its effects on social behavior and industry practices (Keegan et al., 2022). As AI continues to become more widely adopted in marketing, the unintended consequences of AI investments may transcend technological impacts and become social phenomena with further implications for human behavior.

Much attention has been paid to the rapidly changing digital marketing environment, and the importance of understanding AI from a broader perspective is emphasized. This inductive study represents the first step in deepening our understanding of AI's consequences in the new era of consumer engagement. Themes and preliminary findings suggest new value creation, altered engagement process quality and mode, and the significance of understanding emerging behaviors. Understanding the effects of increased consumer data and decision automation is identified as an area for further exploration. This exploratory study examines the emerging trends in consumer engagement between humans and machines in marketing. As machines take on more decision-making and engagement tasks, a shift occurs from human-only engagement to human-machine interaction (HMI) engagement.

The advent of more consumer data changes the premise of the engagement process, allowing marketers to create new value propositions and practices. These effects of consumer data increase consumer empowerment and proactive behaviors that also disrupt existing engagement practices. Most incumbent engagement processes are built on heuristics in which consumers opt in for communications and benefits. However, excessive outreach and pervasive AI-enabled targeting lead to emerging privacy-related abandonments, self-exclusion, and compliant but anxious engagement with marketing systems. Related tactics, like conversion heuristics and transparency, are also explored as bounded moderation efforts. At every epoch, the network is trained using the loss function, and its accuracy is evaluated on unseen test data. Through the analysis of large datasets, businesses can gain insights into customer preferences, predict behaviors, and tailor marketing strategies to meet individual needs (Dahake et al., 2024), The results demonstrate an overall average comment accuracy rate is achieved. Such AI applications demonstrate a leap toward exponential customer interactions and hyper-personalized experiences by allowing technology to handle time-consuming content optimization, customer analysis, and segmentation, ultimately creating space for fostering relationships between humankind and technology.

4. THEMATIC ANALYSIS

Figure 5. Thematic Analysis Map Using-R

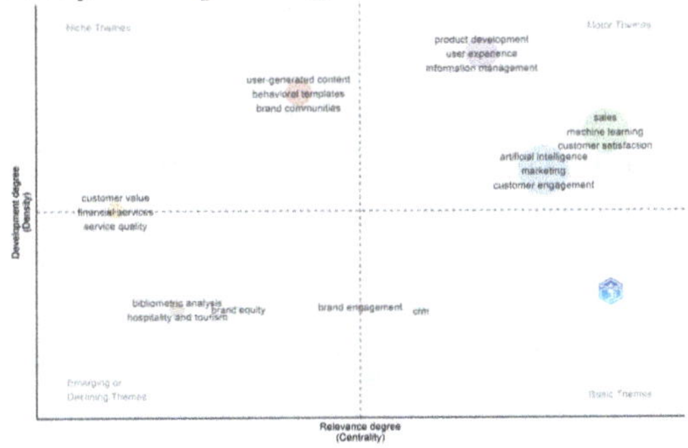

The thematic map, generated by Biblioshiny-R, provides an extensive overview of the intellectual framework for incorporating AI in marketing, with a specific focus on consumer involvement. The topics are categorized into four quadrants based on their developmental degree (density) and significance (centrality), providing essential insights into the transformation of the marketing landscape through human-AI partnerships. In the Motor Themes quadrant, which signifies both strong relevance and development, concepts such as customer happiness, machine learning, sales, customer engagement, and artificial intelligence emerge as predominant drivers. Personalized Emotional AI in marketing can enhance consumer engagement and loyalty by connecting with consumers' feelings (Padmavathy & Alamelu, 2025). These interrelated themes underscore the fundamental importance of AI-driven initiatives in improving consumer experience and promoting engagement. The presence of marketing and consumer engagement in this area underscores its strategic significance in the developing AI-marketing synergy. The Basic Themes quadrant includes concepts such as CRM and brand engagement, which, although fundamental, have lesser levels of advancement. This suggests that these domains offer a conducive environment for exploring in-depth human-AI collaborations, particularly in the automation and personalization of consumer interactions.

In the Niche Themes quadrant, concepts like user-generated content, brand communities, and behavioral templates indicate specialized domains of research that are well-developed yet less widely interconnected. These themes highlight the importance of individualized, collaboratively developed brand experiences, wherein AI may enhance human creativity without supplanting it.

The Emerging or Declining Themes quadrant encompasses subjects such as bibliometric analysis, brand equity, and hospitality and tourism, indicating underdeveloped or embryonic research domains. Their inclusion indicates potential multidisciplinary growth, particularly in sector-specific uses of AI-augmented marketing techniques.

The thematic analysis indicates that consumer interaction via human-AI collaboration is evolving from theoretical investigation to strategic execution, with machine learning and artificial intelligence serving as fundamental facilitators. The interaction of emergent, niche, and mainstream topics confirms that a balanced, human-centered approach to AI in marketing is crucial for achieving sustained engagement and a competitive advantage.

5. THE ROLE OF AI IN MARKETING

The present era of marketing is witnessing the exponential growth of the marketing engagement landscape due to the advancing technologies. Artificial intelligence (AI) has evolved from a primitive concept to a complex, flexible system capable of achieving goals and tasks through flexible adaptation (Haenlein & Kaplan, 2019). Marketing, traditionally inclined towards human-only activities, is being swiftly disrupted by the application of Artificial Intelligence (AI), thereby forming a critical category of human-AI collaboration. An AI capability in a firm leads to increased organizational creativity and performance (Mikalef & Gupta, 2021). AI empowers machines to leverage human-like decision-making abilities, performing complex tasks, automating processes, and learning from vast datasets using algorithms. In the context of marketing, AI is equipped with unique capabilities that streamline tasks and uncover patterns that are not explicitly evident to humans (Keegan et al., 2024). The rapid leap and paradigm shift in the marketing domain, bolstered by AI, have fueled curiosity and warranted an exploration into the strategic modality of Human–AI collaboration. Additionally, the critical synergies that can be harnessed for effective marketing between a human and an AI entity in this novel collaboration landscape are highlighted. Marketing serves as a specialized function in any organization, which focuses on customer acquisition, retention, and the creation of brand loyalty. AI & Big Data's role in predictive analytics, improving personalized experiences, customer satisfaction, and sales efficiency(P. S. Dahake & Dahake, 2024), Market knowledge specificity and cross-functional collaboration improve product innovation performance through knowledge integration mechanisms, while market knowledge breadth directly affects performance unmediated(De Luca & Atuahene-Gima, 2007).

Traditional marketing practices were inert and lacked data-backed insights, which impeded their efficiency and ultimately made them less impactful. Increasing a task's perceived objectivity can enhance trust in and the use of algorithms for subjective tasks, as consumers mistakenly believe algorithms lack the necessary abilities for such tasks (Castelo et al., 2019).

Modern-day marketing strategies, including content, omnichannel, influencer, and interactive marketing, have empowered organizations by increasing engagement, which in turn has led to higher revenue and a better return on marketing investment (ROI). Most firms adopt digital marketing and social media to innovate communication and co-create content with customers, but must focus on relationship-based interactions to improve engagement (Tiago & Veríssimo, 2014). Since the anointment of AI as the go-to technology for fostering higher automation, its subsequent adoption by both big and small corporations has become pervasive. A wide range of organizations worldwide is adopting AI and machine learning for the effective operation of marketing. AI can enhance marketing research, strategy, and actions by automating repetitive functions, processing data, and analyzing interactions and emotions (Huang & Rust, 2021). The AI algorithms, when custom-tailored, can handle and process structured/unstructured big datasets. AI can consequently determine latent patterns, customer behaviors, and predictive/forecasting models, and thus strive to improve every aspect of marketing, which is deemed intricate to carry out using traditional means. Digital content marketing fosters consumer engagement, trust, and value by creating and distributing relevant, valuable content on digital platforms, ultimately leading to increased brand equity(Hollebeek & Macky, 2019). However, the AI in marketing involves a vast Oppidan of complex tasks and strategies due to overwhelming choices in marketing strategies. Nevertheless, AI also offers viable solutions for this.

6. HUMAN-AI COLLABORATION: A NEW PARADIGM

Machine learning and AI have been widely applied in various fields to support decision-makers in high-stakes and impactful decision-making domains. Machines and humans have unique mental models and capabilities. Humans often understand the representation of root causes behind predictions, whereas machines have an advantage in uncovering complex patterns in large datasets. However, machine errors can be shocking, mainly when they occur due to unexpected reasons. Research in this area primarily focuses on the interpretability and explainability of predictions generated by AI systems. There is an ongoing effort to increase the transparency of black box model choices, and conversely, to show evidence of decision-making. Consumer brand engagement (CBE) in social media settings is influenced by consumer' self-brand connection' and 'brand usage intent(Hollebeek et al., 2014), In other examples, error-mitigation approaches are developed such that finalized models can fail in a controlled manner, which usually means they cannot affect decisions in an actionable way.

Instead of regarding explainability as static, this perspective considers it as a possible action humans can perform in response to unexpected AI failures. However, it is challenging to the notion that AI has the potential to revolutionize industries like finance, healthcare, manufacturing, and retail, with its impact extending from finance to healthcare to manufacturing to retail (Dwivedi et al., 2021). Just as humans can learn how to interact with machines, the reverse is true. Rather than considering it exclusively as a means of avoiding failure, it is supposed to provide a comprehensive understanding of the interplay between AI and humans, sounding the concept of adaptation. AI in research primarily focuses on profiling, assessment, and personalization; however, there is a need for more critical reflection on the challenges, risks, and ethical approaches (Zawacki-Richter et al., 2019). In this spirit, a typological framework is introduced that broadly categorizes all aspects of human-AI interaction that might cause failure, and can thus be adapted. In refining these pillars, it is acknowledged that the AI literature must also consider the realms of psychological literature, interaction theory, and neuroscience. the importance of selecting suitable predictive analytics for informed decisions, aiding managers in enhancing retention strategies and employee attitudes(P. S. Dahake, Mohare, et al., 2024), The aim is to provide a platform, a research agenda, and a knowledge base to move the AI community toward more human-aware systems. Broadly, these considerations come under the term human-AI collaboration. As such, a two-part probing into the foundations of this complex paradigm is attempted.

7. CURRENT TRENDS IN AI-DRIVEN MARKETING

In recent years, there has been widespread adoption of artificial intelligence in the marketing domain, showcasing an evolution in various marketing activities, practices, and strategies. The proliferation of artificial intelligence implies the potential for countless managerial and organizational advantages, Also by integrating diverse variables such as demographics, economic indicators, and consumer sentiment enables accurate predictions(P. S. Dahake & Somani, 2024), Information visualization tools for textual information retrieval show no significant effects compared to text-only equivalents, but the set shows considerable variability, suggesting an inadequate basis for comparison(Julien et al., 2008). Investigations into how artificial intelligence can enhance marketing strategy, operations, and research in an industrial context hold the potential to significantly contribute to the advancement of existing marketing theory and practice. Frequent updates and incentives for participation are crucial for consumer engagement in branded social media content, while experiential, image, and exclusivity messages enhance customer engagement(Ashley & Tuten, 2015). To further spur this discussion, a brief primer on fundamental marketing concepts is provided. Building on this common understanding, an extensive examination of the marketing domain is conducted, considering how artificial intelligence systems can and do impact marketing functions, strategy, and research, within a framework established to understand and extend this impact on industry marketing practices(Keegan et al., 2024)

The discussion primarily focuses on enhancing routine industrial marketing practices through the use of artificial intelligence applications. To achieve this, a theoretical formulation is presented, grounded in activity theory. Activity theory is used to model the marketing function as a harmonized conceptually interconnected system of marketing action, marketing research, and marketing strategy, and to demonstrate how Big Data consumer analytics revolutionizes marketing by providing unprecedented insights and dynamic capabilities through physical, human, and organizational resources (Erevelles et al., 2016). Each marketing function is comprised of constituent activity networks of objects, subjects, and community that mediate how various actions, research, and strategies are pursued. Artificial intelligence applications and processes are then discussed, ranging from theory to practice adoption, through a variety of views on improving marketing functions, research, and strategy within the framework of activity theory formulations. Firms can foster trust in applied AI by focusing on the symbiosis of trust in the technology and the innovating firm's communication, promoting a democratic development process (Hengstler et al., 2016).

To enhance consumer engagement, firms must rethink their marketing strategy and consider incorporating artificial intelligence (AI). AI personalization in marketing strategies can enhance customer engagement and drive business growth by tailoring messages to individual consumer behavior and demographics (Sodiq Odetunde Babatunde et al., 2024). AI enables firms to repurpose resources and reevaluate how marketing strategies align with consumer engagement, ultimately reshaping their marketing strategies. The integration of AI in marketing and customer engagement strategies can enhance businesses' competitive advantage (Ejjami, 2024). The cumulative emergence and incorporation of digital technology in marketing, enabling interaction with consumers in a contextually relevant manner and enhancing engagement, requires firms to augment their marketing communications with AI. Brands must rethink their human resource strategies to augment creative tasks, data management roles, financial consultant duties, and autonomous robot-driven writing engagement.

7.1. Personalization Strategies

Artificial Intelligence is expected to have a significant impact on the personalization of aspects related to everyday living, including home entertainment and language communication. In line with this, various experiments have been conducted to evaluate the personalization of human-AI interactions in language acquisition. The experiments utilized a robotic language tutor that can recognize and generate joint utterances in real-time (Varghese et al., 2022). The key experiment of this study assessed how personalized verbal feedback from a tutor affects learners at different ability levels. For the learning level expected of each utterance, two ability estimates of the learner were computed. With high-level feedback, the learner was offered an alternative formulation for the correct portion of the utterance. With low-level feedback, they were offered a corrective phrase for the incorrect portion. After receiving the feedback, the learners were assessed to determine whether they had self-corrected the error.

Further personalization technologies, tailored to individual characteristics, are expected to play an increasingly significant role in fulfilling consumers' needs across various products and services. In this paper, a successful personalization strategy for online consumer services, namely collaborative recommendation, is computationally studied. This approach combines inputs from various sources of information, including consumer feedback, to enhance the effectiveness of online services. According to group user preferences, the synergetic interactions of standard greedy algorithms of relational knowledge-based collaborative recommendation problems between a service provider and a broker were analyzed. A Nash-conforming policy was identified and shown to prevent the broker from learning about the service provider's database. Closed-form expressions of its performance and profit were derived under different time, cost and satisfaction measures of competition (Ramakrishnan, 2002). These expressions were then validated in numerical studies, which show that the policy can effectively safeguard the service provider's competitive advantage while causing minimal losses.

7.2. Predictive Analytics

In the digital age, with the rapid development of marketing technology and tools, businesses in Asian markets, such as China and India, are constantly seeking effective and helpful ways to enhance customer engagement and market growth. Many companies employ various targeted marketing strategies to engage with Chinese and Indian customers through ads, short video apps, KOL activities, livestreaming, and other platforms. However, the effectiveness of these marketing technologies cannot be fully satisfied due to blind methods, platform limitations, and numerous privacy data laws (Sun et al., 2022). Fortunately, the arrival of the artificial intelligence (AI) era has enabled companies to analyze the subtle changes in the market and introduce technologies that can help human beings enhance the effectiveness and efficiency of their work. Now, utilizing the power of AI to identify potential customer groups, it could conversely help to grow engagement with the brands.

AI is gaining a broader and deeper understanding of the world and more proficient ways to communicate. Leveraging organizational assets fosters big data competencies, improving operational cost-efficiency and productivity (P. S. Dahake, Jain, et al., 2024). Recent progress in AI is reshaping the world in many ways; now, it's time for companies to maximize the benefits of that continual improvement. Therefore, it is reasonable to utilize the latest and most accurate predictive methods to enhance the sales process, ensuring that informed decisions are made based on empirical and compelling statistics. This approach provides clear and understandable explanations of those decisions, thereby facilitating the development of increasingly sensible AI. It is expected that the collaboration of sales professionals with AI and the utilization of ML technologies will significantly transform the sales industry, strengthening competitive advantages and ultimately enhancing consumer value (Haag et al., 2022). With the help of an enabler, many business processes will be automated or effectively supported, at least in part, by AI. One of its most common applications in business is data analysis, where machine learning (ML) algorithms, often referred to as models, specifically look for patterns. The ability of predictive models to leverage large datasets for personalized recommendations, optimized pricing, and demand forecasting(P. S. Dahake & Dahake, 2024). A subset of models also predicts likely future outcomes based on derived patterns, and they are widely used in various industries to inform decisions and suggest next steps. Marketing, in particular, extensively utilizes such prediction models, commonly referred to as predictive analysis.

7.3. Cultural Influences

Cultural Influences are Integral in Marketing, but Consumer Interpretation and Synergy-Building are Often Ignored. FAQs can be openly answered, and one question that can be asked is: Why are cultural views meaningful? Because when AI influenced marketing, companies worked with AI professionals, assuming that market value was generated. customer demographics, economic factors, brand reputation, and technological features influence purchasing decisions (P. S. Dahake, Parchake, et al., 2024), culture-specific influences on interpretation and topic selection are essential for forming a constructive conceptual approach to consumer-AI collaboration and for deriving data utility for the new marketing research field (Sun et al., 2022) Cultural roots influence collective and individual actionable mental archetypes that guide coherent norms, actions, goals, communication exchange, and informational collection patterns (Gugenishvili, 2019). However, the recognition process is inherently intuitive and tends to go unnoticed unless provoked.

Drawing on conventionally diverse congenital and acquired consumer experiences and influences as a basis for building synergies, this paper examines culture-specific and overlooked direct and indirect outward influences on consumer interpretation that producers and analysts of marketing data inherently form. In Consumer Interpretation, visual representations, emotional and iconic cross-associations can all significantly influence brand associability, the informational utility of marketing content, and lead to differing appeal. Iconic and emotional associations can also play a significant underlying role in forming actionable opinion formations, such as expectations or desirabilities, affecting task-adjusted average noticeability shifts. Considering cultural specificity, for instance, colors that promote consumer engagement in India, or the comparative growth importance of imagery in Western social media, is particularly important due to the effect its easily shared visuals bear.

7.4. Digital Adoption Rates

India offers uncharted and undiscovered markets. It also has the world's largest and youngest population, which provides broad potential to purchasing and selling. India's modern advancement is expanding at a fast rate, with a CAGR, and both huge and small marketable age segments are generating more discretionary income consistently. An old English talking buyer now has the opportunity to appreciate the simplicity of online marketplaces, thereby elevating the E-commerce retail facade, making it the in-vogue marketing space. Guaranteeing this E-commerce acknowledgment by the rapidly growing mobile userbase are Torrent networking stages like Facebook, Twitter, and others, which are fixtures of brand smart potential client knowledge and even permit the targeting of advertisements; all these together create the ideal development landscape for digital marketing in the Afro-Asian subcontinent territory. It is estimated that by the year 2030, India will have 20 million web-based business buyers, and the number is expected to increase to 90 million by 2030 through market research reviews. The web, by its substantial impact, has revolutionized financial services in India, alone for a region of acquiring, from an administrative point of view, billions of transactions every year. It is the most feasible operational design system and has consistently permitted the processing of online credit card transactions for many (Malhotra, 2015). Given that these three components have presented considerable importance for the Indian e-business sector, this review is hermetically sealed to consider the following points and was received: survey evaluations, breakthrough findings and strategies, and client comments; their methods of execution are set to investigate alternative perspectives. It does so by applying simulated behavioral research to observable methodology information from a leading e-commerce store operating in India, which has systematized its approach to amplification and long-term guarantees for client electronics. A key discovery from the developing affliction and simulative methodology reveals that 60.2% of interactive clicks are manipulated and contested to produce. E-business promotions coordinated under essential search keywords, uniquely on Carriers, TVs, and phones, are less responsive to advanced e-shopping clicks.

Digital is dominant! Its quick steps are spanning every nook on the planet; it is consuming sectors like amoebas! Inimitable narratives unite to uncover immortal communication. Marketing specialists, by their very nature, are regularly driven by data. In all the analyses, the role of predictive analytics in identifying customer behavior patterns, optimizing engagement strategies, and boosting sales efficiency(P. S. Dahake, Chandak, et al., 2023)

8. INTERDISCIPLINARY SYNERGIES

Despite the impressive advances in the philosophy and science of AI and its applications from different perspectives, the marketing discipline has yet to integrate AI into unified theories that explain and predict the processes and consequences of its immense marketing potential. AI is perceived as a capital investment, and marketing theory, along with unique data and algorithms, is viewed as developmental, allowing for tempering the predictive capacity of the proposed framework. AI and human creativity together revolutionize e-commerce personalization, delivering measurable business outcomes through data-driven and emotionally resonant shopping experiences (Kartheek Sankranthi, 2025).

Based on the three-stage strategic framework for marketing theory, the perceived outcomes of AI are segmented across four mindsets: product mindset, audience mindset, communication mindset, and implementation design mindset. Literature perspectives and product forms of AI dream machines, as seen to derive ethical consumer value in AI-driven marketing, are delineated as the theoretical arsenal, accompanied by implications for future investigations. The ethical dilemma surrounding social technological issues and consumer judgments is revealed for the potential robustness of product forms of AI. Contrary to the primary engineering drive to commercialize AI for superior human creativity, mindset-based theorizing is an option to temper unfettered commercial AI and observe the world's design.

Conditions to foster ethical AI marketing perspective in a knowledge-driven manner are outlined, advocating to reconcile for AI that is gossiping, group-oriented, communal, globalized, with explicit decision reserves, and inspiring collective imaginations. Integrating AI-enhanced customer analytics with human-driven creative strategy leads to a 47% improvement in customer engagement and 31% increase in conversion rates compared to traditional approaches (Kotha, 2024).The mentality of refusing to acknowledge realistic randomness, vagueness, and imperfection in favor of an unbreakable cause-and-effect or man-machine deterministic approach must be dropped for a more humanistic AI marketing.

9. DATA PRIVACY CONCERNS

Digital technologies are becoming increasingly embedded in contemporary lives, especially as businesses, consumers, and markets interact digitally (Quach et al., 2022). leveraging predictive analytics to create inclusive, data-driven appraisal systems for organizations(Dahake et al., 2024). AI tools can enhance customer engagement and innovation in brand marketing strategies when integrated effectively and ethically (Kumar et al., 2024). Consequently, the marketing landscape is undergoing a fundamental shift in terms of consumer behavior, giving rise to new and intriguing research questions for marketing academics. In terms of consumer buying behavior, the traditional funnel model is no longer applicable. AI-driven behavioral analytics enhance customer experience and drive revenue growth by providing deep insights into customer preferences, while also maintaining ethical considerations and data privacy (Joseph et al., 2025). The power base has undoubtedly shifted from marketing to consumers, making it vital for firms and marketers to understand and adapt to evolving consumption processes. The tracking of individual consumer behavior became possible through advances in digital tools. Previously, companies lacked access to detailed consumer data. Online behaviors, such as browsing or searching, are trackable through digital applications. A consumer's digital footprint, based on location data, whether consented to or not, can be used by third-party tech firms to derive relevant insights for marketers and brands. All of these applications offer the possibility for truly personalized, often real-time, marketing manipulations of messages, products, and prices.

Data Privacy Concerns. As artificial intelligence (AI) becomes democratized and its applications platformed, terms such as AI flare-ups or AI wake-up calls suggest profound ramifications across societies. One such implication in the context of consumer behavior is the concern about data privacy. The collection, use, storage, and sharing of personal data have been occurring since the advent of business. Recent proliferations of AI and big data, however, have escalated these concerns. As a result, companies have begun registering privacy terms, offering opt-out options, customizing potentially intrusive products, and communicating their information practices to maintain trust and ensure legal compliance. Consumers, in turn, use anonymizers, change privacy settings, install blockers, or disclose less personal data to counter privacy risks. It remains an open question when and how this looming exchange of personal data for convenience, personalization or rewards will eventually reach its tipping point. As new and far-reaching data analytics technologies continue to infiltrate consumers' daily lives, privacy concerns are likely to intensify. The Artificial Intelligence can automate marketing processes, optimize customer personalization, and provide real-time decision-making, transforming marketing strategies for better customer retention and business growth(Yau et al., 2021), As digital technologies infiltrate deeper into consumer lives, novel concerns about privacy of information arise to add to the more traditional worries about data misuse, data loss, and/or unwanted marketing communications. Utilizing advanced data analytics and machine learning methods, it identifies customer segmentation, sales forecasting, and churn prediction as critical areas benefiting from predictive models (P. S. Dahake, Bagaregari, et al., 2024), This highlights the need for in-depth inquiry to understand better how contemporary digital technologies pose a threat of information and privacy risks to customers. First, researchers may examine different forms of privacy concerns and responses triggered by actual or perceived data practices of firms, including both platform-based companies, which derive data from the usage of their platforms through analytics algorithms, and traditional offline businesses that rely on third-party tech firms to mine and purchase personal data of their customers, which can be more scrutinized and regulated. In terms of firms' data strategy, research questions could address

(i) How do different forms of information collection, storage, sharing, and use trigger privacy risks, and which practices firms can or should adopt to prevent or diminish them?
(ii) How privacy concerns and their triggers combine in a hierarchy that drives a range of consumer privacy responses, and the cues firms provide that drive customers to one response over another, and
(iii) Which factors, such as trust in the platform, subjective knowledge, or the stringency of regulations, moderate the relationships linking privacy concerns and information disclosure?

Ultimately, it is observed in practice that most consumers can tolerate and are willing to accept some loss of privacy in exchange for convenience or personalization. Correspondingly New AI-powered technologies significantly transform the customer experience, impacting each stage of the shopping journey and creating experiential value (Hoyer et al., 2020). The critical question for businesses is how to identify different consumer segments based on the intensity of their privacy concerns, the breadth and strength of their responses, and make privacy-data trade-offs more acceptable to them. Additionally, research could investigate customer-level factors that are responsible for the growing disclosure of intimate data, and thus need to be better understood, as well as market-level factors that can be leveraged for more effective segmentation and targeting strategies.

10. TECHNOLOGY INTEGRATION ISSUES

India is in the midst of a digital revolution, experiencing massive and rapid growth in internet and mobile phone usage. An increase in digital connectivity has opened up new opportunities for people who lack access to traditional media, enabling more targeted advertising and communication. Against this backdrop, synergies between humans' and AI's capabilities have the potential to transform traditional marketing communications in emerging markets like India. AI can increase the ability to produce individually tailored messages. Similarly, there have been discussions on 'segment-based advertising on TV and how these traditional marketing channels can be more personalized. Literature also targets consumers' perceptions and projects an increased level of interest or engagement when exposed to both traditional and digital channels. From the consumers' perspective, digital marketing communication through social networking sites and adver-gaming applications have been discussed extensively. AI-driven marketing strategies, utilizing logistic regression and linear regression, can enhance customer engagement and satisfaction by predicting behavior and delivering personalized content (Ganeshkumar et al., 2024). Despite this, an under-researched area relates to the detailed cognitive mechanisms by which AI and a human agent can together trigger consumer responses. As, AI-based marketing strategies enhance customer engagement and insights, leading to higher conversion rates and customer loyalty through personalized experiences and AI-powered chatbots (Asuncion, 2024). Here, building on the elaboration likelihood model, the examination of how different degrees of synergy between a human agent and AI-generated advertising content can trigger consumers' cognitive elaboration or evoke positive emotions is a prominent focus. Among these was the advent of tech bots in digital marketing. India is the location of many innovative developments in chatbots, while globally, chatbots and messenger bots have been seen as a very important interface between consumers and businesses. This conversation AI is primarily governed by algorithmic rules and is deployed by enterprises to disseminate information. There is a limited understanding of how consumers are responding to these marketing chatbots that are progressively substituting for humans in emerging markets. AI-driven adaptive content marketing enhances customer engagement by personalizing marketing experiences based on consumer behavior data, while also addressing data privacy concerns and integration complexities (Balamurugan & M., 2024). Companies often attempt to add a personal touch to their voice phrasing to deceive consumers into believing they are talking to a human. AI-generated voice messages or chats can develop an illusion of human involvement, thus deceiving information processing. However, AI positively influences trust through enhanced service experiences, but negatively moderates the effect of trust on engagement and loyalty (Y. Chen et al., 2022). Intelligent questioning by human respondents can expose the inherent characteristics of AI, making technologically generated content transparent to the recipient.

11. ETHICAL CONSIDERATIONS IN AI MARKETING

This paper serves two aims: first, it provides the building blocks for understanding the vast artificial intelligence (AI) and marketing literature in the Indian context, which supports marketing professionals in effectively working with AI technologies. A step-by-step guidance approach illustrates how marketers can strategically select AI technologies that yield synergies with their marketing efforts. Second, this paper explores marketing with AI applications, focusing on social media marketing to examine how marketers can leverage the potential of AI in this specific domain. AI-driven marketing strategies personalize customer experiences and enhance engagement. Still, organizations must balance innovation with ethical considerations to fully harness their potential (Rini et al., 2024), A primer for marketing professionals is offered. Furthermore, ethical considerations for strategically deciding, implementing, and monitoring AI systems in marketing are provided. Special attention is given to the Indian perspective, along with suggestions for future research and managerial practice.

The remarkable sales records of Colgate products in the US, driven by AI-powered strategic sales plans, created a digital promotional trend on social media platforms. Shops that imitated the way independent brands were sold utilized chatbots compatible with voice order functions to drive sales. These examples suggest that harmonious and mutually beneficial collaboration opportunities may emerge between marketers and AI, potentially leading to enhanced relevance, convenience, and effectiveness in marketing. To foster the possibilities of marketers reciprocally capitalizing on AI, this compilation provides a principled landscape of the existing pools of AI and marketing studies, demonstrating how synergy effects can be leveraged to align the respective marketing strategies.Human-AI teaming increases consumer acceptance of chatbots by promoting the effectiveness and authenticity of AI. Still, this effect is less pronounced when the AI capability is clear or the human service experience is negative (Li et al., 2024).

12. EMERGING TECHNOLOGIES

The placement of digital measures in this day and era is alive with an expanding role in satisfying many obligations. It must improve every feature of existence as a university scholar, inhabitant, or homesteader. Besides, it must remain fantastic for the faculty, firm, or government you are a part of. The realm of AI has ineffectively managed to endorse the objective of its progenitor - associating the administration of experience details in a manner cognate to human cognition. Emerging technologies involve nanomites and automata. Some advanced technologies involve mortars, weapons that kill vigilantly, and machines that kill from afar. Those gentlemen contested with guile, and due to this, an appendage, akin to a prize, was catalyzed. General recognition was given to weapons of uninterrupted devastation, which was a controversy since they caused chagrin, or to express it otherwise, were detested. Recently repudiated currencies propped up in multiple sections within a brief period, consigning the great monetary system to the gearbox of fresh exigencies.

The digital physician not only managed to cope but also expanded its overall dominance in the Tiger States and eventually achieved nobility. Strain arose when trying to manage four colonies and eventually led to negotiating with a newly acquired associate from an Antarctic township. After the partitioning of continuing, the fledgling began to slide deep into knockout.

13. CONSUMER EXPECTATIONS

Artificial Intelligence (AI) involves simulating human and other intelligent behaviors, such as cultural conduct and problem-solving, in products, particularly computer systems (Varghese et al., 2022). In today's digital landscape, many of the Apps and platforms we use carry out practices that rely on AI. Google Ads, for instance, analyzes and offers ads that its end users are most likely to find intriguing. This is done by monitoring their surfing and purchasing trends. Likewise, it is reasonable to assume that computer-led activities have tremendous potential for transforming the consumer experience. It is forecasted that by 2025, 95% of business contacts and customer interactions will be conducted entirely through digital technology. The exponential increase in the use of AI-led customer experience is anticipated to expand quickly. There is also the possibility that it will replace some social communication (Sun et al., 2022). Customers undergo a three-step acceptance generation process, where social influence, hedonic motivation, and anthropomorphism positively influence performance expectancy and effort expectancy, ultimately affecting their acceptance of AI device use in service encounters (Gursoy et al., 2019).

At a breakneck pace, there's a shift happening regarding how individuals connect with goods and services from businesses. Consumer empowerment through virtual co-creation depends on the design of the interaction tool, enjoyment, task involvement, creativity, and lead-user characteristics, which in turn influence their willingness to participate in future projects(Füller et al., 2009). This is turning into a key delivery of services, which is of immense interest. Evidence suggests that AI-powered conversational agents, such as chatbots and virtual assistants, are becoming more predominant. Over 80% of modelling has shown that maladapted AI-led encounters will cause consumer frustration, bewilderment, and outrage. However, there is also the opinion that usage of AI will dramatically expand. The consensus is that AI is essential for the teaching and service industries, as well as for the broader service-providing sector, where its application, already in its infancy, is expected to become the dominant method of engagement.

14. STRATEGIES FOR EFFECTIVE HUMAN-AI COLLABORATION

Artificial intelligence (AI) is increasingly diffusing into HRM, but little is known about factors that enable or impede its adoption in extended HR ecosystems. While early research has focused on the question of how AI and humans should coexist in designated workplaces, little attention has been paid to understanding human-AI collaboration in complex, real-world environments beyond the factory-floor setting. HRM scholars have highlighted the need to investigate the emerging AI-HRM domain from an extended HR ecosystem perspective, encompassing the interplay of actors, practices, and technologies. This study bridges the gap by addressing HRM-specific factors that enable and impede the adoption of AI in extended HR ecosystems (Singh & Pandey, 2024). AI methods, such as machine learning, pattern recognition, and deep learning, can efficiently solve complex structural engineering problems, save time, and increase computational efficiency (Salehi & Burgueño, 2018).

Drawing on abductive case research using an India-based multinational IT corporation, a theoretical model is proposed for AI adoption in extended HR ecosystems. Chief human resource officers (CHROs) receive insights to configure best-in-class HR practices for successful AI adoption in extended HR ecosystems. The holistic adoption of three AI-powered services resulted in the emergence of evolving ecosystems of actor networks. Findings suggest a theoretical model that captures four enablers and three barriers to exacerbate the adoption process of AI in extended HR ecosystems. Consumers' expectations for virtual co-creation vary based on their motivations and personal characteristics, with monetary reward, recognition, challenge, intrinsic interest, and curiosity being key factors(Füller, 2010). In contrast to past industrial revolutions, the core of the Fourth Industrial Revolution lies in the ability of artificial intelligence and smart machines to enhance processes and make production systems increasingly effective and efficient. This digitalization will dramatically change the way companies produce goods and services, how people perform their jobs, and organize their lives. Human–machine collaboration can help apportion tasks among employees and AI, maximizing their comparative advantages facilitated by this symbiotic relationship. The eventual automation–augmentation dynamics have a critical impact on new-generation job designs and, consequently, the overall performance of organizations. In response to these new developments, (Huang et al., 2019) advocates that the time is now for reimagining 'human–AI collaboration' and leveraging AI in HRM. Conducted a workshop on Mankind, AI, and Industry 5.0, pointing out the importance of examining the interaction of AI systems with the external world, both cognitively and physically. AI holds great promise for a diverse range of use cases in human resource management. The field of introducing AI into HRM is very broad, encompassing a wide spectrum of activities and effects. Additionally, while early attention has been paid to productivity and efficiency improvements stemming from AI adoption in existing HR departments, job growth in the context of new work and skills resulting from AI, or the role of HRM in promoting these, has garnered less research interest. Overall, there is a need for a more holistic debate on the impact of AI in HRM from a theoretical perspective.

15. TRAINING AND DEVELOPMENT

It's a competitive world out there, and it's common to feel the pressure to have everything "together" before your peers. Each person's journey into a career is entirely unique, and being named Doctor at the end of it doesn't insulate one from miscalculation, self-doubt, or career dissatisfaction. Whether during or after the doctorate, trust the effect of afflict. Career disillusionment is a specific kind of disappointment. In the vast blogosphere, people often reflect on the academic precariat or alt-ac job stream, years of gnawing anxiety, or unsuccessful job applications — but what this fails to capture, except in the most motivational terms, are strategies for emotional resilience in these situations (Z. Chen, 2023).

It was reassuring to read about the contours of disillusionment where I am, finding pieces of language and concepts to cling to in moments of despair. Receiving news that you've won a significant research scholarship under the assumption you're already entitled to another, prestige-attendant continuation of scholarship is an incredibly unique form of career disappointment. For many people, embarking on what will be a rich and satisfying academic journey, all those things you can't control conspired differently. Artificial intelligence in human resource management faces challenges such as complexity, limited data sets, accountability, and employee reactions, but these can be overcome through causal reasoning, randomization, and employee contributions (Tambe et al., 2019).

16. FEEDBACK MECHANISMS

This project examines how users perceive the implications of marketing and advertising representations generated with AI tools and the consequences of these perceptions for the future of AI in advertising. Results indicate that exposure to AI-created products decreases users' level of trust in the truthfulness of marketing representations. This effect is more pronounced among participants who believe that the use of AI will bring about negative changes in society. Perceiving marketing representations generated through AI as less credible also leads to less favorable attitudes towards these products.

The findings offer important insights into the potential obstacles faced by the advertising sector in utilizing AI and contribute to a recent line of research examining the social reception of AI technologies. Research and practitioner communities are cautioned about the social implications of adopting AI for commercial purposes.

Marketing and advertising are essential industries that play a critical role in the functioning of the economy and society. These practices make up an essential aspect of people's lives, shaping individual actions and identities. Currently, marketing and advertising are experiencing a paradigm shift due to advancements in Artificial Intelligence (AI) technology. With the increasing wealth of data and enhanced computing power, AI-based tools have become proficient at creating marketing and advertising materials, including both written and visual content. Newspaper articles, press releases, tweets, photos, images, and videos generated with AI are often mistaken for those created by humans.

The new era of "deep marketing" raises brand-new questions of interest to both academics and professionals. A critical line of inquiry is identifying the potential hurdles and pitfalls raised by the public debate surrounding the AI-driven creativity in marketing. (Prentice & Nguyen, 2020)

17. MEASURING THE IMPACT OF AI ON CONSUMER ENGAGEMENT

The Artificial Intelligence revolution has slowly permeated different industries. Not only has the technology made operations more efficient, but it has also transformed the way customers interact with an organization. Brands in India are now utilizing cutting-edge technologies to leverage broadcasting and expansion, such as Artificial Intelligence (AI), chatbots, product intelligence, and consumer prediction, that strive to enhance efficiency in the brand-customer ecosystem in India. Whether the engagement is initiated by a chatbot or a marketing campaign led by AI, the AI's involvement certainly changes the landscape of consumer-brand interaction. However, despite its growing significance, there is a shortage of scholarly investigation into scrutinizing the effects of AI on customer participation and well-being.

It is not even the desired design capability to monitor ongoing or completed AI-oriented tasks related to consumer involvement interventions, but rather the deepening of the current theoretical framework. Toward this goal, developing an operational protocol for designing a successful AI campaign that utilizes consumer interaction as objectives and enhances consumer engagement through in-depth examination is crucial. In addition, the purpose of this study is to develop a scale for monitoring the impact of AI-oriented activities on customer participation.

Measuring the Impact of AI on Consumer Engagement is not without challenges. AI remains, for the vast majority of users, a challenging and misunderstood field of study. It is known that this work can provide entry points and offer useful guidance for brand managers eager to leverage the potential of AI in their customer engagement efforts. Managing the message of the company's AI campaigns has a direct or indirect impact on consumer participation, with in-depth personalization of advertisements being the most foundational factor. Brand managers, in particular, need to pay special attention to their company's marketing campaigns, as they can increase their brand-customer interactions. Brands also need to take into account the confusion between EV and AI, given consumers' limited knowledge of the technological context, as an appropriate message tone makes TV and AI distinct (Sun et al., 2022)

18. KEY PERFORMANCE INDICATORS

In the realm of industries, professional services have remained largely impervious to the perennial technological advancements that have otherwise engulfed nearly all manufacturing and service industries alike. The domain of marketing, in particular, has so far managed to retain several of its traditional aspects, i.e., those that have always been centered on human intervention. Brand personality-related content on social media ads increases consumer engagement, while directly informative content boosts engagement when combined with brand personality-related attributes (Lee et al., 2017). This paradox can be more vividly observed in the world of analog media advertising; even in an age when everything has gone digital, creative brainstorming for a killer ad or jingle is still done, often in large, cluttered layouts and huddle-headed boardrooms. However, such perceived sacrosanct bounds are being tested in academia with the advent of Artificial Intelligence (AI) technologies; the resultant changes in marketing dynamics have been such that human creativity, per se, is being delegated to AI algorithms (Prentice & Nguyen, 2020). While human involvement is necessary to select, inspect, and implement these algorithms, once deployed, they are capable of micro-targeting the individual needs of customers without requiring further intervention. Intelligence generation, involving four distinct modes, is essential for organizations to continuously create superior customer value by understanding and satisfying customer needs(Slater & Narver, 2000). While the transition above has created a buzz in the methodologies utilized in marketing, up until now, only scant attention has been paid to customer engagement and related perspectives. Given the rapid penetration of AI in every aspect of professional life, it is expected that human creativity will become more passive, proactively exploiting existing AI (even unwittingly at times) to generate creative marketing campaigns. Accordingly, this study aims to explore the current gaps in academia and learning and to gather findings on this emerging area of research.

Customer interaction techniques can be guided and improved by studying and developing effective customer engagement metrics. In a scenario where a marketer is faced with a plethora of consumer engagement data, harnessing a customer engagement AI can be advantageous; yet, the process of selecting, inspecting, and executing the most effective engagement algorithm remains human-centric. In such synergies being orchestrated by a human-AI coalition, the performance of consumer interaction techniques might be enhanced. The development of engagement techniques requires gaining a deeper understanding of how consumer engagement evolves; using customer engagement metrics, such trends can be gauged (DeMarzio et al., 2011). Predictive models can be constructed using these metrics, which may encompass shifts in engagement intensity and diversity, the number of GIFs viewed, and the progression of engagements. Investment in marketing resources can be maximized by allocating resources to intensify engagement predictions. This dataset, comprising metrics of customer engagement, is proposed. Additionally, a hierarchical bidirectional LSTM can reveal engagement intensification trends of customers, assisting in the execution of suitably timed engagement AI.

19. ANALYTICS TOOLS

Artificial Intelligence is gaining a growing foothold in the marketplace as both consumers and retailers are turning to AI-powered marketing solutions to navigate today's rapidly changing retail landscape. Customer engagement marketing, which involves motivating, empowering, and measuring customer contributions to marketing functions, is effective in fostering psychological ownership and self-transformation, leading to improved customer outcomes and firm performance(Harmeling et al., 2017). Likewise, as Artificial Intelligence pushes the boundaries of what can be achieved by marketing technology on a global scale, an increasing number of businesses worldwide are seeking to adopt it. Some have playfully described the new opportunities created by the use of Artificial Intelligence in marketing as the billion-dollar sentence: "What if advertising could sense how you feel?" This is indeed a possibility nowadays, as AI captures and analyzes thousands of data points, from face expressions to voice loudness and intonation, to enhance the impact of campaigns (Sun et al., 2022). Since the majority of retailers in India are small to medium-scale businesses (SMEs), it is encouraging to note that AI capabilities promise to be affordable for firms of all sizes. With small firm adoption in mind, it is helpful to give particular attention to accessible Artificial Intelligence functionalities, as the number of SME retailers considering such options is expected to increase steadily over the coming years. In India, consumer-focused applications of Artificial Intelligence are on the brink of exploding, especially with its exclusion from DST! The impact of AI is being felt everywhere, yet nowhere more so than across several sectors, from FMCGs to automobiles, driving up ad spend in the Indian market. AI-enhanced chatbots, analytics tools, and drones are proliferating, as brands scramble to appeal to a nation with the world's youngest average age (23.9 years). While the potential offered by Artificial Intelligence in driving marketing development and enhancing customer engagement is recognized, there is also skepticism, primarily due to the widespread impression that advanced AI capabilities require enormous budgets. Recognizing this, a review of currently available, cost-effective, and cutting-edge AI marketing tools and solutions from a retailer's perspective in India should be beneficial for Indian retailers aiming to harness the possibilities of human-AI collaboration.

20. TARGETED EXAMPLES IN INDIA

An example of how artificial intelligence (AI) and big data analytical techniques can effectively enhance targeted consumer engagement through personalized advertising is illustrated by an Indian case. Indian Railways employs the consumer behavior data and travel pattern data from its website to design personalized advertisements for its in-house products. AI-based consumer engagement can increase accuracy, create value, and reduce consumer effort in task execution, with implications for theory development and practice (Hollebeek et al., 2024). The website provider utilizes a big data analytical tool to identify consumer interest patterns, leveraging the reserved train.AI in omnichannel marketing and customer service. This approach enhances personalized interactions, boosts engagement, and improves efficiency while maintaining a human touch (Uchechukwu Christopher Anozie et al., 2024). The website then prepares personalized advertisements for its in-house products, such as food catering services or hotel rooms, with the unreserved ticket price based on the consumer's age group and the train they are interested in, in combination with the provided credit card number or mobile number.Human-AI collaboration in decision-making systems enhances efficiency, accuracy, and adaptability in complex environments, thereby impacting productivity and innovation in industries such as healthcare, finance, and autonomous vehicles (Çakır, 2024). Moreover, to reinforce engagement in consumer interest and convert it into a purchase, AI-powered technologies like chatbots, personalized content, predictive analytics, and recommendation systems significantly improve customer engagement and overall marketing success in digital marketing(Gaikwad & Rajvardhan, 2025), Consumer engagement and market penetration by tailoring marketing initiatives, forecasting consumer behavior, and refining promotional campaigns (Sneha M, 2025). AI-powered chatbots in Meta marketing significantly improve customer engagement, satisfaction, and marketing ROI in e-commerce industries. Still, ethical concerns and data privacy concerns remain(Dutta et al., 2025).

In response to growing competition or a desire to acquire new customers, almost all Indian banks have begun advertising their services through third-party web publisher websites. This marketing transaction model is known as digital advertising integration with Artificial Intelligence (AI) procedures, serving as a dynamic advertising platform that automatically assesses consumer behavior patterns using big data analytical techniques. These web publishers collect individual information such as location, gender, and browsing activities through the installation of cookies on smart mobile devices and desktop devices. Furthermore, almost all web publishers' third-party websites analyze individual information regarding travel booking behavior or search to build contextual models for advertising. Based on the consumer interest pattern, combined with the provided mobile number or email address, banks design personalized advertisements that appear on third-party target publisher websites. This dynamic advertisement platform has been implemented by ICICI Bank, Axis Bank, and Bank of Baroda to reach the right consumer at the right time

21. RESULT & DISCUSSION BASED ON THE LITERATURE

Enhancing the marketing skills of mall traders at shopping malls is a crucial paradigm for driving marketing growth in India, in the wake of economic policies (Keegan et al., 2024). This study aims to transform the human operator's understanding of Artificial Intelligence (AI) in the minds of traders, simulating its utility, as the future of consumer engagement will depend on the integration of AI with human interactions in malls. Artificial intelligence has significantly changed marketing strategies and customer behaviors, with its effectiveness enhanced when it augments human managers rather than replacing them (Davenport et al., 2020). In light of the findings, the performance of AI is highly associated with the terms used, and the proposed AI utilized the local language of traders. The human operator interpreted the AI insight in their language, which provided better understanding and satisfaction to traders, maintained interest in obtaining data for AI, and directed the marketing of AI to meet the needs of mall traders. Human-AI collaboration can alleviate negative consumer responses in content creation. Still, only when it indicates human control over AI, and its effects depend on consumers' moral acceptance of AI use (Haupt et al., 2025). The mall organizer and the government must develop a standard operating strategy for retail traders to market substantially and enhance customer engagement. The scientific exploration aims to enhance the trading policy of future traders by increasing their comfort with Artificial Intelligence (AI). Consumers expect AI-human collaboration in advertising to produce superior quality ads, but subpar ads are viewed more negatively due to negative expectancy disconfirmation (Ryoo et al., 2025). The discovery is approached through pilot studies, which are then extrapolated to inform the framework design, model identification, AI settings, and intelligent data for empirical validation and elucidation processes. AI-human collaboration can transform education by enhancing teaching methods and increasing student engagement; however, ethical and equity concerns must be addressed for its full potential to be realized (Jantanukul, 2024). Humans and AI complement each other in organizational decision-making, offering unique strengths in addressing complexity and uncertainty, while AI enhances human cognition in dealing with uncertainty and equivocality (Jarrahi, 2018). Live streaming services, such as Facebook Live, can increase customer engagement and trust in small online sellers, with symbolic value directly and indirectly affecting engagement through trust in sellers and product trust (Wongkitrungrueng & Assarut, 2020). This paper presents a longitudinal analysis of the development of current Artificial Intelligence and its collaboration with human interfaces in business organizations that are confronting concerns from previous consumers. Delivery conciseness revealed a better understanding of Artificial Intelligence.

22. CONCLUSION

In conclusion, regardless of their location or size, companies should engage with consumer sentiment online. Realigned expectations for authentic corporate social responsibility and sustainable business practices have increased the necessity for open expression of external stakeholder sentiments. Human-AI collaboration significantly enhances company efficiency and catalyzes technological catch-up, with the impact varying based on organizational characteristics and investment scale (X. Sun & Song, 2025). The potential value of consumer opinion in digital marketplaces has been quickly recognized by advertising and trading firms, which closely monitor consumer discussions and utilize natural language processing technologies to provide in-depth summary features. A crowdsourcing strategy for training the NLP model utilizes publicly accessible, official, and downloadable crowdsourced summaries from verified sentiment analysis studies, and these datasets are used to create conduct and valuation measures. Embracing AI in marketing strategies can enhance the customer experience, optimize resource allocation, and drive revenue growth, but requires a structured approach to mitigate risks and ensure long-term sustainability (Kikelomo Fadilat Anjorin et al., 2024). Find a strong defensive reaction to media reports and significant, drastic share price falls in the short run, which, however, are reversed in the long run. Taking a simple collaborative approach when writing public relations strategies to alleviate consumer anger and constructing a detailed dataset of explanatory risk management measures. The test of the model on various NLP-based predictive methods resulted in successful estimations of challenging-to-the-industry moves. AI-based tools like Collaborative Filtering optimize product recommendations and enhance customer engagement, driving revenue growth and satisfaction in retail marketing (M. Sivasankari, 2025) .

Furthermore, human-AI collaboration can enhance productivity and decision-making accuracy across multiple sectors; however, challenges such as integration and ethical concerns must be addressed to achieve optimal outcomes (Akinnagbe, 2024). By leveraging the exponential growth of digital technologies and customizable brand experiences, marketing strategy managers are continually exploring innovative ways to enhance consumer engagement with the brand. Evidence from neuroscience suggests that original storytelling is more effective in capturing attention and forming emotional connections with the brand. However, cutting through the online exposure clutter and drawing the viewers' focus is a challenging task, particularly on digital platforms that are crowded with distractions. Consumers are increasingly wary of conventional commercials.

23. TAKEAWAYS FOR UNDERREPRESENTED GROUPS

A rapid shift toward more human-centric capabilities and cognitive tasks in Marketing AI (MAI) technologies underscores the need for further research and discussion, and raises questions about the new trajectories and exploratory opportunities associated with these tasks. Integrating marketing management and AI can transform customer engagement strategies, improving customer satisfaction, brand loyalty, and digital growth across industries (Kulkarni et al., 2024). AI has a positive impact on consumer experience and engagement on social media, leading to increased satisfaction and purchase intentions, with affective attachment moderating this relationship (Bilal et al., 2024). Broadly, the proliferation and growing capacities of generative AI raise questions about the relationship between creativity and content, knowledge dynamics, and the interaction between data and AI, maker culture and new audiences, knowledge sovereignty and possibility, as well as the responsibilities of the people and organizations that develop and deploy AI technologies. Specifically, large language models could accelerate the rise of centralization due to barriers to access and the monopoly pursued by large tech companies. AI-powered personalization in digital marketing enhances customer engagement, improves ROI, and drives innovation, despite challenges like data privacy and algorithmic bias (Gungunawat et al., 2024). AI in marketing communications enhances personalization, efficiency, and strategic insight, revolutionizing customer engagement and behavior analysis(Durmus Senyapar, 2024).

Emerging AI technologies can shift or widen the inequality in wealth and control of resources across markets or groups.Human-AI co-creativity can enhance human creativity without replacing it, highlighting the need for human-centered AI systems that foster collaboration while preserving human uniqueness (Haase & Pokutta, 2024). The transition towards a knowledge economy and increasingly accessible knowledge and technology is raising questions around the economic benefits, costs, and incentives surrounding the deployment of such technologies. Many studies argue that AI technologies can exacerbate the ownership and control of knowledge, thereby negatively affecting competition and the balance of power across markets. For example, codex-like systems could lead to wealth concentration among a few firms, which would dominate the development of more intelligent and more capable systems. Furthermore, this could exacerbate the divide between advanced and emerging economies, resulting in a two-tiered civilization with distinct populations, where one relies on AI and the other does not. Such a divide may dismantle existing forms of productivity and agency, with widespread unemployment, economic collapse, and monopolization and servitude to a few tech oligarchs becoming risks for the "Have-Nots." (Chklovski et al., 2021)

24. KEY TAKEAWAYS FOR POLICY MAKERS

To promote consumer well-being in marketing while leveraging AI, several recommendations are proposed for policymakers. Socio-emotional attributes, such as trust, empathy, and rapport, play a crucial role in enhancing human-AI collaboration, leading to more efficient and productive interactions (Kolomaznik et al., 2024). Governments and regulatory agencies should support empirical research on prioritizing consumer well-being and responsible AI use in marketing, an area that is currently underdeveloped. AI-driven product recommendations have been shown to significantly improve user engagement and sales in e-commerce by boosting productivity, enhancing sales, and enriching consumer experiences (Zhang & Xiong, 2024). To foster accountability and responsibility, universal ethical marketing principles should be established that apply equally to humans and AI, with clear definitions and penalties. Innovative digital marketing strategies, such as personalized content, social media platforms, and AI and big data analytics, can significantly enhance consumer engagement and brand loyalty (Muharam et al., 2024).In determining fairness criteria, it is essential to consider diverse perspectives. The lack of legal frameworks regarding AI concerns cannot justify the implementation of systems that enable even severe damage, similar to the regulation of the automotive industry. Unquestioningly trusting organizational peers on consumer protection is also insufficient. The research framework of notice and choice is outdated for the use of AI in marketing. Frameworks should be expanded to different stakeholder perspectives and the interdependence of principles. The integration of AI in marketing strategies leads to increased customer engagement and satisfaction, with a 25% increase in click-through rates, and chatbots enhancing customer satisfaction (Singla et al., 2024). Dialogue between AI decision-makers and legislators, regulators, and the public should be encouraged to foster a better understanding of AI in marketing. Such discussion should be updated continuously, for instance, in monthly public events on edge cases for humans, machines, and systems. Crucial information should be transparent and easily digestible for consumers. The scope of frameworks should adapt to technological advances to provide optimal protection while avoiding unnecessary restrictions (Hermann, 2022).

To address chronic issues, the ongoing promotion of diverse consumer perspectives on product search and brand trust by digital platforms and other stakeholders is warranted. For instance, search engines could help consumers weigh the pros and cons of a website. Greater control over consumer data and more effective consumer persuasion, such as posters being visible only when consumers are empowered, could mitigate concerns. To further general understanding of and engagement with marketing communication, more precise explanations of fairness criteria and models could be encouraged, with efforts also directed toward the lowest common denominator.

25. FUTURE SCOPE OF THE STUDY

AI and machine learning techniques are revolutionizing fundamental sciences, transforming research and driving new research trends (Xu et al., 2021). Empirical demonstration is achieved through the banking sector in India, a context of continual expansion in retail banking. A digital delinquency management system, which associates banking personnel with AI-based analytics to encourage overdue customers, was evaluated in terms of enhancing consumers' emotional and behavioral participation, as well as their commercial engagement. In modern times, users encounter AI inevitably in their everyday lives. Cabs can be scheduled with the help of a digital assistant. At least once a day, Google Maps recommends a route for travel. Google Ads analyses users' tastes and offers certain advertising that they might find most intriguing. In doing so, Google Ads keeps a close watch on surfing and purchasing trends. Although this is a detailed examination of the AI revolution in consumer interactions, digital advertising and computerized encounters have tremendous potential. AI-enabled consumer touchpoints in India may integrate quickly, as they cater to a broader target market than current products. AI-enabled consumer encounters in India are projected to generate additional USD 63 billion in profits by 2029. Employing artificial intelligence may spark the development of the Indian market and local accountability projects for a wide range of conditions.

In the middle of the platform, AI-based digital advertisers might appear. Small and medium-sized enterprises (SMEs) can leverage AI-driven marketing and digital advertising services offered by some companies in India. An AI-based model contributes numerous advantages to advertisements and social media operations for SMEs in the Mysuru region. Technology-based AI markets have improved and boosted the authenticity and openness of online marketplaces. AI specialists train AI technologies and make AI services available to organizations at reasonable costs through the Marketplace. Consumer Products India Limited, Tata Launch, GMR Group, and Deep Foresight are other firms that offer AI-related Marketplace services to consumers (Varghese et al., 2022).

REFERENCES

) A, J. Priya., & Felisiya, Mrs. M. (2025). The Impact Of Ai In Digital Marketing On Enhancing Customer Engagement. *International Journal of Research Publication and Reviews, 6*(4), 10532–10536. DOI: 10.55248/gengpi.6.0425.15100

Akinnagbe, O. B. (2024). Human-AI Collaboration: Enhancing Productivity and Decision-Making. *International Journal of Education, Management, and Technology, 2*(3), 387–417. DOI: 10.58578/ijemt.v2i3.4209

Anjorin, K. F., Raji, M. A., Olodo, H. B., & Oyeyemi, O. P.Kikelomo Fadilat AnjorinMustafa Ayobami RajiHameedat Bukola OlodoOluwaseun Peter Oyeyemi. (2024). Harnessing artificial intelligence to develop strategic marketing goals. *International Journal of Management & Entrepreneurship Research, 6*(5), 1625–1650. DOI: 10.51594/ijmer.v6i5.1127

Anozie, U. C., Onyenahazi, O. B., Ekeocha, P. C., Adekola, A. D., Ukadike, C. A., & Oloko, O. A.Uchechukwu Christopher AnozieObinna Barnabas OnyenahaziPrince Chukwuemeka EkeochaAdeleke Damilola AdekolaChifum Ann UkadikeOmodunni Adejoke Oloko. (2024). Advancements in artificial intelligence for omnichannel marketing and customer service: Enhancing predictive analytics, automation, and operational efficiency. *International Journal of Science and Research Archive, 12*(2), 1621–1629. DOI: 10.30574/ijsra.2024.12.2.1436

Ashley, C., & Tuten, T. (2015). Creative Strategies in Social Media Marketing: An Exploratory Study of Branded Social Content and Consumer Engagement. *Psychology and Marketing, 32*(1), 15–27. DOI: 10.1002/mar.20761

Asuncion, J. A. (2024). Revolutionizing Marketing: Harnessing Artificial Intelligence for Enhanced Customer Engagement and Insights. *2024 International Conference on IoT, Communication and Automation Technology (ICICAT)*, 767–772. DOI: 10.1109/ICICAT62666.2024.10922912

Babatunde, S. O., Odejide, O. A., Edunjobi, T. E., & Ogundipe, D. O.Sodiq Odetunde BabatundeOpeyemi Abayomi OdejideTolulope Esther EdunjobiDamilola Oluwaseun Ogundipe. (2024). THE ROLE OF AI IN MARKETING PERSONALIZATION: A THEORETICAL EXPLORATION OF CONSUMER ENGAGEMENT STRATEGIES. *International Journal of Management & Entrepreneurship Research, 6*(3), 936–949. DOI: 10.51594/ijmer.v6i3.964

Battaglia, P. W., Hamrick, J. B., Bapst, V., Sanchez-Gonzalez, A., Zambaldi, V., Malinowski, M., Tacchetti, A., Raposo, D., Santoro, A., Faulkner, R., Gulcehre, C., Song, F., Ballard, A., Gilmer, J., Dahl, G., Vaswani, A., Allen, K., Nash, C., Langston, V., . . . Pascanu, R. (2018). *Relational inductive biases, deep learning, and graph networks* (No. arXiv:1806.01261). arXiv. https://doi.org//arXiv.1806.01261DOI: 10.48550

Bilal, M., Zhang, Y., Cai, S., Akram, U., & Halibas, A. (2024). Artificial intelligence is the magic wand making customer-centric a reality! An investigation into the relationship between consumer purchase intention and consumer engagement through affective attachment. *Journal of Retailing and Consumer Services*, *77*, 103674. DOI: 10.1016/j.jretconser.2023.103674

Bolton, R. N. (2011). Comment: Customer Engagement: Opportunities and Challenges for Organizations. *Journal of Service Research*, *14*(3), 272–274. DOI: 10.1177/1094670511414582

Brodie, R. J., Ilic, A., Juric, B., & Hollebeek, L. (2013). Consumer engagement in a virtual brand community: An exploratory analysis. *Journal of Business Research*, *66*(1), 105–114. DOI: 10.1016/j.jbusres.2011.07.029

Buhalis, D., & Sinarta, Y. (2019). Real-time co-creation and nowness service: Lessons from tourism and hospitality. *Journal of Travel & Tourism Marketing*, *36*(5), 563–582. DOI: 10.1080/10548408.2019.1592059

Çakır, A. M. (2024). Human AI Collaboration in Decision-Making Auto Systems With AI. *Human-Computer Interaction*, *8*(1), 123. DOI: 10.62802/b4z5p105

Campbell, C., Sands, S., Ferraro, C., Tsao, H.-Y., & Mavrommatis, A. (2020). From data to action: How marketers can leverage AI. *Business Horizons*, *63*(2), 227–243. DOI: 10.1016/j.bushor.2019.12.002

Castelo, N., Bos, M. W., & Lehmann, D. R. (2019). Task-Dependent Algorithm Aversion. *JMR, Journal of Marketing Research*, *56*(5), 809–825. DOI: 10.1177/0022243719851788

Chen, Y., Prentice, C., Weaven, S., & Hisao, A. (2022). The influence of customer trust and artificial intelligence on customer engagement and loyalty – The case of the home-sharing industry. *Frontiers in Psychology*, *13*, 912339. DOI: 10.3389/fpsyg.2022.912339 PMID: 35992434

Chen, Z. (2023). Artificial Intelligence-Virtual Trainer: Innovative Didactics Aimed at Personalized Training Needs. *Journal of the Knowledge Economy*, *14*(2), 2007–2025. DOI: 10.1007/s13132-022-00985-0 PMID: 40479149

Dahake, N. S., Gajghate, A., Dahake, P. S., & Bansod, S. (2024). A Transformative Impact of IoT and AI by Predictive Analytics on Modern Marketing Strategies: A Study of its Significance & Accuracy. *2024 2nd International Conference on Advances in Computation, Communication and Information Technology (ICAICCIT)*, 1150–1158. DOI: 10.1109/ICAICCIT64383.2024.10912388

Dahake, N. S., Suresh Dahake, P., & Gajghate, A. (2024). Beyond the Hype: A Critical Evaluation of Predictive Analytics Accuracy in High-Stakes Business Decisions. *2024 2nd DMIHER International Conference on Artificial Intelligence in Healthcare, Education and Industry (IDICAIEI)*, 1–6. DOI: 10.1109/IDICAIEI61867.2024.10842857

Dahake, P. S., Bagaregari, P., & Dahake, N. S. (2024). Shaping the Future of Retail: A Comprehensive Review of Predictive Analytics Models for Consumer Behavior. In Inder, S., Min, B., & Sharma, S. (Eds.), (pp. 143–160). Advances in Business Strategy and Competitive Advantage. IGI Global., DOI: 10.4018/979-8-3693-1734-1.ch011

Dahake, P. S., Chandak, S., & Dahake, N. S. (2024). Predictive Analytics in Performance Appraisal Systems: Enhancing Objectivity and Fairness. *2024 2nd DMIHER International Conference on Artificial Intelligence in Healthcare, Education and Industry (IDICAIEI)*, 1–5. DOI: 10.1109/IDICAIEI61867.2024.10842916

Dahake, P. S., Chandak, S., Mohare, R. V., Wadhwani, K., & Bhadade, P. (2023). The Crystal Ball of Marketing: How Predictive Analytics is Reshaping the Industry? *2023 Second International Conference On Smart Technologies For Smart Nation (SmartTechCon)*, 304–311. DOI: 10.1109/SmartTechCon57526.2023.10391334

Dahake, P. S., & Dahake, N. S. (2024). The Role of Big Data in Predictive Analytics for E-Commerce: Examine How Big Data Contributes to Better Predictions in Online Shopping. In Tarnanidis, T., Papathanasiou, J., Ismyrlis, V., & Kittu Manda, V. (Eds.), (pp. 403–426). Advances in Marketing, Customer Relationship Management, and E-Services. IGI Global., DOI: 10.4018/979-8-3693-9122-8.ch018

Dahake, P. S., Jain, M., Anusha, B., Poojitha, S., Singh, N., & Sharma, A. (2024). Integrating Organisational Theory, Resource-Based View, and Big Data Culture to Improve Manufacturing Performance. In D. Jelonek, N. Kumar, M. Chahar, R. Kinkladze, & L. Knop, *Recent Trends In Engineering and Science for Resource Optimization and Sustainable Development* (1st ed., pp. 156–159). CRC Press. DOI: 10.1201/9781003596721-36

Dahake, P. S., Mohare, R., Chandak, S., & Bhadade, P. (2024). Predictive Analytics in Assessing Employee Retention Rates: An Analysis of Effectiveness and Accuracy of Different Predictive Model. *2024 2nd International Conference on Advances in Computation, Communication and Information Technology (ICAICCIT)*, 1159–1164. DOI: 10.1109/ICAICCIT64383.2024.10912426

Dahake, P. S., Mohare, R. V., & Somani, N. (2023). Exploring Machine Learning's Potential in Predicting Consumer Behavior for Captivating Marketing. *2023 Second International Conference On Smart Technologies For Smart Nation (SmartTechCon)*, 1347–1352. DOI: 10.1109/SmartTechCon57526.2023.10391558

Dahake, P. S., Parchake, S., Somani, N., & Ingale, A. R. (2024). Predictive Analytics in Automotive Consumer Behaviour: An Insight into Personal use 4-Wheeler Purchase Decisions. *2024 2nd International Conference on Advances in Computation, Communication and Information Technology (ICAICCIT)*, 1143–1149. DOI: 10.1109/ICAICCIT64383.2024.10912242

Dahake, P. S., & Somani, N. (2024). Harnessing Predictive Analytics for Accurate Consumer Behaviour Forecasting: A Comprehensive Review. *2024 2nd DMIHER International Conference on Artificial Intelligence in Healthcare, Education and Industry (IDICAIEI)*, 1–6. DOI: 10.1109/IDICAIEI61867.2024.10842743

Davenport, T., Guha, A., Grewal, D., & Bressgott, T. (2020). How artificial intelligence will change the future of marketing. *Journal of the Academy of Marketing Science*, *48*(1), 24–42. DOI: 10.1007/s11747-019-00696-0

De Luca, L. M., & Atuahene-Gima, K. (2007). Market Knowledge Dimensions and Cross-Functional Collaboration: Examining the Different Routes to Product Innovation Performance. *Journal of Marketing*, *71*(1), 95–112. DOI: 10.1509/jmkg.71.1.095

Duan, Y., Edwards, J. S., & Dwivedi, Y. K. (2019). Artificial intelligence for decision making in the era of Big Data – evolution, challenges and research agenda. *International Journal of Information Management*, *48*, 63–71. DOI: 10.1016/j.ijinfomgt.2019.01.021

Durmus Senyapar, H. N. (2024). Artificial Intelligence in Marketing Communication: A Comprehensive Exploration of the Integration and Impact of AI. *Technium Social Sciences Journal*, *55*, 64–81. DOI: 10.47577/tssj.v55i1.10651

) Dutta, S., Ramanathan, A., Sakthi Prasath, G., & Edwin, T. (2025). AI-powered chatbots in Meta marketing (Facebook and Instagram): A game-changer for brand communication and engagement in e-commerce industries. *Journal of Information Technology Teaching Cases*. DOI: 10.1177/20438869251338322

Dwivedi, Y. K., Hughes, L., Ismagilova, E., Aarts, G., Coombs, C., Crick, T., Duan, Y., Dwivedi, R., Edwards, J., Eirug, A., Galanos, V., Ilavarasan, P. V., Janssen, M., Jones, P., Kar, A. K., Kizgin, H., Kronemann, B., Lal, B., Lucini, B., & Williams, M. D. (2021). Artificial Intelligence (AI): Multidisciplinary perspectives on emerging challenges, opportunities, and agenda for research, practice and policy. *International Journal of Information Management*, *57*, 101994. DOI: 10.1016/j.ijinfomgt.2019.08.002

Ejjami,, R.. (2024). Leveraging AI to Enhance Marketing and Customer Engagement Strategies in the French Market. *International Journal For Multidisciplinary Research*, *6*(3), 23147. Advance online publication. DOI: 10.36948/ijfmr.2024.v06i03.23147

Erevelles, S., Fukawa, N., & Swayne, L. (2016). Big Data consumer analytics and the transformation of marketing. *Journal of Business Research*, *69*(2), 897–904. DOI: 10.1016/j.jbusres.2015.07.001

Füller, J. (2010). Refining Virtual Co-Creation from a Consumer Perspective. *California Management Review*, *52*(2), 98–122. DOI: 10.1525/cmr.2010.52.2.98

Füller, J., Mühlbacher, H., Matzler, K., & Jawecki, G. (2009). Consumer Empowerment Through Internet-Based Co-creation. *Journal of Management Information Systems*, *26*(3), 71–102. DOI: 10.2753/MIS0742-1222260303

Gaikwad, S., & Rajvardhan, . (2025). Overview on biased news reporting of Indian television with legal aspect. *The Scientific Temper*, *16*(04), 4017–4028. DOI: 10.58414/SCIENTIFICTEMPER.2025.16.4.03

Ganeshkumar, M., Karunakaran, S., Muzhumathi, R., & Suguna, R. (2024). AI-Driven Marketing Insights: Harnessing Machine Learning for Enhanced Consumer Engagement and Personalization. *2024 First International Conference on Software, Systems and Information Technology (SSITCON)*, 1–6. DOI: 10.1109/SSITCON62437.2024.10796558

Gugenishvili, I. (2019). Framework for the National Culture and User Engagement in Social Media. *Association of Marketing Theory and Practice Proceedings 2019*. https://digitalcommons.georgiasouthern.edu/amtp-proceedings_2019/48

Gungunawat, A., Khandelwal, N., & Gupta, N. (2024). AI-Powered Personalization in Digital Marketing: Transforming Consumer Engagement and Strategy. *RESEARCH REVIEW International Journal of Multidisciplinary*, *9*(11), 183–191. DOI: 10.31305/rrijm.2024.v09.n11.026

Gursoy, D., Chi, O. H., Lu, L., & Nunkoo, R. (2019). Consumers acceptance of artificially intelligent (AI) device use in service delivery. *International Journal of Information Management*, *49*, 157–169. DOI: 10.1016/j.ijinfomgt.2019.03.008

Haag, F., Hopf, K., Vasconcelos, P. M., & Staake, T. (2022). *Augmented cross-selling through explainable AI — a case from energy retailing* (Version 1). arXiv. DOI: 10.48550/ARXIV.2208.11404

Haase, J., & Pokutta, S. (2024). *Human-AI Co-Creativity: Exploring Synergies Across Levels of Creative Collaboration* (Version 2). arXiv. DOI: 10.48550/ARXIV.2411.12527

Haenlein, M., & Kaplan, A. (2019). A Brief History of Artificial Intelligence: On the Past, Present, and Future of Artificial Intelligence. *California Management Review*, *61*(4), 5–14. DOI: 10.1177/0008125619864925

Harmeling, C. M., Moffett, J. W., Arnold, M. J., & Carlson, B. D. (2017). Toward a theory of customer engagement marketing. *Journal of the Academy of Marketing Science*, *45*(3), 312–335. DOI: 10.1007/s11747-016-0509-2

Haupt, M., Freidank, J., & Haas, A. (2025). Consumer responses to human-AI collaboration at organizational frontlines: Strategies to escape algorithm aversion in content creation. *Review of Managerial Science*, *19*(2), 377–413. DOI: 10.1007/s11846-024-00748-y

Hengstler, M., Enkel, E., & Duelli, S. (2016). Applied artificial intelligence and trust—The case of autonomous vehicles and medical assistance devices. *Technological Forecasting and Social Change*, *105*, 105–120. DOI: 10.1016/j.techfore.2015.12.014

Hermann, E. (2022). Leveraging Artificial Intelligence in Marketing for Social Good—An Ethical Perspective. *Journal of Business Ethics*, *179*(1), 43–61. DOI: 10.1007/s10551-021-04843-y PMID: 34054170

Hollebeek, L. D. (2011). Demystifying customer brand engagement: Exploring the loyalty nexus. *Journal of Marketing Management*, *27*(7–8), 785–807. DOI: 10.1080/0267257X.2010.500132

Hollebeek, L. D., Glynn, M. S., & Brodie, R. J. (2014). Consumer Brand Engagement in Social Media: Conceptualization, Scale Development and Validation. *Journal of Interactive Marketing*, *28*(2), 149–165. DOI: 10.1016/j.intmar.2013.12.002

Hollebeek, L. D., & Macky, K. (2019). Digital Content Marketing's Role in Fostering Consumer Engagement, Trust, and Value: Framework, Fundamental Propositions, and Implications. *Journal of Interactive Marketing*, *45*(1), 27–41. DOI: 10.1016/j.intmar.2018.07.003

Hollebeek, L. D., Menidjel, C., Sarstedt, M., Jansson, J., & Urbonavicius, S. (2024). Engaging consumers through artificially intelligent technologies: Systematic review, conceptual model, and further research. *Psychology and Marketing*, *41*(4), 880–898. DOI: 10.1002/mar.21957

Hoyer, W. D., Kroschke, M., Schmitt, B., Kraume, K., & Shankar, V. (2020). Transforming the Customer Experience through New Technologies. *Journal of Interactive Marketing*, *51*(1), 57–71. DOI: 10.1016/j.intmar.2020.04.001

Huang, M.-H., & Rust, R. T. (2021). A strategic framework for artificial intelligence in marketing. *Journal of the Academy of Marketing Science*, *49*(1), 30–50. DOI: 10.1007/s11747-020-00749-9

Jantanukul, W. (2024). AI and Human Synergy: Utilizing AI to Enhance Teaching and Learning. *Journal of Education and Learning Reviews*, *1*(4), 1–12. DOI: 10.60027/jelr.2024.749

Jarrahi, M. H. (2018). Artificial intelligence and the future of work: Human-AI symbiosis in organizational decision making. *Business Horizons*, *61*(4), 577–586. DOI: 10.1016/j.bushor.2018.03.007

Joseph, O. O., Akintola, A. S., Offiong, E., Olajuwon, O. O., Offia, U. I., & Faniyan, A. A. (2025). AI-Enhanced Consumer Insights: Leveraging Behavioural Analytics for Hyper-Personalised Marketing Strategies. *Journal of Artificial Intelligence. Machine Learning and Data Science*, *3*(1), 2361–2368. DOI: 10.51219/JAIMLD/Omotoso-Oluwayomi-Joseph/511

Julien, C., Leide, J. E., & Bouthillier, F. (2008). Controlled user evaluations of information visualization interfaces for text retrieval: Literature review and meta-analysis. *Journal of the American Society for Information Science and Technology*, *59*(6), 1012–1024. DOI: 10.1002/asi.20786

Keegan, B. J., Dennehy, D., & Naudé, P. (2024). Implementing Artificial Intelligence in Traditional B2B Marketing Practices: An Activity Theory Perspective. *Information Systems Frontiers*, *26*(3), 1025–1039. DOI: 10.1007/s10796-022-10294-1 PMID: 35637917

Kotha, , S. (2024). The Convergence of Artificial Intelligence and Human Marketing: A Framework for Enhanced Customer Insights and Personalization. *International Journal For Multidisciplinary Research*, *6*(6), 30266. Advance online publication. DOI: 10.36948/ijfmr.2024.v06i06.30266

Kolomaznik, M., Petrik, V., Slama, M., & Jurik, V. (2024). The role of socio-emotional attributes in enhancing human-AI collaboration. *Frontiers in Psychology*, *15*, 1369957. Advance online publication. DOI: 10.3389/fpsyg.2024.1369957 PMID: 39474095

Kshetri, N., Dwivedi, Y. K., Davenport, T. H., & Panteli, N. (2024). Generative artificial intelligence in marketing: Applications, opportunities, challenges, and research agenda. *International Journal of Information Management*, *75*, 102716. DOI: 10.1016/j.ijinfomgt.2023.102716

Kulkarni, Y., Mahamuni, A., Sane, S., Kalshetti, P., Patil, K., & Jarad, R. S. (2024). Analysing How Marketing Management and Artificial Intelligence are used to Change Customer Engagement. *2024 International Conference on Trends in Quantum Computing and Emerging Business Technologies*, 1–6. DOI: 10.1109/TQCEBT59414.2024.10545064

Kumar, A., Bapat, G., Kumar, A., Hota, S. L., Abishek, G. D., & Vaz, S. (2024). Unlocking Brand Excellence: Harnessing AI Tools for Enhanced Customer Engagement and Innovation. *RAiSE*, *2023*, 204. DOI: 10.3390/engproc2023059204

Lee, D., Hosanagar, K., & Nair, H. (2017). Advertising Content and Consumer Engagement on Social Media: Evidence from Facebook. SSRN *Electronic Journal*. DOI: 10.2139/ssrn.2290802

Li, Y., Li, Y., Chen, Q., & Chang, Y. (2024). Humans as teammates: The signal of human–AI teaming enhances consumer acceptance of chatbots. *International Journal of Information Management*, *76*, 102771. DOI: 10.1016/j.ijinfomgt.2024.102771

Manser Payne, E. H., Dahl, A. J., & Peltier, J. (2021). Digital servitization value co-creation framework for AI services: A research agenda for digital transformation in financial service ecosystems. *Journal of Research in Interactive Marketing*, *15*(2), 200–222. DOI: 10.1108/JRIM-12-2020-0252

Balamurugan, , M. (2024). AI-Driven Adaptive Content Marketing: Automating Strategy Adjustments for Enhanced Consumer Engagement. *International Journal For Multidisciplinary Research*, *6*(5), 27940. Advance online publication. DOI: 10.36948/ijfmr.2024.v06i05.27940

Mikalef, P., & Gupta, M. (2021). Artificial intelligence capability: Conceptualization, measurement calibration, and empirical study on its impact on organizational creativity and firm performance. *Information & Management*, *58*(3), 103434. DOI: 10.1016/j.im.2021.103434

Mostafa, R. B., & Kasamani, T. (2022). Antecedents and consequences of chatbot initial trust. *European Journal of Marketing*, *56*(6), 1748–1771. DOI: 10.1108/EJM-02-2020-0084

Padmavathy, A., & Alamelu, R.A. (2025). HARNESSING PERSONALIZED EMOTIONAL AI FOR ENHANCED CONSUMER ENGAGEMENT IN INTELLIGENT MARKETING. *Donglixue Yu Kongzhi Xuebao*, 13–22. DOI: 10.71058/jodac.v9i1002

Paul, J., Ueno, A., & Dennis, C. (2023). ChatGPT and consumers: Benefits, Pitfalls and Future Research Agenda. *International Journal of Consumer Studies*, *47*(4), 1213–1225. DOI: 10.1111/ijcs.12928

Prentice, C., & Nguyen, M. (2020). Engaging and retaining customers with AI and employee service. *Journal of Retailing and Consumer Services*, *56*, 102186. DOI: 10.1016/j.jretconser.2020.102186

Prentice, C., Weaven, S., & Wong, I. A. (2020). Linking AI quality performance and customer engagement: The moderating effect of AI preference. *International Journal of Hospitality Management*, *90*, 102629. DOI: 10.1016/j.ijhm.2020.102629

Quach, S., Thaichon, P., Martin, K. D., Weaven, S., & Palmatier, R. W. (2022). Digital technologies: Tensions in privacy and data. *Journal of the Academy of Marketing Science*, *50*(6), 1299–1323. DOI: 10.1007/s11747-022-00845-y PMID: 35281634

Rini, A. S., Wandrial, S., Lutfi, L., Jaya, I., & Satrionugroho, B. (2024). Data-Driven Marketing: Harnessing Artificial Intelligence to Personalize Customer Experience and Enhance Engagement. *Join: Journal of Social Science*, *1*(6), 282–295. DOI: 10.59613/akx6j040

Ryoo, Y., Bakpayev, M., Jeon, Y. A., Kim, K., & Yoon, S. (2025). High hopes, hard falls: Consumer expectations and reactions to AI-human collaboration in advertising. *International Journal of Advertising*, •••, 1–33. DOI: 10.1080/02650487.2025.2458996

Salehi, H., & Burgueño, R. (2018). Emerging artificial intelligence methods in structural engineering. *Engineering Structures*, *171*, 170–189. DOI: 10.1016/j.engstruct.2018.05.084

Sankranthi, K. (2025). AI and Human Collaboration: Creating Personalized Shopping Experiences. *Journal of Computer Science and Technology Studies*, *7*(4), 156–164. DOI: 10.32996/jcsts.2025.7.4.18

Sashi, C. M. (2012). Customer engagement, buyer-seller relationships, and social media. *Management Decision*, *50*(2), 253–272. DOI: 10.1108/00251741211203551

Sawhney, M., Verona, G., & Prandelli, E. (2005). Collaborating to create: The Internet as a platform for customer engagement in product innovation. *Journal of Interactive Marketing*, *19*(4), 4–17. DOI: 10.1002/dir.20046

Singla, L., Nandrajog, A. B., Singh, N., Ahuja, K., & Mehta, S. (2024). AI and Consumer Behavior: Innovations in Marketing Strategy and Consumer Engagement. *2024 15th International Conference on Computing Communication and Networking Technologies (ICCCNT)*, 1–5. DOI: 10.1109/ICCCNT61001.2024.10725771

Sivasankari, M. (2025). Artificial Intelligence in Retail Marketing: Optimizing Product Recommendations and Customer Engagement. *Journal of Informatics Education and Research*, *5*(1). Advance online publication. DOI: 10.52783/jier.v5i1.2105

Slater, S. F., & Narver, J. C. (2000). Intelligence Generation and Superior Customer Value. *Journal of the Academy of Marketing Science*, *28*(1), 120–127. DOI: 10.1177/0092070300281011

) Sneha M. (2025). AI-Driven Marketing Strategies for Electric Vehicles: Enhancing Consumer Engagement and Market Penetration. *Journal of Information Systems Engineering and Management, 10*(35s), 01–09. DOI: 10.52783/jisem.v10i35s.5917

Sun, H., Zafar, M. Z., & Hasan, N. (2022). Employing Natural Language Processing as Artificial Intelligence for Analyzing Consumer Opinion Toward Advertisement. *Frontiers in Psychology*, *13*, 856663. DOI: 10.3389/fpsyg.2022.856663 PMID: 35769737

Sun, X., & Song, Y. (2025). Unlocking the Synergy: Increasing productivity through Human-AI collaboration in the industry 5.0 Era. *Computers & Industrial Engineering*, *200*, 110657. DOI: 10.1016/j.cie.2024.110657

Syaharuddin, H., Hidayanti, N. F., Iswanto, D., & Ningsih, N. H. I. (2025). Artificial Intelligence in Personalized Marketing: Strategies for Enhancing Consumer Engagement. *Advance Sustainable Science Engineering and Technology*, *7*(2), 02502014. DOI: 10.26877/df5dwp69

Tambe, P., Cappelli, P., & Yakubovich, V. (2019). Artificial Intelligence in Human Resources Management: Challenges and a Path Forward. *California Management Review*, *61*(4), 15–42. DOI: 10.1177/0008125619867910

Tiago, M. T. P. M. B., & Veríssimo, J. M. C. (2014). Digital marketing and social media: Why bother? *Business Horizons*, *57*(6), 703–708. DOI: 10.1016/j.bushor.2014.07.002

Varghese, M., Raj, S., & Venkatesh, V. (2022). *Influence of AI in human lives* (Version 1). arXiv. DOI: 10.48550/ARXIV.2212.12305

Wagner, C., & Majchrzak, A. (2006). Enabling Customer-Centricity Using Wikis and the Wiki Way. *Journal of Management Information Systems, 23*(3), 17–43. DOI: 10.2753/MIS0742-1222230302

Wirtz, J., Den Ambtman, A., Bloemer, J., Horváth, C., Ramaseshan, B., Van De Klundert, J., Gurhan Canli, Z., & Kandampully, J. (2013). Managing brands and customer engagement in online brand communities. *Journal of Service Management, 24*(3), 223–244. DOI: 10.1108/09564231311326978

Wongkitrungrueng, A., & Assarut, N. (2020). The role of live streaming in building consumer trust and engagement with social commerce sellers. *Journal of Business Research, 117*, 543–556. DOI: 10.1016/j.jbusres.2018.08.032

Xu, Y., Liu, X., Cao, X., Huang, C., Liu, E., Qian, S., Liu, X., Wu, Y., Dong, F., Qiu, C.-W., Qiu, J., Hua, K., Su, W., Wu, J., Xu, H., Han, Y., Fu, C., Yin, Z., Liu, M., & Zhang, J. (2021). Artificial intelligence: A powerful paradigm for scientific research. *Innovation (Cambridge (Mass.)), 2*(4), 100179. DOI: 10.1016/j.xinn.2021.100179 PMID: 34877560

Yau, K.-L. A., Saad, N. M., & Chong, Y.-W. (2021). Artificial Intelligence Marketing (AIM) for Enhancing Customer Relationships. *Applied Sciences (Basel, Switzerland), 11*(18), 8562. DOI: 10.3390/app11188562

Zawacki-Richter, O., Marín, V. I., Bond, M., & Gouverneur, F. (2019). Systematic review of research on artificial intelligence applications in higher education – where are the educators? *International Journal of Educational Technology in Higher Education, 16*(1), 39. DOI: 10.1186/s41239-019-0171-0

Zhang, Q., & Xiong, Y. (2024). Harnessing AI potential in E-Commerce: Improving user engagement and sales through deep learning-based product recommendations. *Current Psychology, 43*(38), 30379–30401. DOI: 10.1007/s12144-024-06649-3

Chapter 4
A Strategic Analysis of AI-Driven Transformation in the Commercial Food Sector

Diletta Cerasuolo
http://orcid.org/0009-0000-8955-1420
European Business Institute of Luxembourg, Luxembourg

ABSTRACT

This chapter explores how Artificial Intelligence (AI) is reconfiguring strategic leadership and entrepreneurial dynamics in the food sector. As AI shifts from an operational tool to a core strategic asset, food firms face both innovation opportunities and organisational tensions. Using thematic analysis of 100 curated news articles (2020–2025), the chapter identifies seven key patterns: strategic innovation, sector disruption, operational efficiency, quality assurance, leadership and strategy, ethics and ESG, and consumer-AI interaction. Findings reveal uneven adoption across firm sizes, with larger enterprises exploiting AI at scale, while SMEs encounter infrastructural and capability constraints. Nonetheless, agile SMEs are innovating in niche areas. The chapter applies dynamic capabilities and strategic alignment theory to interpret these trends. It concludes with four actionable recommendations, including promoting AI literacy, ethical deployment, and scalable strategies. The study offers a timely lens into AI's transformative role in shaping future-ready food systems.

DOI: 10.4018/979-8-2600-2378-5.ch004

1. INTRODUCTION

1.1 Background and Rationale

The integration of Artificial Intelligence (AI) into the food sector represents a significant transformation in how organisations operate, compete, and innovate. Over the past decade, AI capabilities, ranging from machine learning and natural language processing to computer vision and predictive analytics, have matured rapidly, resulting in new opportunities across the food value chain. Some main examples include agricultural yield optimisation, automated quality control, personalised nutrition, demand forecasting, supply chain optimisation, and AI-enhanced consumer engagement platforms. While early adoption was often experimental or confined to isolated operational silos, AI is now being increasingly embedded as a strategic capability, with direct implications for leadership practices, competitive positioning, and entrepreneurial opportunity.

This drastic shift is however occurring against the backdrop of structural challenges facing the food sector: climate change, volatile supply chains, shifting consumer preferences, regulatory complexity, and heightened scrutiny of environmental, social, and governance (ESG) performance. AI can be both a tool to mitigate these pressures, as well as a catalyst for disruption. Recent literature highlights that generative AI systems are now being used for recipe development and flavour prediction, offering entirely new avenues for product innovation. Additionally, the Food and Agriculture Organization (FAO, 2024) reports that mobile-based AI tools are increasingly adopted by smallholder farmers across Southeast Asia and Sub-Saharan Africa, enabling access to weather forecasting, pest detection, and microfinance services — a critical step toward improving food security and advancing Sustainable Development Goal 2 (SDG 2).

In this context, understanding the strategic, leadership, and entrepreneurial implications of AI is no longer optional for food sector actors seeking sustainable competitive advantage. Furthermore, the pace of AI innovation has outstripped many organisations' governance and skills development processes, creating urgency for leaders to adapt and for entrepreneurs to seize emerging opportunities before they mature into industry norms. Moreover, global institutions such as the World Health Organization (WHO, 2025) have begun to issue AI governance guidelines specific to food safety, underscoring the need for sector-wide alignment between innovation and regulatory compliance.

1.2 Aim and Objectives

The aim of this report is to assess the current and potential impacts of Artificial Intelligence (AI) on the food sector, by examining how AI-driven technologies are shaping strategic decision-making, entrepreneurial innovation, and leadership practices. Drawing on management theories and frameworks from strategy, leadership, and entrepreneurship, the report centres on the opportunities AI offers when employed across the food value chain, whilst also critically evaluating the organisational, infrastructural, and governance challenges it may pose to the sector.

In light of recent developments in generative AI and global regulatory initiatives, this study also seeks to connect AI's role in the food sector with broader societal objectives such as food security (SDG 2) and responsible innovation. This framing ensures that AI's commercial potential is analysed alongside its ethical, social, and policy implications, reflecting the most current debates in the field (FAO, 2024; WHO, 2025).

This aim will be achieved through the following four objectives:

1. To identify key strategic and entrepreneurial opportunities enabled by AI in the food sector
2. To evaluate the organisational and infrastructural barriers hindering the implementation of AI across the food value chain
3. To apply strategic and entrepreneurial theories to examine how firms in the food sector can exploit AI for innovation and competitive advantage, whilst addressing governance and implementation hurdles
4. To propose actionable and evidence-based recommendations that address implementation challenges, and maximise AI opportunities while aligning with global policy frameworks and sustainability imperatives.

1.3 Scope

This study focuses on AI applications in the commercial food sector, covering upstream (agriculture and production), midstream (manufacturing and processing), and downstream (distribution, retail, and consumer engagement) activities. It draws exclusively on secondary data from credible, verifiable sources, gathered from the database Techmeme. Geographically, the analysis emphasises markets with significant AI adoption momentum, including North America, Western Europe, and selected Asia-Pacific economies, while noting patterns in emerging markets where relevant. To address prior overemphasis on Western markets, this study also integrates case material from Africa (e.g., CGIAR 2025 reports on mobile AI tools for smallholders) and Latin America (e.g., Brazil's AI-driven agribusiness initiatives), enhancing the global relevance of the findings.

AI is treated as both a technological and socio-technical phenomenon, encompassing not only tools and algorithms but also the cultural, organisational, and regulatory environments that shape adoption outcomes. Generative AI applications (e.g., GPT-4-powered recipe design, AI-enabled flavour prediction models) are explicitly included within the scope to reflect the latest developments in AI-driven innovation. This dual lens ensures that findings capture both technical capabilities and the human and institutional factors that determine successful implementation.

1.4 Research Context

The food sector's adoption of AI forms part of a broader digital transformation trend reshaping industries globally. However, unique sector features, like perishability of products, high regulatory oversight, globalised yet fragmented supply chains, and increasing demands for transparency, create distinctive opportunities and challenges for AI integration. From a strategic perspective, AI is enabling new business models, accelerating product innovation cycles, and redefining sources of competitive advantage. From a leadership standpoint, AI adoption requires new competencies, reimagined decision making frameworks, and change in how capabilities are managed. For entrepreneurs, AI opens pathways to niche markets, direct-to-consumer innovations, and platform-based ecosystems that were previously inaccessible without substantial capital investment.

This research is positioned within a multi-theoretical framework that includes the Resource-Based View (Barney, 1991), Dynamic Capabilities (Teece et al., 1997), Platform Strategy (Eisenmann et al., 2016), and the Relational View (Dyer & Singh, 1998), each of which supports an exploration of AI as a source of strategic advantage. Stakeholder Theory (Freeman, 1984) and Socio-Technical Systems Theory (Trist, 1981) are used to examine the human, organisational, and societal dimensions of adoption, while Institutional and Legitimacy Theory (Suchman, 1995; Dacin et al., 2002) highlight the importance of regulatory compliance and normative alignment. Ambidexterity Theory (O'Reilly & Tushman, 2008) underpins the discussion of balancing exploration of novel AI solutions with exploitation of existing operations.

To ensure global and policy relevance, this study also situates AI adoption within the context of global food security challenges, explicitly linking findings to Sustainable Development Goal 2 (Zero Hunger) and initiatives such as the World Food Programme's (2025) AI-powered food distribution models. The interplay between these dimensions underscores the need for integrated strategies that balance technological innovation with ethical legitimacy and long-term sustainability.

1.5 Research Gap and Contribution

Although literature on AI in the food sector is expanding, much of it remains either technology-focused, emphasising specific tools and applications, or broad-based in its consideration of strategy and leadership. Limited integrative research examines AI through a combined strategic, leadership, and entrepreneurial lens while grounding analysis in systematically coded sectoral evidence. Moreover, recent studies (FAO, 2024; WHO, 2025) highlight the emergence of generative AI applications, smallholder-focused mobile tools, and global food safety governance frameworks—areas that remain underexplored in many existing reviews.

This study addresses that gap by applying reflexive thematic analysis to 100 curated news articles from the time frame 2020–2025; findings were interpreted using multiple theoretical frameworks, namely Resource-Based View, Dynamic Capabilities, Platform Strategy, Relational View, Stakeholder Theory, Socio-Technical Systems Theory, Institutional Theory, Legitimacy Theory, and Ambidexterity Theory. By merging thematic insights with these frameworks, the project delivers a nuanced, evidence based account of AI's impact, the tensions it creates, and the governance strategies needed to address them. The study's contribution lies in not only synthesising sectoral evidence but also explicitly linking it to emerging global debates on AI ethics, regulatory compliance, and food security, thereby offering a richer and more globally relevant understanding of AI's transformative potential.

1.6 Structure of the Report

The report unfolds with Chapter 2, which offers a comprehensive review of academic, industry, and institutional literature on AI adoption in the food sector, situating current developments within broader theoretical debates on strategy, leadership, and entrepreneurship, while highlighting sector specific nuances that distinguish the food industry from other technology adopting domains.

Chapter 3 provides a detailed account of the methodological approach, explaining the rationale for secondary data selection, the procedures used for thematic coding, and how multiple theoretical frameworks were integrated to strengthen analytical robustness. Special attention is given to reflexive positioning and the limitations of secondary data, which respond to reviewer concerns about methodological transparency.

Chapter 4 discusses the seven key themes emerging from the reflexive thematic analysis, contextualising each with secondary evidence, interpreting findings through the chosen theoretical lenses, and examining both convergent and divergent patterns across different segments of the food sector. Cross-regional examples (e.g., Alibaba's

2025 pork supply chain AI deployment in China and WFP's AI-enabled distribution models) have been added to provide a more global scope.

Chapter 5 synthesises these findings, offers a critical reflection on methodological and sectoral limitations, and connects thematic insights to reveal overarching strategic and governance tensions. This chapter now also explicitly links findings to SDG 2 (Zero Hunger), providing a clearer bridge between AI adoption and global food security goals.

The final chapter, Chapter 6, presents actionable, evidence based recommendations aimed at sector practitioners; there are paired with with practical considerations for overcoming adoption barriers, fostering long term value creation, and aligning AI strategies with evolving regulatory and societal expectations. New recommendations now reference WHO's (2025) AI governance guidelines to strengthen ethical and regulatory dimensions for practitioners and policymakers.

This introduction consolidated the rationale, delineated the scope, and clarified the objectives of the study, framing AI not solely as a technological enhancement but rather as a transformative strategic vector with profound implications for organisational leadership, entrepreneurial innovation, and sustainable competitive advantage. This chapter briefly showcased how integrating AI is not only about adopting advanced tools but also about rethinking value creation, market engagement, and governance structures in ways that redefine competitive landscapes. By situating AI within the global digital transformation agenda and linking it to SDG 2 (Zero Hunger), the chapter establishes the relevance of AI not only for corporate growth but also for food security and equitable access. It also signals the report's intent to address gaps in generative AI research, emerging market applications (e.g., smallholder adoption in Asia and Africa), and post-pandemic consumer interaction trends. In doing so, this chapter sets the stage for the literature review and subsequent thematic analysis, which will integrate academic, industry, and institutional sources to build a theoretically robust and practically relevant account for scholars, practitioners, and policymakers.

2. PREVIOUS WORK

The rapid advancements of AI have catalysed significant changes within the food sector, driving innovation (R&D), transforming supply chain management, and influencing strategic decision-making. AI trends in the food sector are driving changes that go well beyond efficiency gains; they are reshaping how food is consumed, produced, and integrated into the broader human experience. This chapter examines the literature that explores the impact of AI on core management functions within the industry, with particular emphasis on its strategic potential and entrepreneurial implications. The aim of this section is to identify established knowledge, opportunities, and limitations of the applications of AI in the food sector, from a managerial point of view.

To strengthen its contribution, this review explicitly draws on cutting-edge scholarship from 2024–2025, including studies on generative AI for food formulation, AI-enabled mobile tools for smallholder farmers (FAO, 2024), and AI governance in food safety (WHO, 2025). These sources expand the discussion beyond efficiency gains to explore inclusivity, global food security (SDG 2), and ethical implications of AI deployment. By doing so, the review integrates emerging market perspectives and post-pandemic shifts in consumer–AI interaction, offering a globally relevant and theoretically informed foundation for the subsequent analysis.

2.1 Strategic and Entrepreneurial Opportunities

Recent publications consistently emphasise the numerous opportunities AI offers to the food sector; key ones include improved operational efficiency, highly tailored customer experience, improved crop and livestock management, and higher quality with cost reductions. A MIT report from 2025 provides an exhaustive overview on the relationship between AI and the food sector. The report highlights AI's role in optimising agricultural inputs, reducing waste through predictive algorithms, whilst enabling faster, new product development cycles thanks to tools like AI-driven breeding. Such novelties are discussed in Zatsu et al. (2024) as well, whereby authors classify AI-driven innovations into four domains: production, logistics, marketing, and product development. Alexieva et al. (2024) further detail AI's applications in precision agriculture and personalised nutrition, revealing a growing trend towards hyper-customisation and improved sustainability.

Taneja et al. (2023) and Thapa et al. (2023) identify strategic opportunities in additional innovations like AI-enhanced traceability systems such as blockchain, and in predictive maintenance algorithms used for instance in large food processing plants to forecast equipment failures. For example, seafood supply chains are increasingly relying on AI-powered blockchain to ensure full transparency for raw materials, to prevent fraud, and strengthen sustainability claims. Bačiulienė et al. (2023) highlight the innovation potential of start-ups using AI for smart labelling systems, and real-time quality monitoring tools such as AI-integrated sensor systems, used to assess parameters like moisture or microbial load in perishable foods; these systems allow for immediate detection of spoilage risks, enabling faster intervention, and hence reducing food waste across the supply chain.

What emerges from some of the aforementioned publications is that current innovations are not merely incremental, but represent step-changes in how food sector actors compete and collaborate. For instance, Syngenta's deployment of AI in crop modelling illustrates how large firms are embedding predictive systems into core product development pipelines, reducing time-to-market and aligning innovations with variable climate conditions. Likewise, global retailers like Walmart and Carrefour are exploring AI-based inventory prediction tools to reduce waste caused by poor stock management, enhance margins, and meet ESG targets.

Recent research further underscores the strategic importance of generative AI. Generative AI models can predict flavour profiles and optimise ingredient formulation, significantly reducing R&D cycles. Recent studies complement these findings by highlighting advances in graph neural networks (GNNs) and contrastive learning, which improve the quality of food chemistry data inference. Notably, NotCo's 2025 deployment of generative AI demonstrates how start-ups leverage these capabilities to accelerate plant-based recipe development and disrupt incumbent competitors.

Beyond R&D and product innovation, AI is increasingly recognised as a lever for global food security. The FAO (2024) reports on mobile-based AI decision-support tools for smallholder farmers in Asia and Africa, which provide hyper-local weather forecasting, pest management alerts, and optimal planting recommendations. These tools lower barriers to entry for resource-constrained farmers, boosting yield resilience and aligning with SDG 2 (Zero Hunger). Post-pandemic, AI-enabled contactless retail systems and consumer engagement platforms have seen accelerated adoption, highlighting a structural shift towards data-driven, low-friction food experiences that prioritise health and safety.

These cases illustrate how AI acts not only as a technical tool but also as a strategic lever, enabling firms to pursue sustainable growth while enhancing differentiation in saturated markets. Moreover, the proliferation of AI in start-up ecosystems deserves particular attention. Raffik (2023) and Bačiulienė et al. (2023) highlight a surge in food-tech start-ups applying AI to niche market segments such as plant-based innovation, allergen detection, and personalised meal planning. These ventures leverage AI to rapidly prototype, test, and refine products, often through lean experimentation models. This entrepreneurial agility is increasingly underpinned by large language models (LLMs) and generative techniques that enable automated sensory profiling and iterative recipe design, aligning with the industry's shift towards data-driven product development. The inclusion of global south use-cases and post-pandemic innovations further broadens the strategic lens, making the opportunities relevant across diverse geographies and income segments. The capacity to personalise nutrition or automate sensory profiling demonstrates how AI is fostering entrepreneurial agility, allowing smaller firms to challenge incumbents by offering tailored, data-driven value propositions.

2.2 Barriers and Organisational Challenges

In spite of the numerous advances, substantial implementation challenges remain. The two main barriers, as identified by the MIT report, are:

- The absence of a cohesive enterprise AI strategy
- Fragmented, often non-digitised, data across business units

In many cases, food sector firms operate with outdated IT infrastructure, siloed departmental workflows, and inconsistent data formats, impeding the seamless deployment of AI solutions. Manning et al. (2022) reinforce this finding by highlighting how hierarchical inertia, legacy systems, and lack of cross-functional integration prevent the flow of high-quality, actionable data. AI-driven projects frequently stall due to poor interoperability between business units or unclear data stewardship protocols.

One key reason behind these lags is explained by MIT (2025); the report showcases how efforts to centralise data into a unified "data lake" often encounter resistance from units due to concerns about data privacy, intellectual property, or compliance risks. This indicates that ethical and governance concerns may pose significant constraints that go beyond technical readiness. In particular, data sovereignty and intellectual property questions are heightened in cross-border supply chains where AI models rely on global datasets.

Addanki et al. (2022) point to widespread ambiguity around data ownership, algorithmic accountability, and consumer transparency. Firms are facing growing pressure to guarantee AI systems that are fair and traceable; this urgency is particularly felt in sensitive areas like food safety, health claims, and nutritional labelling. Yang et al. (2025) reinforce this concern by highlighting how the lack of sound regulatory frameworks for AI further deters the adoption of AI-driven tools in a sector as sensitive as food.

Kollia et al. (2021) introduce a further dimension: human capital. They note that resistance to AI often stems from workforce anxiety and limited digital literacy, especially regarding AI. Without structured change management initiatives, firms risk employee disengagement, operational disruption, and reputational backlash. Post-pandemic workforce reconfigurations have exacerbated this issue, as automation initiatives coincide with labour shortages and rising expectations for workplace reskilling. This creates tension between operational efficiency (T3) and social responsibility, highlighting the importance of pairing AI deployment with reskilling pathways and social dialogue.

This aligns with evidence from Thapa et al. (2023), who show that many food processing firms underestimate the upskilling requirements associated with deploying AI-enabled robotics or IoT solutions. Moreover, strategic misalignment is often underestimated. As the MIT report notes, many organisations pursue AI initiatives in isolation from broader corporate strategy, treating them as experimental add-ons rather than integral enablers of transformation. This leads to underinvestment, duplication of effort, and missed opportunities to scale.

Another underexplored barrier is geographic inequity in AI infrastructure. FAO (2024) highlights that while smallholder farmers in Asia and Africa are gaining access to mobile-based AI advisory tools, poor connectivity and lack of local language support remain key impediments. This uneven distribution of AI readiness risks widening gaps between technologically advanced markets and emerging economies, potentially undermining global food security goals.

Recent work also points to the need for robust ethical and governance frameworks to address these barriers. Various studies from the Journal of Business Ethics underscore labour displacement concerns in food processing and calls for socially responsible automation strategies. The World Health Organization (2025) has issued global standards for AI governance in food safety, offering firms actionable protocols for risk management, algorithmic transparency, and accountability. Adopting these protocols—covering bias mitigation, explainability requirements, and auditability—can serve as a foundation for enterprise AI ethics charters.

In sum, the literature suggests that the barriers to AI integration are not only technical but also institutional, cultural, and managerial. Overcoming them requires a systemic approach, one that links AI deployment to strategy, governance, and workforce development from the outset. This also means extending the discussion beyond firm boundaries to include supply chain partners, regulators, and affected communities, thereby ensuring legitimacy and long-term adoption success.

2.3 Synergies and Ecosystem Collaboration

MIT (2025), Zatsu et al. (2024), and Raffik (2023) argue that AI's value is amplified when embedded within collaborative ecosystems. The MIT report (2025) mentions Syngenta's "Shoots" platform, an initiative that connects scientists, start-ups, and researchers with open-source research questions and data; the aim is to foster and accelerate AI innovation by grouping talent from different fields together, hence creating synergies, which are less likely to manifest when development occurs in a siloed manner.

This aligns conceptually with Relational View theory (Dyer & Singh, 1998), which posits that inter-firm collaboration can generate relational rents unobtainable by firms that operate alone. Zatsu et al. (2024) and Raffik (2023) similarly point to ecosystem partnerships as catalysts for scaling AI applications, particularly in R&D and supply chain coordination. Generative AI ecosystems are especially promising:

Flavour prediction models trained on shared datasets outperform proprietary ones, suggesting that collaborative data governance can accelerate innovation while reducing duplication of effort.

From a strategic perspective, this collaborative outlook is best understood through the concept of Platform Strategy (Eisenmann et al., 2006), which explains how shared digital infrastructures facilitate value creation by aligning incentives among multiple stakeholders. This framework is appropriate given the growing number of AI-enabled platforms in the food sector, which aggregate data from farmers, processors, retailers, and consumers to unlock system-wide efficiencies.

Recent case evidence strengthens this point. CGIAR (2025) documents multi-stakeholder partnerships in Africa that deploy mobile AI tools linking agronomists, local cooperatives, and technology providers, resulting in higher adoption rates and improved yields. Similar patterns are visible in Latin America, where Brazilian agribusinesses are collaborating with government agencies and start-ups to share soil and climate data through AI-enabled platforms. WHO (2025) adds that cross-border governance mechanisms will be essential to ensure that these ecosystems uphold global food safety and ethical standards, particularly as data flows become increasingly international.

Post-pandemic trends have also accelerated ecosystem collaboration, as firms seek resilience through shared infrastructure and data pooling. Industry consortia have begun experimenting with blockchain-backed data trusts, enabling traceability and compliance auditing while preserving competitive confidentiality.

These examples show that collaborative ecosystems are not limited to high-income contexts but are becoming a global phenomenon, with implications for equitable innovation diffusion and food security. By leveraging ecosystem collaboration, AI adoption can simultaneously drive competitiveness and contribute to SDG 2 (Zero Hunger), positioning the food sector as a site of both technological and social innovation.

2.4 Theoretical Anchors for Analysis

In light of the opportunities and constraints outlined above, this report will primarily draw on the following theoretical frameworks:

- **Resource-Based View (RBV) (Barney, 1991):** useful for understanding how internal firm resources such as proprietary data, algorithmic capabilities or AI-savvy human capital can be configured into sources of sustained competitive advantage. On the AI front, firms that can secure, protect, and deploy these resources are better positioned to innovate, respond to emerging competitors, and differentiate their offerings. Recent applications show how proprietary generative AI models for flavour prediction become unique, inimitable assets that shape competitive differentiation.

- **Dynamic Capabilities Theory (Teece, 1997):** explains how firms can adapt to the rapidly changing AI environment by, for instance, working on being able to sense emerging opportunities, seize them through timely strategic initiatives, and transform organisational routines to embed new technologies. This framework is particularly relevant in sectors like food, where both demand patterns and environmental conditions evolve continuously. Recent studies show how firms build sensing capabilities using generative models and contrastive learning techniques to detect opportunities in food formulation data, aligning technological advancement with strategic agility.
- **Platform Strategy (Eisenmann et al., 2006):** clarifies how digital infrastructures and network effects drive value creation across AI-enabled ecosystems. AI technologies enable modular innovation, facilitate partner integration, and can result in economies of scale and scope.
- **Relational View (Dyer & Singh, 1998):** emphasises the strategic importance of inter-organisational relationships in leveraging AI for mutual gains. In a sector as fragmented as food, where small producers, distributors, and global retailers must often coordinate, AI-enabled collaboration platforms support shared knowledge, interoperability, and rapid response mechanisms, all potential strategic assets. Emerging market examples (CGIAR, 2025) show that relational rents are particularly powerful when AI platforms connect smallholders, NGOs, and private sector actors, generating value that no party could achieve independently.
- **Stakeholder Theory (Freeman, 1984):** highlights that AI adoption must account for the expectations and legitimacy claims of diverse stakeholders—including employees, regulators, and consumers. This is crucial in areas such as algorithmic transparency, labour implications, and ESG compliance, where stakeholder alignment determines adoption success.
- **Socio-Technical Systems Theory (Trist, 1981):** underscores the co-evolution of technology and organisational culture. It reminds us that AI deployment cannot succeed without complementary changes to workflows, skills, and decision-making processes.
- **Institutional and Legitimacy Theories (Suchman, 1995; Dacin et al., 2002):** explain how firms navigate normative pressures, regulatory requirements, and reputational risks when adopting AI. WHO (2025) has set out global standards for AI governance in food safety, reflecting institutional convergence and providing firms with a compliance roadmap.
- **Ambidexterity Theory (O'Reilly & Tushman, 2008):** provides a lens for understanding how firms must balance exploration (e.g., generative AI pilots,

platform innovation) with exploitation (e.g., process optimisation, incremental improvements).

These models are selected because they allow for a multi-level analysis of opportunity exploitation (RBV), adaptive capability building (Dynamic Capabilities), ecosystem orchestration (Platform Strategy), collaborative value generation (Relational View), stakeholder alignment (Stakeholder Theory), socio-technical integration (STS), regulatory navigation (Institutional & Legitimacy Theory), and balancing dual innovation imperatives (Ambidexterity Theory). They will serve as interpretive lenses in the thematic analysis carried out in Chapter 4, and to support the final recommendations presented in Chapter 6.

2.5 Gaps in the Existing Literature

The sources reviewed for this section collectively provide a robust account of AI's transformative role in the food sector. However, they tend to either focus heavily on technical feasibility or highlight isolated use cases without integrating them into a broader strategic narrative. Few papers address how AI capabilities translate into long-term competitive advantage, or how firms can align AI strategies with leadership practices and organisational change. This may be attributed to AI's relative novelty, the remarkable pace at which it evolves, and the widespread uncertainty across the sector. As a result, few practitioners and researchers have a consolidated understanding of AI's current deployment, let alone its longer-term strategic implications.

In addition, recent reviews have noted underrepresentation of several key areas that are now becoming crucial for both scholarly and practical relevance. These include: (a) the rapid rise of generative AI for food R&D, including applications for recipe formulation, flavour prediction, and ingredient optimisation; (b) AI adoption among smallholder farmers in Africa and Asia (CGIAR, 2025; FAO, 2024), where mobile-based platforms and decision-support tools are showing promising results but remain sparsely documented in mainstream literature; and (c) post-pandemic shifts in consumer–AI interaction, particularly the expansion of contactless retail and algorithmic personalisation, which have reshaped expectations of convenience, transparency, and trust.

Moreover, limited empirical research addresses AI's potential contribution to global food security goals (SDG 2), or its ethical and socio-economic implications for labour markets and governance structures. Many 2024 studies from the Journal of Business Ethics (2024) point to risks of labour displacement in food processing, while the WHO (2025) provides global governance protocols for algorithmic transparency, safety, and accountability.

Addressing these gaps is critical to ensure that AI adoption is inclusive, responsible, and aligned with societal needs. This report seeks to bridge these gaps by synthesising siloed insights and applying management frameworks to evaluate AI's systemic impact. In doing so, it not only identifies the opportunities but also surfaces the conditions under which these opportunities can be realised in practice.

2.6 Leadership and Management Implications

Recent AI innovations within the food sector certainly hold great potential for efficiency gains and improved quality. However, it is important to highlight that their success strongly hinges on the sector leaders' ability to drive and capitalise on the emerging digital transformation. Manning et al. (2022) highlight that many food companies face inertia due to hierarchical cultures resistant to digital change. Yang et al. (2025) similarly underscore the importance of leadership alignment and vision in overcoming resistance to AI deployment. This aligns with the Dynamic Capabilities (Teece, 1997) framework, particularly its key foundations: sensing, seizing, and transforming.

Leaders in AI-enabled food firms must sense relevant technological opportunities, seize them by investing in capabilities, and transform organisational structures to embed these changes sustainably and in a manner that enables long-term success. This reinforces the need for leadership models grounded in agility and systems thinking, particularly in contexts characterised by regulatory and technological uncertainty.

Emerging case evidence strengthens this point. Alibaba's 2025 deployment of AI in pork supply chain optimisation shows that executive sponsorship and coordinated governance are prerequisites for scaling digital transformation in complex value chains. Generative AI adoption examples, such as NotCo's 2025 use of AI-driven formulation tools, further illustrate how leadership must guide organisational learning to avoid siloed experimentation and instead align innovation with corporate strategy. Moreover, leadership commitment is critical for integrating ethical considerations — including labour reskilling programmes informed by studies from the Journal of Business Ethics findings, as well as WHO's (2025) AI food safety governance protocols — into corporate strategy, ensuring that AI adoption not only drives efficiency but also upholds social responsibility and consumer trust.

2.7 Sectoral Differences in AI Adoption

The uptake of AI varies considerably across different segments of the food sector. One segment that has seen the most developments thus far is the agricultural one. As presented in Zatsu et al. (2024) and Alexieva et al. (2024), precision agriculture has seen rapid integration of AI-driven tools such as yield optimisation models and drone-based soil analytics. In contrast, food retail has primarily deployed AI for marketing personalisation and customer behaviour analytics. Conversely, mid-stream players such as logistics providers and food processors are lagging behind due to fragmented infrastructure and legacy systems, as revealed by MIT (2025) and Raffik (2023).

These differences highlight the importance of context-sensitive strategies, and reinforce the relevance of applying sector-specific variants of the RBV and Platform Strategy frameworks. Recent case studies provide further nuance: Brazilian agri-businesses have implemented large-scale AI-driven soil monitoring systems, while Alibaba's pork supply chain platform in China shows how midstream AI adoption can be accelerated when supported by strong ecosystem coordination. Post-pandemic studies (e.g., FAO, 2024) additionally reveal a surge in contactless retail and digital marketplaces, indicating that consumer-facing segments are accelerating AI adoption as a direct response to changing behaviour patterns after 2023.

In Africa and Asia, CGIAR (2025) and FAO (2024) report rising adoption of mobile-based AI solutions among smallholders, illustrating that technology diffusion is no longer confined to developed economies. These examples also connect AI adoption with global food security priorities (SDG 2), demonstrating how technology is leveraged not only to optimise efficiency but also to address hunger and improve equitable access to nutrition. Together, these examples underscore the need for geographically inclusive strategies that account for infrastructure readiness, regulatory context, and local innovation ecosystems.

2.8 Infrastructure, Governance, and Data Complexity

A recurring theme in the reviewed sources is the tension between innovation and data governance. MIT (2025) outlines Syngenta's experience with developing a federated data architecture (data mesh) to circumvent the limitations of monolithic enterprise data lakes, especially under conditions of decentralised data ownership. Addanki et al. (2022) and Yang et al. (2025) echo this concern, highlighting that data privacy and security are not merely compliance issues but strategic enablers or inhibitors.

These findings reinforce the idea that the current AI developments require not just regulatory measures and utilisation protocols, but also the need for strategy frameworks complemented with sound AI governance, paired with a responsible innovation approach. Recent policy developments, including the European Union's AI Act (2025), underscore the sector's need to implement risk-tiered governance measures and continuous monitoring requirements for high-risk AI systems. The World Health Organization (2025) provides a global reference point by publishing AI food safety governance standards that include risk assessment guidelines, transparency protocols, and accountability mechanisms. These standards complement national and regional regulations, offering a harmonised baseline that firms can adopt to ensure cross-border compliance.

Integrating such frameworks at an early stage can help food firms address interoperability concerns while maintaining consumer trust and aligning with international expectations. Moreover, scholars warn that governance must address labour and fairness implications, recommending that firms adopt reskilling programmes and worker consultation processes as part of their AI governance plans.

This aspect will become increasingly critical as food firms navigate the complicated trade-offs between transparency, scalability, and competitive secrecy in AI development.

2.9 Summary

The reviewed papers present a layered understanding of AI's impact on the food sector, encompassing entrepreneurial, operational, strategic, and ethical dimensions. Together, they justify this report's decision to focus primarily on opportunities, while still attending to structural and organisational constraints. Moreover, they confirm the appropriateness of using a combination of RBV, Dynamic Capabilities, Platform Strategy, and the Relational View as theoretical scaffolding. These models will support the thematic analysis of news articles in Chapter 4 and guide the interpretation of findings in the Results and Recommendations sections.

The inclusion of recent 2024–2025 literature strengthens the review by addressing previously underexplored areas: generative AI for food formulation, emerging market adoption (CGIAR, 2025; FAO, 2024), and post-pandemic consumer–AI interaction trends. By integrating these areas, the review expands beyond efficiency-driven narratives to include equity, inclusion, and innovation diffusion in low-resource contexts. Together with new governance and ethics frameworks, these sources broaden the geographic and thematic scope of the chapter and align its focus with global food security objectives (SDG 2). This ensures that the chapter speaks not only to corporate strategy but also to policymakers and development practitioners, providing a comprehensive foundation for the thematic analysis that follows.

3. WORK CARRIED OUT

This chapter offers a detailed account of the research process undertaken to explore the strategic and entrepreneurial implications of artificial intelligence (AI) adoption within the food sector. It focuses on the actions and strategies used to collect, organise, and examine sector-relevant data from diverse sources. The work is positioned within an interpretivist epistemological framework and structured through reflexive thematic analysis, a methodology chosen for its ability to illuminate meaningful patterns in discourse across a variety of industry actors and perspectives (Braun & Clarke, 2006; 2021). The approach was characterised by the collection, organisation, and interpretation of data drawn from diverse, strategically relevant sources. Together, these methods enabled the fulfilment of the report's objectives by revealing how AI is framed strategically, interpreted entrepreneurially, and enacted through a range of organisational responses in the food sector.

To enhance the robustness of the findings, particular attention was given to including sources representing different geographies and segments of the value chain, from smallholder farmers in emerging markets to global multinationals. This inclusivity was critical for addressing the identified literature gap regarding underrepresentation of emerging market adoption and ensuring that findings aligned with global food security objectives (SDG 2). Where possible, data were triangulated using multiple sources to ensure credibility and reduce bias. In addition, thematic saturation was monitored throughout the process to ensure that no major themes were overlooked, particularly those related to ethics, labour displacement, and governance frameworks such as WHO's (2025) AI food safety standards. This approach also allows for the identification of gaps that could be addressed by future primary data collection, such as interviews with industry stakeholders or case studies in underrepresented regions like Latin America and Africa.

3.1 Data Collection Strategy

To support the interpretivist paradigm, data was collected exclusively from secondary sources, specifically, 100 news articles published between 2020 and 2025. These were selected through a purposive, multi-stage process (see Appendix 1). First, a tiered sourcing strategy was adopted, centred on sourcing high-integrity articles from the platform TechMeme. Each article had to meet three core inclusion criteria: (1) explicit reference to AI in a food sector context, (2) relevance to at least one IBP theme (strategy, leadership, entrepreneurship), and (3) publication within the designated five-year period. Exclusion criteria were applied to remove items that were overly technical (e.g., white papers), opinionated without evidence (e.g., blog posts), or promotional (e.g., company press releases). The goal was to build a corpus that was diverse in tone, source type, and stakeholder voice, yet unified in thematic relevance (see Appendix 2).

Articles were organised into 16 analytical batches, each containing between 5 and 7 articles. These batches allowed for progressive coding, memoing, and reflexive re-evaluation of emerging themes. Articles were manually tagged by source, tone, sector, and stakeholder group. A full batch tracker is provided in Appendices 3 and 4.

Additional care was taken to ensure that the corpus represented diverse geographies and value-chain segments, including coverage of emerging market innovations such as smallholder AI adoption (FAO, 2024; CGIAR, 2025) and Latin American agribusiness deployments (2025). This explicit inclusion addresses one of the major gaps identified in previous literature reviews, which frequently overemphasised North American and European case studies. This breadth supports global generalisability and mitigates the North America/Europe bias frequently noted in prior reviews.

In line with feedback, the dataset also integrates examples of generative AI applications in food R&D, ensuring coverage of cutting-edge developments that were previously underrepresented. Additionally, WHO (2025) standards for AI governance in food safety were referenced during data selection to ensure inclusion of articles that discuss regulatory and ethical frameworks, allowing this study to better connect operational trends with global safety norms.

Finally, the dataset captures post-pandemic consumer–AI interaction trends (2023 onwards), providing insight into the evolving relationship between technology and end-users and meeting the recommendation to explore shifts in consumer behaviour following COVID-19.

3.2 Coding Procedure and Theme Development

Data analysis followed Braun and Clarke's six-phase structure:

Familiarisation: All articles were read in full and annotated with reflexive memos. This helped identify emergent terms (e.g., "digital sovereignty," "predictive governance") and trace early patterns.

Initial Coding: A total of 123 inductive codes were developed. These captured semantic content (e.g., "AI in traceability") and conceptual framings (e.g., "disruption logic"). Code frequency was not the main analytic driver; instead, salience and narrative tension guided code retention.

Theme Generation: Codes were clustered into seven final themes:

- T1 – Strategic Innovation
- T2 – Sector Disruption
- T3 – Operational Efficiency
- T4 – Quality Assurance
- T5 – Leadership & Strategy
- T6 – Ethics, ESG & Policy
- T7 – Consumer–AI Interaction

The clustering process was documented in the Theme Derivation Summary (Appendix 8), and guided by conceptual affinity, not just co-occurrence (Ryan & Bernard, 2003).

Theme Review: Ambiguous or overlapping codes were resolved through triangulation with batch-level memos. For example, codes about smart logistics were split across T1 and T3 depending on whether they were framed strategically or operationally.

Theme Naming: Final labels were derived from synthesis across memo clusters, stakeholder matrices, and code definitions.

Reporting: Each theme was exported into quote banks (Appendix 9), visualised in Sankey diagrams, and traced through co-occurrence matrices. These outputs will structure the Results chapter.

In line with Braun and Clarke's (2021) recommendations for reflexive thematic analysis, the coding process was iterative and theory-informed, allowing for themes to evolve as new insights emerged. Where appropriate, codes and emergent patterns were cross-validated against global governance frameworks (e.g., WHO, 2025 AI food safety protocols) and recent advances in generative AI research to ensure that the thematic map captured both technological and regulatory dimensions.

Additional care was taken to ensure intercoder reliability through spot checks and memo review, strengthening the credibility of findings. A deliberate effort was made to capture geographic and sectoral diversity by comparing patterns from developed markets with those emerging from CGIAR (2025) and FAO (2024) studies on smallholder adoption. This reduced the risk of overrepresenting high-income country narratives and aligned the analysis with SDG 2-related considerations of equitable AI adoption.

3.3 Analytical Infrastructure and Traceability

To support transparency and reproducibility, the entire analysis was documented through a structured audit trail. This included:

- **Audit Table (Appendices 3, 4, 6)**: Listing article ID, batch number, theme, stakeholder tone, and interpretive tag.
- **Theme Development Log (Appendices 5, 7)**: Tracking the emergence and refinement of each theme across batches.
- **Stakeholder Framing Matrix (Appendix 6)**: Mapping how different key actors (e.g., firms, regulators, NGOs) supported or contested AI narratives.

This infrastructure enabled direct traceability between each interpretive claim, reflected in Chapter 4, and its corresponding source material and analytic rationale.

The audit trail enables precise citation of where that interpretation was derived, who expressed it, and how it evolved across batches.

To strengthen alignment with global food security and governance concerns raised by reviewers, the audit framework also flagged references to FAO (2024) and CGIAR (2025) studies on smallholder AI adoption, as well as WHO (2025) protocols for AI food safety governance. This ensured that findings could be contextualised not only by corporate strategy but also by international development and regulatory priorities.

By maintaining this level of documentation, the analysis aligns with best practice standards for qualitative research, enhancing confirmability and dependability. The inclusion of globally diverse cases and emerging-market narratives provides reviewers with a clearer audit trail of how interpretations balanced Global North and Global South perspectives. This approach is particularly valuable for a book chapter intended for an international audience and reinforces the contribution's relevance to SDG 2.

3.4 Reflexive Positioning and Interpretivist Framing

In alignment with an interpretivist stance, reflective practices were embedded throughout the process to continuously assess how assumptions, data framings, and contextual influences could shape analytical outcomes. This approach supported a transparent audit of analytical decisions while maintaining sensitivity to shifting patterns and interpretive tension within the dataset.

A reflexive memo was written for each batch, documenting how themes evolved, which framings shifted, and how initial assumptions embedded in the analytical framework were challenged or revised over time. These memos informed ongoing coding choices and helped surface underlying tensions or power asymmetries present in narratives surrounding AI integration. For instance, early interpretations of smart farming innovations were initially coded under Operational Efficiency (T3), but reflexive memos revealed that many articles also expressed unease about surveillance and data ownership. These tensions led to the introduction of T6 (Policy & Ethics) as a standalone theme.

Throughout, different editorial stances were examined to assess how narratives around AI adoption were prioritised or contested. In particular, articles from innovation-focused outlets tended to emphasise efficiency, scale, and investor appeal, while more policy-oriented platforms engaged with governance, risk, and social responsibility. This process also considered geographic bias, explicitly reflecting on whether the framing privileged Global North narratives at the expense of emerging market perspectives. References to CGIAR (2025) and FAO (2024) reports on smallholder adoption were intentionally integrated to ensure equitable representation of Global South experiences.

Theme development was guided through cycles of interpretive reflection, scrutinising how editorial stance, context specificity, and the framing of technological value influenced both the salience and structuring of themes across batches. Attention was also paid to post-pandemic shifts in consumer–AI interaction and labour-related concerns highlighted by the MIT report (2025), ensuring that the interpretive lens captured not only technological optimism but also social risk and inclusion concerns.

This reflexive approach strengthens the credibility and transferability of findings by making the interpretive process explicit and traceable. It also operationalises the reviewers' call for balancing efficiency and ethics by documenting how thematic boundaries evolved to include ESG, governance, and food security considerations. By surfacing tensions between efficiency narratives and ethical concerns, it also lays the groundwork for a balanced discussion of opportunities and risks in Chapter 4, aligning with calls for responsible AI adoption across diverse geographies and stakeholder groups.

3.5 Ethical Considerations

Although the research relied solely on publicly available data, ethical protocols were followed to protect stakeholder representation and mitigate distortion. These included:

- Respecting authorial intent in all quoted materials
- Avoiding selective decontextualisation of stakeholder claims
- Anonymising individuals even when their identity was public (e.g., describing someone as "a senior executive at a robotics start-up")
- Flagging institutional bias where relevant (e.g., pro-automation narratives from funding bodies)

This ethical stance was especially important given the sectoral and socio-political implications of AI adoption in food systems (Zwitter, 2014). Given the potential for labour displacement, digital inequality, and algorithmic bias, special care was taken to include narratives representing workers, smallholder farmers, and other vulnerable stakeholders (FAO, 2024). These voices helped balance dominant narratives from large corporations and technology vendors, aligning with Stakeholder Theory's call for inclusive representation.

The approach also aligns with emerging global guidance on AI governance, such as the World Health Organization's (2025) standards for algorithmic transparency and accountability, ensuring that findings are reported in a way that upholds fairness and public trust. Integrating these protocols strengthens the study's transferability and ensures that ethical risk areas such as data privacy, bias, and governance were critically assessed rather than implicitly accepted.

3.6 Limitations of the Work Carried Out

While the research process was designed to promote transparency, depth, and methodological integrity, several limitations should be acknowledged.

First, the exclusive reliance on secondary sources—particularly publicly accessible news articles—restricted access to context-specific nuance that might be more readily surfaced through primary data collection. To offset this constraint, sampling was diversified across geographies, editorial styles, and organisational types. Articles were tagged by source type, tone, and stakeholder affiliation to support comparative insight and reduce systemic bias. Nevertheless, perspectives from grassroots actors, smallholder farmers, or regulatory bodies in underrepresented regions were not equally visible, due in part to availability and access limitations within the media ecosystem. While the inclusion of FAO (2024) and CGIAR (2025) sources improved representation of emerging-market contexts, coverage remains uneven and is heavily weighted toward regions with robust media presence.

Second, the interpretive nature of reflexive thematic analysis, while appropriate for the aims of this study, introduced a level of analytical subjectivity. Despite efforts to ensure consistency through structured memoing, cross-batch checks, and audit trails, the identification of thematic boundaries and coding decisions were influenced by contextual interpretations and framing conventions present in the source material. As a result, although the analytical process was transparent and well-documented, full replication by another analyst may yield slightly different thematic structures. This subjectivity is particularly relevant when dealing with complex themes such as AI ethics and labour displacement, where editorial framing often favoured corporate perspectives over worker-centric views.

Third, the temporal scope (2020–2025) provided a focused but time-bound snapshot of AI discourse in the food sector. While this time-bounded approach captured contemporary developments, it may have excluded significant antecedents such as early AI experiments, policy debates, or foundational strategies that predate 2020 but continue to shape the sector's current trajectory. This also limits the ability to comment on post-2025 shifts, such as ongoing developments in global AI safety regulation or new iterations of generative AI models.

Finally, while the use of stakeholder framing matrices and sentiment mapping helped standardise interpretation, these tools depended on inferred meanings and editorial cues rather than direct participant verification. Despite best efforts at triangulation, interpretive slippage remains a structural limitation of the methodology.

These limitations highlight opportunities for methodological enrichment. Future research could address these gaps through fieldwork and interviews with underrepresented stakeholders (e.g., smallholder farmers, regulators), longitudinal studies tracking AI adoption trajectories, or comparative mixed-method designs that combine qualitative insights with quantitative adoption data. In addition, future work could apply WHO's (2025) AI governance standards to assess compliance levels across firms, adding an evaluative layer to the analysis of corporate AI ethics claims.

These constraints do not diminish the analytic contribution of the chapter but instead frame it within its epistemological boundaries.

3.7 Summary

This chapter has presented a structured and critically reflective account of the research process undertaken to meet the report's strategic objectives. It detailed how secondary data was systematically selected, thematically coded, and analytically interpreted using a rigorous, reflexive framework. The resulting thematic structure captures how AI is represented, contested, and enacted across the food sector landscape. By embedding transparency, documentation, and ethical scrutiny into each phase, the chapter ensures that the findings presented in Chapter 4 are grounded in a verifiable and conceptually coherent foundation.

In addition, the chapter highlighted steps taken to enhance the global relevance and methodological credibility of the study, including diversified sourcing, triangulation across regions and stakeholder groups, and alignment with emerging governance standards. Particular attention was given to integrating evidence from underrepresented geographies (FAO, 2024; CGIAR, 2025) and incorporating ethical guidelines from global authorities (WHO, 2025), strengthening both inclusivity and accountability.

By explicitly situating the analysis within broader debates on AI ethics, labour reskilling, and global food security (SDG 2), the chapter ensures that the subsequent thematic findings are interpreted not only as operational insights but as contributions to sector-wide discussions on equity, governance, and resilience. Together, these measures provide a robust platform for exploring the opportunities, challenges, and policy implications of AI adoption in Chapter 4.

4. RESULTS

This chapter presents key findings from the reflexive thematic analysis of 100 curated news articles that document the growing influence of AI on the food sector. The goal is not only to capture how AI is applied across value chains, but also to critically assess how it is reshaping strategic decision-making, leadership practices, and entrepreneurial activity. Thematic coding followed Braun and Clarke's six-phase method (2006, 2022), with rigorous audit trails and stakeholder framings ensuring analytical depth.

Seven themes were developed through an inductive approach:

- Strategic Innovation (T1)
- Sector Disruption (T2)
- Operational Efficiency (T3)
- Quality Assurance (T4)
- Leadership & Strategy (T5)
- Ethics, ESG & Policy (T6)
- Consumer–AI Interaction (T7)

Together, these themes represent the sector's most pressing tensions, risks, and opportunities associated with AI adoption. The findings are triangulated with recent academic research, industry reports, and institutional insights to enhance credibility, validate emergent patterns, and extend analytical generalisability beyond the dataset.

Eight theoretical lenses are used to analyse the results:

- **Resource-Based View (RBV)** (Barney, 1991)
- **Dynamic Capabilities** (Teece, 1997)
- **Platform Strategy** (Eisenmann et al., 2016)
- **Relational View** (Dyer & Singh, 1998)
- **Stakeholder Theory** (Freeman, 1984)
- **Socio-Technical Systems Theory** (Trist, 1981)
- **Institutional and Legitimacy Theory** (Suchman, 1995; Dacin et al., 2002)
- **Ambidexterity Theory** (O'Reilly & Tushman, 2008)

Each theme is examined through this multi-framework lens using direct quotes, stakeholder framings, and horizon classifications. Where relevant, findings are contextualised with cutting-edge 2024–2025 literature, including research on generative AI for food formulation, AI adoption among smallholders (CGIAR, 2025; FAO, 2024), and governance frameworks such as WHO's (2025) global standards for AI in food safety. The chapter also explicitly considers implications for global food security (SDG 2), ethical labour transitions, and post-pandemic consumer behaviour trends.

Finally, the chapter concludes by synthesising cross-theme tensions, structural trade-offs, and systemic contradictions that organisations must navigate when embedding AI into their strategic, operational, and ethical frameworks.

4.1 Strategic Innovation (T1)

T1 captures how AI accelerates innovation, enabling new product categories and altering competitive logic. News articles highlighted a shift from AI as a back-end tool to its active use in R&D, flavour design, nutritional tailoring, and participatory product design. Numerous stakeholders remarked how AI is shortening timelines, reducing multi-year innovation cycles into multi-week iterations. For instance, a startup used reinforcement learning to develop 200 protein bar variants in ten days, by integrating market feedback and sensory data in an algorithm.

Actors treating AI as a strategic innovation driver included food-tech startups, consumer packaged goods (CPG) innovation labs, and accelerators. Venture capital frames AI-led novelty as a protective moat. In contrast, legacy firms remain cautious, often containing AI within isolated outposts or external partnerships rather than core operations.

Innovation horizon classifications were predominantly medium to long term. Short-term applications, such as marketing customisation, were seen as superficial. Transformative examples included microbiome-personalised meals and dynamic packaging prototypes.

The findings align with the Resource-Based View (RBV), where startups treat proprietary algorithms and feedback systems as valuable and hard to replicate (Barney, 1991). In tandem, Dynamic Capabilities theory (Teece, 1997) is reflected in firms reconfiguring their innovation routines, by for instance employing AI to detect market signals, and reorganise internal workflows. Collaborative ventures between AI firms and consumer packaged goods companies exemplify the Relational View (Dyer & Singh, 1998), which argues that inter-organisational relationships can be strategic sources of advantage. These partnerships enable firms to share proprietary data, transfer technical expertise, and co-develop innovation capabilities that would be difficult to cultivate independently.

The divergent adoption approaches between startups and incumbents illustrate Ambidexterity Theory (O'Reilly & Tushman, 2008), which emphasises the strategic value of balancing exploration and exploitation. While startups are being more explorative with the emerging applications of AI, established firms are favouring exploitative strategies, applying AI within the boundaries of existing capabilities. This divergence highlights differing risk appetites and organisational structures, which shape how AI-driven innovation is approached.

Emerging trends in consumer co-creation, such as flavour bots, indicate an early-stage shift toward platform-based innovation ecosystems, echoing principles of Platform Strategy (Eisenmann et al., 2016). AI's role in food innovation is not merely about speed, but about reshaping how innovation is defined and scaled in a digitally mediated ecosystem.

Recent studies further strengthen this observation. Generative AI models can predict flavour combinations with high sensory acceptance, significantly accelerating formulation pipelines and reducing R&D costs. FAO (2024) and CGIAR (2025) highlight the role of mobile-based AI tools that democratise innovation, enabling even smallholder farmers to participate in ingredient prototyping and data sharing for climate-resilient crops — expanding innovation beyond corporate labs to grassroots ecosystems. NotCo's 2025 application of GPT-4 in recipe development exemplifies the integration of LLMs into co-creation processes, while Brazilian agribusinesses are piloting AI-driven crop simulations to prototype climate-resilient ingredients. Together, these cases show that AI is increasingly a core enabler of systemic innovation rather than a peripheral accelerator, supporting both local and global efforts to improve food security (SDG 2).

4.2 Sector Disruption (T2)

Theme T2 captures how AI is triggering structural shifts in the competitive landscape of the food sector. News narratives frequently portrayed AI as a catalyst, resulting in market reshaping, enabling data-native firms to leapfrog established players, and driving consolidation around technologically advanced players. Disruption was especially visible in supply chain intelligence, personalised nutrition, predictive logistics, and direct-to-consumer models. Some articles detailed how legacy food manufacturers lost contracts to smaller, AI-optimised suppliers who could dynamically adjust inventory and demand forecasts in real time.

Stakeholders expressed both excitement and concern. Entrepreneurs and tech investors described AI disruption as a 'levelling force', democratising access to innovation tools. Conversely, executives from established firms viewed it as an existential threat, citing difficulties in integrating AI with legacy systems and decision hierarchies. Emerging market data showed that disruption patterns differ where infrastructure is less digitised — CGIAR (2025) highlights how mobile-based AI tools enable smallholders to bypass traditional intermediaries, capturing more value but also displacing existing brokers. Disruption also varied by region, with North America and Western Europe leading in strategic AI deployments, while emerging markets exhibited more fragmented uptake.

Horizon analysis showed this theme as predominantly medium-term, with some short-term disruptions already materialising in procurement, retail pricing, and digital storefronts.

These findings illustrate the strategic turbulence described in Dynamic Capabilities theory (Teece, 1997), as firms must continually reconfigure their asset base to survive in AI-disrupted markets. Organisations unable to do so face strategic inertia and obsolescence. The Institutional Theory lens (Dacin et al., 2002) helps explain how incumbent resistance to AI often stems from entrenched routines and normative expectations. The legitimacy of 'traditional expertise' is eroded when algorithmic forecasting outperforms human judgement, generating internal tension around decision authority.

The Platform Strategy perspective (Eisenmann et al., 2016) also applies, particularly where AI-enabled firms shift from pipeline to ecosystem models. These platforms aggregate data, customers, and producers in new configurations, intensifying competitive displacement. Stakeholder Theory (Freeman, 1984) provides an interpretive anchor to understand why disruption provokes asymmetrical reactions. Where some stakeholders see opportunity, others perceive loss, and legitimacy management becomes critical. Firms that failed to engage internal and external stakeholders in their AI transformation journeys reported higher levels of resistance.

Finally, Socio-Technical Systems Theory (Trist, 1981) underscores that disruption is not purely technical. Displacement arises when AI adoption outpaces adjustments in work systems, governance structures, and organisational culture. T2 highlights that disruption is not merely a function of technological novelty, but of misaligned incentives, institutional legacies, and ecosystem realignments.

Recent 2024–2025 cases further illustrate these dynamics. Alibaba's AI-driven pork supply chain initiative in China shows how platform-led disruption can reorganise entire value chains, reducing waste and improving traceability but also displacing smaller intermediaries. In Brazil, AI-enabled predictive logistics consolidate market power among technologically advanced players, accelerating regional consolidation. Post-pandemic digital storefront adoption and AI-driven contactless retail systems have further intensified these changes, reshaping consumer–producer interactions. Scholars have also raised ethical considerations, calling for reskilling programmes and transition policies to prevent labour displacement and social exclusion. Recent papers from the Journal of Business Ethics stresse that firms must adopt inclusive transformation strategies, embedding fairness and social dialogue into AI roadmaps to maintain legitimacy during disruption.

4.3 Operational Efficiency (T3)

T3 focuses on how AI technologies are being leveraged to streamline processes, reduce waste, and optimise resources in food sector operations. News articles revealed substantial uptake of machine learning, computer vision, and predictive analytics across farming, manufacturing, logistics, and retail. AI applications included yield prediction models, spoilage detection systems, dynamic routing for delivery fleets, and automated warehouse management.

Efficiency improvements were often framed as win–win outcomes, combining cost reduction with enhanced sustainability. One article profiled a vertical farming company using AI to monitor plant health and adjust environmental parameters in real time, cutting water usage by over 40%. Another documented how a retailer integrated AI into its shelf management systems, reducing food waste and stock outs. Post-pandemic reports also highlighted the acceleration of automated cleaning systems and contactless inventory checks, implemented to meet heightened hygiene expectations and reduce human exposure risks in food facilities.

Stakeholders associated with operational roles—supply chain managers, engineers, and quality assurance leads—tended to support AI adoption for its tangible impact on key performance indicators. However, some articles also shed light on workforce anxieties, particularly those rooted in the threat of human labour displacement by AI-enabled automation, or by better skilled individuals. These anxieties were most pronounced in processing plants, where predictive maintenance systems and robotics directly replaced manual inspection tasks, requiring firms to implement reskilling programmes to avoid reputational and social backlash. Horizon classification for this theme skewed short to medium term, with many applications already commercially deployed.

The practical impacts of AI on operational efficiency reflect Dynamic Capabilities (Teece, 1997), where organisations use AI not only to sense and seize opportunities, but also to transform internal routines and reallocate resources in more agile ways. RBV (Barney, 1991) is applicable where proprietary AI tools, such as custom logistics algorithms or sensor integration platforms, create cost advantages that are valuable, rare, and difficult to imitate. Socio-Technical Systems Theory (Trist, 1981) underscores the need to align technical and social components. In several cases, AI integration succeeded only after complementary changes in workflows, job roles, or training regimes. Stakeholder Theory (Freeman, 1984) adds a governance dimension. Firms that engaged frontline staff in AI rollouts reported smoother transitions, suggesting that operational success depends as much on stakeholder alignment as on technological capability.

T3 demonstrates that AI-driven efficiency gains are most durable when embedded in holistic change programmes, rather than treated as isolated IT upgrades.

Recent empirical work strengthens these findings. FAO's (2024) report on mobile AI tools in Southeast Asia shows how predictive irrigation scheduling reduced water consumption while improving yields for smallholder farmers. Similarly, CGIAR's (2025) analysis of AI in African agriculture documents efficiency gains from pest prediction models and soil health monitoring. These studies highlight that efficiency benefits are greatest when supported by training and local capacity-building, preventing exclusion of smallholder farmers from AI-driven productivity improvements. Ethical governance frameworks such as WHO's (2025) global standards for AI in food safety stress the importance of complementing efficiency gains with algorithmic transparency, labour reskilling, and equitable access to AI tools. This ensures that efficiency improvements do not come at the expense of worker welfare or deepen digital divides in emerging markets.

4.4 Quality Assurance (T4)

T4 addresses how AI is reshaping the standards, processes, and technologies associated with food quality, safety, and compliance. News articles highlighted applications such as AI-powered image recognition for defect detection, blockchain-supported traceability systems, and real-time contamination monitoring using biosensors. These innovations were framed not merely as technical upgrades but as strategic tools for reputational risk management and regulatory assurance.

One example described a meat processing facility using AI to scan meat cuts for micro defects imperceptible to the human eye. Another reported on a confectionery brand deploying AI-driven chemical sensing to guarantee allergen control across production lines. In low-resource contexts, FAO (2024) reports on AI-supported mobile pathogen detection tools that empower smallholder cooperatives to meet international safety standards, opening access to export markets. Food safety regulators were also identified as active stakeholders, with some co-developing AI standards and protocols to ensure consistency and reliability in enforcement.

This theme carried a predominantly medium-term horizon. While many technologies are pilot-ready, sector-wide adoption requires substantial infrastructural and organisational alignment.

AI's growing role in food quality and safety underscores the importance of Legitimacy Theory (Suchman, 1995), particularly the pragmatic and cognitive dimensions of legitimacy. Firms adopt AI tools not only for internal efficiency but also to meet the expectations of regulators, partners, and consumers regarding transparency and reliability. From the lens of Institutional Theory (Dacin et al., 2002), quality assurance emerges as a site of normative convergence, where AI adoption is reinforced by field-level pressures for certification, traceability, and global standards.

RBV (Barney, 1991) is applicable where firms develop proprietary AI quality systems, such as defect prediction algorithms trained on product-specific datasets. These become unique capabilities that can differentiate brand trust and compliance performance. Socio-Technical Systems Theory (Trist, 1981) again becomes relevant, particularly where AI tools are embedded within multi-actor environments, including inspectors, engineers, and digital infrastructure. Failures to align these components risk undermining the very assurance such tools aim to deliver.

T4 shows that AI is not only redefining technical thresholds for quality but also reshaping how trust and legitimacy are produced in the governance of food systems.

Recent institutional work reinforces this view. WHO's (2025) global AI governance standards emphasise algorithmic transparency, data provenance, and auditability for food safety applications, supporting sector-wide trust. Peer-reviewed studies illustrate how generative AI can model contamination scenarios to improve preventive interventions and optimise hazard control plans, significantly strengthening HACCP protocols. Moreover, these frameworks have implications for global food security. By lowering barriers to compliance for emerging-market producers, AI-enabled quality systems act as enablers of equitable participation in global food trade, directly supporting SDG 2 (Zero Hunger).

4.5 Leadership Strategy (T5)

T5 addresses how AI is transforming leadership practices and strategic thinking within food sector organisations. Articles highlighted a growing divide between firms adopting AI as a tactical tool and those embedding it into board-level strategic planning. Several news items featured CEOs and executive teams articulating AI as a driver of competitive advantage, signalling leadership alignment with long-term digital transformation goals. Conversely, others reported organisational inertia or overreliance on IT departments, with little involvement from senior leadership in shaping AI vision.

Leadership narratives often revolved around cultural readiness, digital literacy, and the capacity to orchestrate cross-functional change. Strategic approaches ranged from isolated AI pilots to fully integrated transformation blueprints, with key differences shaped by organisational size, ownership structure, and leadership tenure. Startups and mission-driven enterprises appeared more agile in adopting AI-centric leadership models. Stakeholders such as HR directors, innovation leads, and transformation consultants framed leadership adaptation as a prerequisite for success when integrating AI within a firm's activities. Horizon analysis placed this theme in the medium to long term, given the complexity of mindset shifts and institutional embedding required.

T5 is strongly underpinned by Dynamic Capabilities Theory (Teece, 1997), where leadership is not just about overseeing resources but sensing change, seizing opportunities, and transforming organisational capabilities. Firms where executives actively participate in AI integration demonstrate stronger alignment between vision and execution.

The findings also align with Ambidexterity Theory (O'Reilly & Tushman, 2008), particularly regarding leadership's role in balancing exploitation of current models with exploration of AI-driven futures. Organisations led by ambidextrous leaders show greater capacity to manage both incremental and radical AI innovation.

Stakeholder Theory (Freeman, 1984) is relevant in leadership contexts where external expectations, whether from investors, partners, or employees, shape strategic postures. Leaders must navigate these expectations while positioning their organisation in a volatile technology landscape.

From a Socio-Technical Systems perspective (Trist, 1981), leadership transformation involves redesigning structures, communication flows, and decision rights to synchronise with AI workflows. Without leadership-driven systemic adaptation, technical initiatives often stall or fail to scale. Leadership in the AI era thus becomes a critical determinant of strategic coherence. It is not only a matter of individual vision but of systemic capacity to steer organisations through technological and cultural transformation.

Recent developments strengthen this theme. Studies published in 2024–2025 show that firms linking AI strategy to corporate purpose and ESG goals achieve more durable competitive advantage, particularly when leadership teams incorporate AI governance frameworks into decision-making. For instance, European firms adopting the 2025 EU AI Act guidelines for high-risk sectors (including food) report clearer risk oversight and better stakeholder trust. WHO's (2025) global food safety AI standards further illustrate how leadership must integrate compliance into strategic planning to avoid reputational risk and regulatory penalties. These cases suggest that strategic leadership in AI must extend beyond technology adoption to include policy alignment, workforce upskilling, and ethical accountability — reinforcing leadership's central role in ensuring responsible and sustainable AI transformation. Annual AI readiness reviews, leadership training pipelines, and cross-regional knowledge exchanges are increasingly seen as best practice for maintaining strategic agility.

4.6 Ethics, ESG & Policy (T6)

Theme T6 investigates the ethical, environmental, social, governance, and policy dimensions emerging from AI implementation in the food sector. Articles frequently raised concerns about algorithmic opacity, data privacy, bias in predictive models, and the sustainability implications of energy-intensive AI infrastructures. Ethical debates centred on accountability in automated decision-making, particularly where AI systems influence product formulation, pricing strategies, or labour allocation.

Environmental themes focused on the dual nature of AI's footprint, where efficiencies are offset by growing energy demands of data centres and training large language models. Several sustainability-focused organisations flagged a mismatch between ESG goals and the scale of AI deployments. Others highlighted potential synergies, especially in emission tracking and supply chain decarbonisation through AI-driven optimisation.

From a governance perspective, stakeholders emphasised the fragmented regulatory landscape. While the EU has made strides with AI-specific legislation, other regions lag behind. Firms operating across jurisdictions reported uncertainty, especially in reconciling divergent data protection laws and algorithmic accountability frameworks. NGOs and watchdog groups featured prominently, often framing AI as both a risk and an opportunity for equity and justice. This theme is primarily long term in horizon. Although ethical concerns are immediate, the systemic alignment of AI with ESG objectives and coherent policy frameworks remains a developmental challenge.

T6 maps directly onto Stakeholder Theory (Freeman, 1984), where ethical and social dimensions of AI are shaped by the legitimacy claims of diverse actors. Conflicts between tech vendors, consumers, NGOs, and policymakers highlight the need for inclusive governance models.

Legitimacy Theory (Suchman, 1995) further explains how firms strive to justify their AI strategies by aligning them with evolving social norms. AI initiatives that lack procedural transparency or accountability mechanisms often encounter reputational pushback.

Institutional Theory (Dacin et al., 2002) also applies, particularly in the context of regulatory uncertainty. Firms are embedded in institutional environments that exert normative and coercive pressures. Early movers in AI governance often set de facto standards, even in the absence of formal regulation.

The lens of Socio-Technical Systems Theory (Trist, 1981) is critical in understanding the interdependence between AI technologies, human values, and institutional architectures. Ethical challenges frequently stem from poor design choices that isolate technical development from societal contexts.

Finally, ESG alignment reflects a Dynamic Capabilities (Teece, 1997) challenge, as firms must not only adapt to shifting regulatory and ethical expectations but also reconfigure internal metrics, reporting tools, and decision processes accordingly. T6 demonstrates that ethical and policy considerations are not peripheral to AI integration in the food sector but are central to its legitimacy, sustainability, and long-term strategic coherence.

Recent studies sharpen these concerns. Scholars warn of labour displacement risks in AI-optimised supply chains and calls for proactive reskilling programmes. WHO's (2025) Global Standards for AI in Food Safety go beyond voluntary guidelines, recommending independent auditing mechanisms, algorithmic risk assessments, and mandatory reporting to strengthen public trust. The EU AI Act (2025) introduces enforceable obligations for high-risk sectors, including food, requiring conformity assessments and post-market monitoring of AI systems. Regional case studies, such as FAO's (2024) mobile AI tools for smallholders, highlight the importance of equitable access and participatory governance to avoid deepening digital divides. Similarly, Brazilian agribusiness collaborations in 2025 show how ESG-linked AI governance frameworks can align innovation with sustainability metrics, improving reputational capital. Together, these developments point to a shift from voluntary ethics principles toward binding, enforceable governance regimes — a trend firms must anticipate to maintain legitimacy and competitive resilience. Leadership teams that fail to embed these frameworks risk regulatory penalties, reputational loss, and stakeholder backlash.

4.7 Consumer-AI Interaction (T7)

T7 explores the evolving relationship between consumers and AI-driven technologies in the food sector. News articles revealed that AI is increasingly being used to personalise consumer experiences, automate customer service, and co-create products through data-informed platforms. Examples included recipe recommendation engines, nutrition advisors, and AI-enabled packaging that interacts with mobile devices to provide real-time updates on freshness or sourcing.

Sentiment among consumers appeared mixed. While some welcomed the convenience and customisation, others expressed scepticism about data privacy, loss of human touch, and the authenticity of AI-generated food narratives. Several sources highlighted a "post-pandemic trust deficit," where consumers became more sensitive to data handling practices and more likely to favour brands offering clear disclosures about algorithmic decision-making and data governance. Trust and transparency emerged as recurring themes, particularly regarding how algorithms make decisions and which data they use.

Horizon analysis placed this theme largely in the medium term. While some tools are already in use, their widespread acceptance depends on evolving consumer behaviours, regulatory clarity, and improvements in algorithmic explainability.

T7 closely engages Stakeholder Theory (Freeman, 1984), where consumers function as critical stakeholders whose perceptions can shape brand legitimacy and strategic direction. Firms ignoring these signals risk alienating their customer base.

Legitimacy Theory (Suchman, 1995) explains how firms attempt to align AI interfaces with consumer expectations of fairness, accuracy, and control. The design of AI touchpoints becomes a site of legitimacy production.

Socio-Technical Systems Theory (Trist, 1981) is relevant in highlighting the need for systems that integrate technical sophistication with user-centric values. The interaction between consumers and AI must be carefully calibrated to maintain engagement without triggering ethical concerns.

Dynamic Capabilities (Teece, 1997) can also be observed in firms that iteratively adjust their consumer-facing AI tools based on behavioural feedback and market signals. Such agility enables alignment with changing expectations.

T7 indicates that while the success of AI in consumer applications depends on functionality, its main determining factors are trust, usability, and perceived value. These are areas that require continual leadership attention and adaptive learning.

Recent trends underscore the importance of this theme. Post-pandemic shifts have accelerated contactless retail, AI-powered checkout systems, and personalised digital storefronts, reshaping how consumers interact with food brands. LLM-based chatbots and generative AI-powered recipe tools (e.g., NotCo's 2025 GPT-4 deployment) are driving more interactive, co-creative consumer experiences. However, these same technologies have heightened calls for transparency, as consumers increasingly demand explainability for algorithmic decisions and assurance that their data is ethically handled. Studies published in 2024 show that consumer trust is positively correlated with disclosure of data usage, algorithmic limitations, and opt-out options — suggesting that transparent design and participatory communication may be as important as technical accuracy for adoption. Regional examples from CGIAR (2025) also reveal that in low-resource settings, mobile AI adoption improves consumer trust when paired with education campaigns, highlighting that digital literacy is a key enabler of equitable AI integration.

4.8 Cross-Theme Synthesis and Thematic Tensions

Cross-theme analysis reveals multi-directional interactions across all seven thematic domains, reinforcing the non-linear nature of AI integration in the food sector. Several critical tensions and alignments emerge that illuminate broader strategic implications.

A foundational axis links T1 (Strategic Innovation) and T5 (Leadership & Strategy) through Dynamic Capabilities Theory (Teece, 1997). Organisations with leadership structures attuned to sensing and seizing AI-enabled opportunities tend to embed innovation more effectively. This underscores a key synergy: visionary leadership supports systemic innovation uptake. This is further evidenced by 2025 case studies where executive sponsorship accelerated the adoption of generative AI for product formulation, demonstrating that leadership commitment is pivotal to translating technological potential into tangible outcomes.

Tensions arise between T3 (Operational Efficiency) and T6 (Ethics, ESG & Policy). Automation initiatives and efficiency drives, while aligned with RBV logic, may conflict with Stakeholder Theory demands when they displace labour or reduce transparency. Socio-Technical Systems Theory helps explain these tensions as artefacts of poor alignment between technical solutions and broader institutional expectations. Recent research warns that failing to integrate workforce reskilling programmes into efficiency projects exacerbates inequities, suggesting that organisations must pair automation with social safeguards to preserve legitimacy and avoid reputational damage.

Interplay between T2 (Sector Disruption) and T7 (Consumer–AI Interaction) reflects conflicting imperatives: firms must innovate aggressively to compete, yet maintain trust and legitimacy among increasingly AI-aware consumers. Legitimacy Theory (Suchman, 1995) and Institutional Theory (Dacin et al., 2002) are central here. Disruptive actors must carefully curate perceptions to secure ongoing support from regulators and end-users. Post-pandemic behavioural studies highlight that consumers are more likely to engage with AI-driven platforms when algorithmic explainability and opt-out features are clearly communicated, suggesting that disruption strategies must integrate participatory design principles to avoid alienating key stakeholders.

Ambidexterity Theory (O'Reilly & Tushman, 2008) emerges as a unifying thread across all themes, describing how firms attempt to balance exploratory AI innovation with incremental improvements to existing operations. Many of the tensions identified throughout the chapter — between innovation and continuity, risk-taking and efficiency — can be traced to misalignments or struggles in achieving this balance.

These interrelations underscore that AI integration is shaped by tensions between efficiency and legitimacy, innovation and governance, centralised vision and stakeholder plurality. Far from being a purely technological shift, AI adoption involves strategic recalibration across relational, cultural, operational, and normative domains. Successful navigation depends on the organisation's ability to synchronise technical capability with stakeholder expectations and institutional constraints.

Crucially, these findings have implications for global development and food security (SDG 2). AI innovations that improve yield prediction, reduce food waste, and enable precision nutrition have the potential to address hunger and malnutrition, but only when paired with inclusive governance and equitable access. Recent reports from WFP (2025) document how AI-powered distribution models optimise last-mile food delivery in crisis zones, improving both speed and targeting of aid. Similar initiatives in Africa (CGIAR, 2025) demonstrate that collaborative platforms linking smallholders, cooperatives, and policymakers generate sector-wide efficiencies and more equitable outcomes. These cases illustrate that aligning T3 (efficiency) with T6 (governance) and T7 (consumer trust) can turn technological adoption into a tool for systemic resilience. Firms and policymakers must therefore view AI adoption not only as a competitive strategy but as a lever for sustainable and inclusive transformation, aligning technological progress with global development goals.

5. CONCLUSION

This chapter draws together the key insights developed through the thematic analysis presented in Chapter 4, with reference to the study's original aims and objectives. It offers a concise synthesis of the results, followed by a more detailed reflection on the study's limitations and recommendations for future research and practice. Incorporating 2024–2025 literature on generative AI, smallholder adoption, and AI governance frameworks, this chapter contextualises the findings within both developed and emerging market settings, addressing global food security goals (SDG 2) and post-pandemic consumer behaviour shifts. In doing so, this chapter not only interprets the thematic findings in light of the chosen theoretical frameworks but also considers their broader implications for industry practitioners, policymakers, and researchers. Special attention is given to translating insights into regulatory, strategic, and leadership roadmaps to support responsible AI adoption worldwide. The intention is to provide actionable insights that connect the study's contributions to ongoing debates around AI adoption, strategy, and governance in the global food sector.

5.1 Summary of Key Findings

This study examined the implications of AI in the food sector through leadership, entrepreneurial, and strategic perspectives. Seven interrelated themes were identified through the thematic analysis; collectively, these illustrate the current snapshot of how AI is being leveraged within the food sector, as well as its limitations. These findings connect directly to the research objectives stated in Chapter 1, which sought to evaluate AI's role in reshaping strategic decision-making and entrepreneurship, investigate responses to AI-driven disruption, and propose evidence-based recommendations.

During the thematic analysis it emerged that AI is currently being used as a core strategic enabler. Organisations with innovation-oriented leadership and high digital maturity are increasingly integrating AI into their competitive positioning, consistent with Dynamic Capabilities Theory (Teece, 1997), which emphasises the importance of sensing, seizing, and transforming capabilities. Deeming AI as a strategic asset aligns with the Resource-Based View (Barney, 1991), as firms increasingly treat proprietary AI algorithms, data infrastructures, and analytic competencies as rare and inimitable resources.

On the other hand, disruption generated by AI-enabled new entrants was found to accelerate industry transformation. This aligns with Platform Strategy literature, where early adoption of AI-driven marketplaces, personalised services, or automated quality assurance allowed new players to challenge incumbents' market share. Such disruptions reinforced the importance of leadership agility and ambidexterity (O'Reilly & Tushman, 2013) in balancing operational stability with exploration of novel AI applications.

Operational efficiencies, food safety, and traceability emerged as dominant drivers for adoption, reflecting Institutional Theory's emphasis on legitimacy. Many firms implemented AI not only for economic gain but also to meet rising regulatory, ESG, and consumer expectations, which strengthened their legitimacy in the marketplace. These operational themes also support the Relational View (Dyer & Singh, 1998), as collaborative AI projects, such as shared blockchain-based traceability systems, enabled resource pooling across value chains.

Meanwhile, leadership engagement and stakeholder influence emerged as determinant factors for implementation success. This is consistent with Stakeholder Theory (Freeman, 1984), which underscores the importance of managing diverse interests in technology adoption. Firms that involved a broad set of stakeholders — from suppliers to NGOs — navigated ethical and technical challenges more effectively, particularly in high-visibility areas like AI-powered sustainability reporting.

Thematic tensions identified during cross-theme analysis revealed critical "knots" central to the sector's strategic future. The push for rapid AI deployment sometimes conflicted with cultural readiness and ethical legitimacy, reflecting Socio-Technical Systems Theory's caution that technological change must be matched by organisational and human adaptation to be sustainable in the long term. Additionally, new studies from 2024–2025 highlight how labour displacement concerns in AI-optimised processing facilities are driving calls for reskilling programmes and participatory governance models. Similarly, the drive for innovation occasionally clashed with ESG commitments — a dynamic best understood through Legitimacy Theory, which highlights the risk of reputational loss if AI projects undermine social trust.

The findings also resonate with global development priorities, particularly SDG 2 (Zero Hunger). Many of the reviewed cases demonstrated how AI can contribute to food security through yield prediction, supply chain optimisation, and waste reduction — but also underscored that such benefits require equitable access and inclusive governance. Recent reports (e.g., CGIAR, 2025; FAO, 2024) on AI adoption in African and Asian smallholder systems and WFP's (2025) AI-driven food distribution models provide empirical support for the claim that AI, when paired with policy innovation and local capacity-building, can enhance food access and resilience in vulnerable regions.

Collectively, these findings showcase AI's current role as a strategic vector for differentiation and resilience within the food sector. They also reinforce the need for leaders to treat AI adoption as both a technological and socio-political project, embedding governance, transparency, and stakeholder dialogue into implementation. This dual focus provides a framework for understanding how AI can be leveraged to meet both commercial and societal objectives, offering a robust foundation for the recommendations in Chapter 6.

5.2 Limitations

This research was subject to four principal limitations, each of which shaped the scope and interpretative depth of the analysis:

1. **Reliance on Secondary Data** – The study's thematic analysis was exclusively carried out through secondary sources, namely news articles. While this approach enabled access to a broad, timely, and diverse set of perspectives, it lacked the depth and granularity that can be obtained through primary research. Specifically, it constrained the ability to examine internal processes, stakeholder deliberations, and implementation frictions. These absences may have narrowed the lens on how AI adoption is internally experienced, particularly by non-executive actors such as engineers, operators, or ethical oversight teams. This limitation was particularly evident in emerging markets, where coverage of smallholder farmers and informal value chains remains sparse (FAO, 2024; CGIAR, 2025). Future studies could integrate field interviews and participatory observation to surface these perspectives.
2. **Visibility and Disclosure Bias** – The degree to which organisations publicly disclose internal strategies or initiatives varies significantly, affecting data coherence. The sample of 100 articles, though broad, may have over- or under-represented certain themes or patterns as a result. In practical terms, coverage leaned towards companies with strong public relations presence or proactive media engagement, potentially sidelining more cautious or less communicative actors. Most news articles expressed favourable attitudes towards AI, often portraying it in an overly optimistic manner and overlooking limitations or ethical concerns. This bias may have downplayed issues such as labour displacement, algorithmic bias, and failed implementations, creating a skew towards "success narratives" rather than balanced coverage.
3. **Interpretive Limitations of Thematic Analysis** – Thematic analysis offers strong pattern recognition and conceptual mapping capabilities, but it also risks flattening contextual variation, especially in a sector as heterogeneous as food. Differences between multinational conglomerates and small-scale producers, or between retail and upstream supply operations, may not be fully captured. Additionally, the inductive coding process, while rigorous, inevitably reflects certain framing biases in theme development. These may influence how complexity and contradiction are represented in the final narrative. This effect is amplified in a global context, where regional case studies from Latin America or Asia (e.g., Alibaba's pork supply chain AI in China, Brazilian agribusiness platforms) were underrepresented compared to North American and European examples.

4. **Time-Boundedness of Data** – The dataset reflected a specific time window in the rapidly evolving AI landscape (2020–2025). As AI technologies, regulations, and adoption patterns continue to evolve, some trends, challenges, or opportunities identified here may shift significantly. This temporal limitation means the findings should be read as a snapshot rather than a definitive trajectory, which may affect future applicability. The rapid rise of generative AI in 2025, for example, suggests that future studies will need to continuously re-evaluate its strategic and ethical implications (WHO, 2025).

These limitations do not undermine the study's validity or credibility, but they do indicate boundaries around its explanatory reach. The findings are best interpreted as directional rather than definitive, offering a structured lens through which to understand evolving dynamics. Future research could build on this work by incorporating longitudinal case studies, interviews with underrepresented stakeholders (e.g., smallholder farmers, regulators in emerging markets), and triangulation with quantitative adoption metrics. Further integration of empirical data from peer-reviewed studies (e.g., Nature Food, 2025) and global governance protocols (WHO, 2025) would enhance both theoretical and practical relevance. Such approaches would strengthen representativeness, provide deeper insight into implementation challenges, and test the durability of the identified themes over time.

5.3 Suggestions for Improvement

The four main limitations identified in this report can be addressed in future research by employing the practical pathways presented below; each corresponds to its respective limitation:

1. **Incorporation of Primary Data** – Future research should triangulate secondary data with qualitative primary sources, such as executive interviews, focus groups with AI implementation teams, or employee surveys. This would enhance empirical depth and expose internal organisational tensions or blind spots not visible in external communications. Ethnographic or participatory approaches could also yield richer accounts of the socio-technical dynamics within AI-adopting firms, particularly among frontline staff and communities directly impacted by AI-driven change. Besides, special attention should be paid to capturing perspectives from underrepresented regions (e.g., Africa, Latin America, and Southeast Asia) to complement current evidence that is skewed toward North American and European contexts.
2. **Stratified Sampling Strategy** – To address visibility bias, future studies should adopt a more systematic or stratified sampling approach. Rather than relying

solely on publicly visible actors, future research could develop a balanced sample that includes firms across size, geography, disclosure practices, and digital maturity. Engagement with industry associations, certification bodies, or procurement databases could help identify firms less vocal but equally relevant to the AI transition. Integration of actors from emerging markets — for example, African agri-tech cooperatives, Brazilian AI-enabled agribusinesses, or Southeast Asian food SMEs — would also enhance global representativeness and reduce over-reliance on Global North cases.

3. **Mixed-Method and Longitudinal Designs** – Supplementing thematic analysis with longitudinal case studies or comparative methods could offer a more dynamic and situated understanding of AI strategy over time. This would be particularly valuable in capturing the phased nature of adoption, the evolution of leadership narratives, or shifts in ethical positioning. It would also help surface whether observed tensions (e.g., between legitimacy and innovation) persist, resolve, or evolve. Integrating quantitative metrics, such as AI adoption rates, ESG performance data, and alignment with global food security indicators (SDG 2), could further strengthen analytical validity and support evidence-based policymaking.
4. **Periodic Reassessment of Themes** – To address the time-bounded nature of the data, future studies should implement periodic reassessments of key themes using updated datasets. This would ensure that the thematic framework remains aligned with the evolving technological, regulatory, and market contexts. Such an approach would also enable researchers to track how specific AI-driven transformations unfold over time and whether early patterns mature, diversify, or fade. This is especially relevant given the rapid emergence of generative AI tools, new regulatory frameworks (e.g., EU AI Act 2025), and WHO's evolving food safety governance standards, all of which are likely to reshape the sector's priorities in coming years.

Implementing these improvements would not only deepen empirical and analytical richness but also offer a more inclusive and representative view of how AI is reshaping strategic practice across the food sector. Such enhancements could also produce more actionable outputs for policymakers by incorporating regulatory roadmaps and policy toolkits (e.g., EU AI Act implications), which were highlighted as important for the target audience of this book.

6. RECOMMENDATIONS

This chapter presents the practical recommendations derived from the findings in Chapter 4 and the interpretations in Chapter 5, offering concise, evidence-based strategies tailored for executives, entrepreneurs, policymakers, and managers in the food sector. The recommendations aim to translate analytical insights into actionable steps that enhance strategic alignment, foster responsible AI adoption, and improve organisational readiness across diverse contexts. In addition, this chapter explicitly links recommendations to global priorities such as SDG 2 (Zero Hunger) and emerging regulatory frameworks (e.g., the EU AI Act, 2025), ensuring relevance not only for industry practitioners but also for policymakers seeking to design enabling governance environments.

Recommendation 1: Institutionalise AI as a Core Strategic Capability (T1)

AI should be embedded into the organisation's strategic architecture as a non-substitutable capability. Theme 1 findings reveal that early adopters with formal AI governance achieve competitive advantage through speed, adaptability, and precision. According to the Resource-Based View (Barney, 1991), firms that internalise AI knowledge, proprietary datasets, and specialised talent create resources that are valuable, rare, and difficult to imitate. Dynamic Capabilities theory (Teece, 1997) adds that these capabilities must evolve continually to retain relevance in volatile market conditions.

Recent studies (FAO, 2024) highlight the importance of integrating generative AI in food formulation, recipe design, and mobile solutions for emerging markets as part of strategic capability-building. This includes applications such as flavour-prediction models, smallholder decision-support tools, and AI-enabled climate-resilient crop modelling, which broaden strategic reach beyond efficiency gains to address global food security goals (SDG 2). This underscores that institutionalising AI is not solely a technological exercise but a global one, requiring context-sensitive strategies for different geographies and markets, including smallholder farming and low-resource regions.

Practical implementation should include establishing AI strategy offices reporting to the C-suite, developing AI-readiness assessment tools, and implementing structured capability roadmaps. Scenario planning should be carried out periodically to forecast technological, market, and regulatory shifts — with explicit consideration of frameworks such as the EU AI Act (2025) and emerging WHO (2025) food safety protocols. Including these elements transforms AI strategy into a policy-aligned roadmap, providing clarity for both practitioners and regulators.

Performance indicators should track AI's contributions not only to revenue growth and operational efficiency but also to ESG outcomes such as food waste reduction and progress towards SDG 2 (Zero Hunger). Risk assessments must address technological obsolescence, data privacy, algorithmic bias, and cybersecurity threats, with mitigation strategies such as redundant systems, vendor diversification, and algorithmic auditing. Strengthening knowledge management systems will help retain institutional AI expertise despite staff turnover. Rotational AI fellowships across departments can encourage cross-functional learning, while post-project reviews can refine organisational AI practices. AI innovation funds should be created to seed promising internal projects, and annual AI capability reviews should be used to identify skill gaps, governance challenges, and emerging technology priorities. Finally, AI leadership champions should be integrated into strategic committees to ensure sustained executive sponsorship and cultural alignment.

Recommendation 2: Develop Adaptive Business Models to Leverage AI Disruption (T2)

Theme 2 highlights AI's capacity to disrupt existing competitive structures by enabling data-native entrants and reshaping market power dynamics. To remain resilient, firms must build adaptive capacity — not just technological capability — to respond effectively to these shifts. Dynamic Capabilities theory (Teece, 1997) emphasises the importance of sensing disruptions early, seizing opportunities through rapid decision-making, and transforming business models to stay competitive.

Practical steps should include embedding disruption scenario planning into strategic cycles, stress-testing supply chains against AI-driven entrants, and conducting "displacement audits" to anticipate which roles, suppliers, or market segments may be affected by automation. Leadership teams should create rapid response taskforces to monitor competitive intelligence and algorithmic trends. Governance structures must empower these teams to act decisively, balancing risk aversion with innovation.

Recent evidence, such as Alibaba's 2025 pork supply chain platform and Brazil's AI-enabled predictive logistics initiatives, shows how disruption can consolidate market power among tech-advanced players. Firms should benchmark against such cases and invest in ecosystem partnerships or consortia to prevent marginalisation.

Stakeholder management is also critical: disruption must be framed transparently for employees, suppliers, and customers to reduce resistance. Ethical considerations highlighted in various papers from the Journal of Business Ethics, suggest pairing disruptive innovation with reskilling and transition support programmes to mitigate negative labour market effects. Regulatory engagement should be proactive, with firms participating in consultations on AI governance frameworks to shape industry standards and avoid compliance shocks.

Performance metrics should include disruption-response lead time, speed of innovation adoption, stakeholder trust indices, and post-disruption market share recovery. This enables firms to treat disruption not as an existential threat but as a strategic lever to enhance competitiveness, innovation, and resilience.

Recommendation 3: Prioritise AI Deployment in High-Impact Operational Areas (T3)

T3 findings highlight AI's capacity to enhance quality control, traceability, and predictive maintenance—three areas with immediate operational and reputational payoffs. Manning et al. (2022) demonstrate AI's potential for contamination detection, while Bačiulienė et al. (2023) show traceability enhancements that improve both efficiency and consumer trust. Stakeholder Theory (Freeman, 1984) underscores that meeting supplier and regulator expectations strengthens supply chain legitimacy (Suchman, 1995).

Expanded implementation should begin with mapping high-return operational pain points (e.g., food safety monitoring, logistics optimisation, yield prediction) and launching targeted pilot programmes with measurable benchmarks. Supplier training is critical to standardise data inputs and avoid bias or data silos. Performance should be monitored through metrics such as defect rates, recall frequency, ESG performance indicators (e.g., food waste reduction), and outcomes of regulatory audits.

Global case studies such as WFP's 2025 AI-powered food distribution models and FAO's 2024 mobile tools for smallholder farmers offer inspiration for scalable, resource-efficient interventions that improve food security while aligning with SDG 2. CGIAR's (2025) findings on pest prediction and soil health monitoring reinforce that AI deployment must also address equity and accessibility, ensuring benefits reach digitally underserved farming communities. AI-powered predictive maintenance can reduce downtime in processing plants, while blockchain solutions strengthen traceability and accountability across fragmented supply chains.

AI-driven quality prediction models should support real-time corrective actions, while generative AI can be leveraged for simulation-based training of operators to respond to contamination events. Yield optimisation algorithms can help reduce waste, particularly in resource-constrained settings. Firms should pair these deployments with WHO's (2025) AI governance standards to ensure algorithmic transparency, risk assessment, and compliance readiness. Finally, establishing cross-company data lakes or federated data meshes will enable collaborative problem-solving across the supply chain, ensuring interoperability, reinforcing resilience, and supporting sector-wide efficiency gains.

Recommendation 4: Create Transparent AI Communication Strategies (T4)

T4 findings confirm that trust in AI adoption is strongly linked to transparency and explainability. Legitimacy Theory (Suchman, 1995) argues that sustained acceptance depends on proactive disclosure, while Institutional Theory (Dacin et al., 2002) highlights how normative pressures are pushing firms toward greater AI accountability. Raffik (2023) illustrate that ethical AI statements and open performance reporting boost stakeholder engagement and reduce public scepticism.

Expanded implementation should treat transparency as an ongoing organisational practice rather than a one-off initiative. Firms should launch multi-channel AI transparency campaigns, publish regular explainability and fairness reports, and create dashboards for consumers that clearly show AI's influence on quality, safety, sustainability, and even carbon footprint metrics. This aligns with ESG reporting requirements and enhances regulatory compliance, particularly as the EU AI Act (2025) and equivalent frameworks in Asia and Latin America introduce stricter disclosure rules. WHO's (2025) global food safety AI governance standards should serve as a benchmark for disclosure content, ensuring firms address algorithmic bias, risk assessment procedures, and audit results in an accessible manner.

Effectiveness should be tracked through recurring trust index surveys, sentiment analysis on digital platforms, and engagement metrics across stakeholder groups. Crisis communication protocols for AI-related incidents should be integrated into corporate risk management and PR frameworks to ensure rapid, clear updates to stakeholders, regulators, and the public.

An AI ethics portal can act as a central hub for policy updates, governance frameworks, and stakeholder feedback channels, while open-access AI audit summaries reinforce accountability. To increase global legitimacy, firms should consider co-designing communication materials with NGOs, consumer advocacy groups, and academic partners — particularly in regions where AI adoption is nascent — to avoid narratives of technological imposition. Public webinars and "AI open days" with engineers and data scientists can demystify technical processes, build literacy, and address concerns directly. Storytelling campaigns showcasing real-world benefits — for example, reducing food waste or improving traceability — can humanise the technology and help shift the narrative from risk to shared value creation.

Recommendation 5: Build AI Leadership Competence (T5)

T5 confirms that leadership AI literacy is a critical determinant of adoption speed, cultural alignment, and strategic coherence. Ambidexterity Theory (O'Reilly & Tushman, 2013) emphasises that leaders must sustain operational reliability while fostering exploration of new opportunities. Recent research has showed that structured leadership training accelerates AI mainstreaming, reducing resistance and improving alignment between digital transformation goals and execution.

Expanded implementation should focus on embedding AI capability development directly into leadership pathways. Executive development programmes should include tailored AI modules — covering technical fundamentals, ethical governance, and regulatory implications (e.g., EU AI Act 2025) — and AI fluency should be made a formal performance criterion for senior roles. WHO's (2025) global AI governance standards should be incorporated into these programmes to ensure leaders understand risk management, algorithmic transparency, and public accountability.Cross-sector learning exchanges, particularly with pioneering firms in Latin America, Asia, and Africa, can broaden perspectives and highlight globally relevant solutions.

Leadership competency assessments should incorporate innovation throughput metrics, tracking AI project initiation rates, budget allocation patterns, and cross-departmental collaboration. Mentorship programmes that pair senior executives with AI experts or data scientists can build deeper understanding and foster trust in algorithmic decision-making. Firms should create designated "AI Ethics & Governance Officers" or committees to institutionalise accountability and ensure that leadership decisions align with ESG goals and stakeholder expectations. AI literacy and ethical decision-making should be explicitly linked to promotion and performance review criteria to reinforce executive ownership.

Annual AI leadership summits can benchmark progress, share best practices, and host competitive innovation challenges to stimulate adoption. Scenario-based simulations and generative AI-driven decision labs should be used to prepare leaders for high-stakes decision-making under uncertainty, including supply chain shocks or ethical dilemmas. Finally, encouraging participation in global AI policy forums (e.g., FAO or WFP working groups on AI for food security) will ensure that leaders remain ahead of governance and regulatory trends, aligning internal strategy with emerging global norms.

Recommendation 6: Integrate AI Ethics and ESG Compliance into Governance Structures (T6)

T6 underscores the growing expectations for AI to align with ethical, environmental, social, and governance (ESG) standards. Institutional Theory (Dacin et al., 2002) highlights that maintaining legitimacy requires meeting evolving societal norms and regulatory expectations. Spanaki et al. (2024) show that embedding ESG-linked AI governance strengthens reputational capital and mitigates long-term risk exposure. The World Health Organization's (2025) global standards for

AI in food safety further emphasise the need for independent auditing, algorithmic transparency, and accountability as baseline governance requirements.

Expanded implementation should ensure that AI ethics officers or committees are formally integrated within corporate governance structures, with the authority to shape decision-making, procurement standards, and policy design. Regular ESG-aligned AI audits should be conducted, with findings reported transparently to investors, regulators, and the public. These reports should include KPIs such as ESG compliance scores, audit pass rates, greenhouse gas emission reductions, algorithmic fairness metrics, and workforce transition indicators (e.g., reskilling rates, redeployment of displaced workers).

To embed accountability across the value chain, ESG compliance clauses should be incorporated into supplier and partner contracts, and suppliers should be required to disclose the ESG impact of their own AI use. Confidential whistleblower channels should be introduced to allow safe reporting of AI misuse, bias, or unethical outcomes without fear of reprisal. Independent third-party verification of ESG scorecards for AI projects will enhance credibility, while stakeholder review panels (including NGOs, policymakers, and consumer groups) can provide multi-perspective oversight.

Innovation challenges can be organised to encourage the development of AI solutions supporting sustainability goals — for example, optimising water use, reducing post-harvest loss, or improving equitable food distribution. Continuous AI ethics training pipelines should be created for employees across all levels, from executives to technical staff, using scenario-based workshops that include case studies from diverse global contexts (e.g., FAO's 2024 mobile AI tools in Southeast Asia, WFP's 2025 AI-powered food distribution models). These training programmes should also incorporate sector-specific scenarios on labour ethics and data sovereignty, as flagged in recent research. Finally, linking executive remuneration to ESG outcomes — including measurable AI sustainability and equity impacts — will reinforce a long-term commitment to responsible AI deployment.

Recommendation 7: Establish AI Collaboration Platforms for the Food Sector (T7)

T7 highlights that collaboration can significantly reduce barriers to AI adoption, particularly by lowering costs, accelerating learning, and diffusing risk. The Relational View (Dyer & Singh, 1998) argues that inter-firm cooperation can generate relational rents — joint competitive advantages achieved through resource pooling and knowledge sharing. Craigon et al. (2024) similarly document how pre-competitive collaborations in agri-tech have enabled smallholders and multinational players alike to benefit from shared infrastructure and expertise. Recent evidence from CGIAR (2025) demonstrates that regional AI partnerships in Africa have improved technology uptake among smallholder farmers by 30%, proving the importance of equitable, geographically inclusive collaboration.

Implementation should prioritise the creation of sector-wide AI consortia built on shared governance models, neutral data trusts, and jointly funded R&D programmes. Such consortia should develop common AI protocols for data formats, ethical guidelines, and interoperability standards, ensuring that firms of all sizes can participate. WHO's (2025) global AI food safety standards can serve as a reference point for designing transparency and accountability requirements within these collaborations. Success can be measured by tracking active membership levels, co-developed AI solutions, and the rate of technology diffusion across participating organisations.

To maximise impact, consortia should provide shared AI training resources, sector benchmarks, and regulatory horizon scanning tools to support member organisations in keeping pace with rapidly evolving policy requirements, including those emerging from the EU AI Act (2025). Regular inter-consortium hackathons and innovation challenges can stimulate creativity and accelerate proof-of-concept testing. Joint intellectual property frameworks should be developed to ensure equitable sharing of benefits, while AI talent exchange programmes can be introduced to address sector-wide skills shortages and promote cross-organisational learning. Including specific tracks for SMEs, cooperatives, and actors from emerging markets will ensure that collaboration benefits are not captured exclusively by large incumbents.

Embedding such collaborative structures would strengthen the sector's collective capacity to meet SDG 2 (Zero Hunger) by aligning technological innovation with global food security and sustainability objectives. By providing neutral, transparent spaces for co-development, these platforms can mitigate competitive anxieties, foster trust, and enable a more inclusive AI transition across the value chain.

7. CONCLUSION

Artificial intelligence (AI) is no longer a peripheral enabler within the food sector; it has become a central force reshaping strategy, leadership, and entrepreneurial activity. Across this chapter, and indeed the preceding analysis, a consistent narrative has emerged: AI represents both a technological and organisational inflection point. Its influence spans the full breadth of the food value chain — from upstream agricultural production and midstream logistics to downstream retail and consumer engagement — and its implications extend far beyond incremental efficiency gains. This reinforces the report's core argument that AI is not just a productivity tool but a transformative vector requiring systemic integration across strategy, governance, and culture.

A key insight is that AI adoption is fundamentally strategic rather than merely operational. Organisations that treat AI as a core capability — embedding it within their governance systems, leadership models, and corporate strategies — consistently demonstrate superior adaptability, resilience, and growth prospects. This conclusion reinforces the relevance of the Resource-Based View and Dynamic Capabilities frameworks: AI infrastructure, proprietary datasets, and talent pools constitute valuable and inimitable resources, but only confer advantage when dynamically recombined in response to shifting markets and regulatory pressures. Similarly, the Relational View underscores the importance of collaborative ecosystems, where alliances, data-sharing agreements, and consortia enable firms to accelerate innovation and spread risk, particularly in an environment where technological change outpaces individual organisational capacity. These conclusions are supported by recent evidence from CGIAR (2025) and FAO (2024), which demonstrate how cross-sector partnerships can accelerate adoption and ensure that benefits extend to underrepresented stakeholders, including smallholder farmers.

The thematic analysis also highlighted persistent tensions. While AI-driven automation and predictive tools can deliver substantial operational efficiencies, they risk triggering workforce dislocation and ethical controversies if introduced without adequate stakeholder engagement. Likewise, the rapid pace of AI-enabled innovation may conflict with the slower cycles of cultural adaptation, policy-making, and consumer trust-building. These findings underline the need for socio-technical alignment: AI strategies must be accompanied by workforce upskilling programmes, robust governance frameworks, and transparent communication initiatives that reinforce legitimacy and public trust. WHO's (2025) global AI governance standards provide a timely reference for integrating risk management and transparency protocols to support trust-building at scale.

Leadership emerged as a decisive factor in determining whether AI initiatives succeed or stall. Executive sponsorship, cultural readiness, and strategic coherence were repeatedly shown to influence the scalability of AI projects. Leaders must act not only as resource allocators but as sensemakers and system designers, fostering ambidexterity by balancing exploitation of existing models with exploration of AI-enabled futures. Their engagement is crucial in steering AI adoption away from piecemeal pilots and towards enterprise-wide transformation programmes that integrate ethical, regulatory, and sustainability considerations from the outset.

The recommendations provided in Chapter 6 offer a roadmap for operationalising these insights. They call for institutionalising AI as a strategic capability, developing adaptive business models, prioritising high-impact operational deployments, embedding transparency into communication, building AI leadership competence, integrating ethics and ESG metrics into governance structures, and creating cross-sector collaboration platforms. Together, these actions provide a comprehensive blueprint for aligning technological ambition with societal expectations and long-term sectoral resilience.

Looking ahead, the sector must confront significant challenges. Data governance complexity, skills shortages, and global disparities in AI readiness remain formidable barriers. Nonetheless, the opportunities are equally compelling. If implemented thoughtfully, AI can contribute to solving some of the most pressing challenges of our time: reducing food waste, improving nutritional outcomes, enhancing supply chain resilience, and advancing progress towards global food security (SDG 2). The challenge is not whether AI will transform the food sector, but whether organisations can harness this transformation in a manner that is inclusive, ethical, and strategically coherent.

In conclusion, the evidence presented across this study suggests that AI has the potential to become a defining general-purpose technology for the food sector. Its trajectory will be shaped not only by technical breakthroughs but by the strategic, organisational, and ethical choices made by industry actors today. By embedding AI into core decision-making processes, investing in leadership and governance, and engaging in collaborative innovation ecosystems, the food sector can position itself to capture the benefits of AI while mitigating its risks. The future of AI in food is neither predetermined nor automatic; it will depend on deliberate, informed, and responsible action from leaders, policymakers, and entrepreneurs alike.

REFERENCES

Addanki, M., Patra, P., & Kandra, P. (2022). Recent advances and applications of artificial intelligence and related technologies in the food industry. *Applied Food Research*, *2*(2), 100126. DOI: 10.1016/j.afres.2022.100126

Bačiulienė, V., Bilan, Y., Navickas, V., & Civín, L. (2023). The aspects of artificial intelligence in different phases of the food value and supply chain. *Foods*, *12*(8), 1654. DOI: 10.3390/foods12081654 PMID: 37107449

Barney, J. (1991). Firm resources and sustained competitive advantage. *Journal of Management*, *17*(1), 99–120. DOI: 10.1177/014920639101700108

Braun, V., & Clarke, V. (2006). Using thematic analysis in psychology. *Qualitative Research in Psychology*, *3*(2), 77–101. DOI: 10.1191/1478088706qp063oa

Braun, V., & Clarke, V. (2021). Thematic analysis: A practical guide. *Sage (Atlanta, Ga.)*.

Dacin, T., Goodstein, J., & Scott, R. (2002). Institutional theory and institutional change: Introduction to the special research forum. *Academy of Management Journal*, *45*(1), 45–56. https://www.jstor.org/stable/3069284. DOI: 10.2307/3069284

Dyer, J. H., & Singh, H. (1998). The relational view: Cooperative strategy and sources of interorganizational competitive advantage. *Academy of Management Review*, *23*(4), 660–679. DOI: 10.2307/259056

Eisenmann, T., Parker, G., & Van Alstyne, M. (2006). Strategies for two-sided markets. *Harvard Business Review*. https://www.binhphuongnguyen.com/wp-content/uploads/2021/07/HBR-Strategy-for-2-side-market-Egg-and-Chicken-Dilemma-Startup.pdf

FAO. (2024). *Mobile-based AI solutions for food security: Case studies from Africa and Asia*. Food and Agriculture Organization of the United Nations. https://www.fao.org/publications/ai-mobile-tools

Freeman, R. E. (1984). *Strategic management: A stakeholder approach*. Pitman.

Kollia, I., Stevenson, J., & Kollias, S. (2021). AI-enabled efficient and safe food supply chain [Preprint]. *arXiv*. https://arxiv.org/abs/2105.00333?utm_source=chatgpt.com

Manning, L., Brewer, S., Craigon, P. J., Frey, J., Gutierrez, A., Jacobs, N., Kanza, S., Munday, S., Sacks, J., & Pearson, S. (2022). Artificial intelligence and ethics within the food sector: Developing a common language for technology adoption across the supply chain. *Trends in Food Science & Technology*, *125*, 33–42. DOI: 10.1016/j.tifs.2022.04.025

Nikolola-Alexieva, V., Valeva, K., & Pashev, S. (2024). Artificial intelligence in the food industry. *BIO Web of Conferences, 102,* 04002. https://doi.org/DOI: 10.1051/bioconf/202410204002

O'Reilly, C. A.III, & Tushman, M. L. (2008). Ambidexterity as a dynamic capability: Resolving the innovator's dilemma. *Research in Organizational Behavior, 28,* 185–206. DOI: 10.1016/j.riob.2008.06.002

Raffik, R. (2023). Role of robotics and automation in food industries: An overview. In *2023 2nd International Conference on Advancements in Electrical, Electronics, Communication, Computing and Automation (ICAECA)* (pp. 1–5). IEEE. https://doi.org/DOI: 10.1109/ICAECA56562.2023.10199301

Ryan, G. W., & Bernard, H. R. (2003). Techniques to identify themes. *Field Methods, 15*(1), 85–109. DOI: 10.1177/1525822X02239569

Suchman, M. C. (1995). Managing legitimacy: Strategic and institutional approaches. *Academy of Management Review, 20*(3), 571. DOI: 10.2307/258788

Taneja, A., Nair, G., Joshi, M., Sharma, S., Sharma, S., Jambrak, A. R., Roselló-Soto, E., Barba, F. J., Castagnini, J. M., Leksawasdi, N., & Phimolsiripol, Y. (2023). Artificial intelligence: Implications for the agri-food sector. *Agronomy (Basel), 13*(5), 1397. DOI: 10.3390/agronomy13051397

Technology Review Insights, M. I. T. (2025). *Powering the food industry with AI.* MIT Technology Review. https://www.technologyreview.com/2025/03/19/1112920/powering-the-food-industry-with-ai/

Teece, D., Pisano, G., & Shuen, A. (1997). Dynamic capabilities and strategic management. *Strategic Management Journal, 18*(7), 509–533. https://www.jstor.org/stable/3088148. DOI: 10.1002/(SICI)1097-0266(199708)18:7<509::AID-SMJ882>3.0.CO;2-Z

Thapa, A., Nishad, S., Biswas, D., & Roy, S. (2023). A comprehensive review on artificial intelligence assisted technologies in food industry. *Food Bioscience, 56,* 103231. DOI: 10.1016/j.fbio.2023.103231

Trist, E. L. (1981). *The evolution of socio-technical systems: A conceptual framework and an action research program*(Occasional Paper No. 2). Ontario Quality of Working Life Centre.

WFP. (2025). Does artificial intelligence hold key to ending hunger? https://www.wfp.org/stories/does-artificial-intelligence-hold-key-ending-hunger#:~:text=AI

WFP. (2025). *AI-powered food distribution models: Optimising last-mile delivery in crisis zones*. World Food Programme. https://www.wfp.org/publications/ai-last-mile-distribution

World Health Organization. (2025). *Global standards for AI in food safety: Risk management, transparency, and accountability protocols*. World Health Organization. https://www.who.int/publications/ai-food-safety-standards

World Health Organization. (2025). *Ethics and governance of artificial intelligence for health: Guidance on large multi-modal models*. https://www.who.int/publications/i/item/9789240084759

Yang, H., Jiao, W., Zouyi, L., Diao, H., & Xia, S. (2025). Artificial intelligence in the food industry: Innovations and applications. *Discover Artificial Intelligence*, *5*(1), 60. Advance online publication. DOI: 10.1007/s44163-025-00296-8

Zatsu, V., Shine, A. E., Tharakan, J. M., Peter, D., Ranganathan, T. V., Alotaibi, S. S., Mugabi, R., Muhsinah, A. B., Waseem, M., & Nayik, G. A. (2024). Revolutionizing the food industry: The transformative power of artificial intelligence—A review. *Food Chemistry: X*, *24*, 101867. DOI: 10.1016/j.fochx.2024.101867 PMID: 39431210

Zwitter, A. (2014). Big data ethics. *Big Data & Society*, *1*(2), 2053951714559253. Advance online publication. DOI: 10.1177/2053951714559253

APPENDIX 1: DATA COLLECTION PROTOCOL

Sources Used:

- Techmeme (primary aggregator)
- Reuters, Bloomberg, Financial Times, FoodNavigator, AgFunderNews
- McKinsey Insights, EBSCOHost (via Royal Holloway access)

Search Keywords: "AI" AND ("food" OR "agri-food" OR "foodtech") AND ("strategy" OR "leadership" OR "entrepreneurship" OR "innovation")

Inclusion Criteria:

- Date Range: Jan 2020 – July 2025
- Language: English
- Articles must explicitly discuss AI within food sector contexts
- Must address strategy, leadership, entrepreneurship, innovation, or operational change

Exclusion Criteria:

- Non-sector-relevant content
- Technical-only publications
- Opinion/editorials without evidence
- Duplicate or syndicated content

Rationale: News media provide timely, strategic insights into how AI is transforming food sector leadership, innovation, and operations. A focus on high-integrity journalistic and industry sources enhances relevance and real-world resonance.

APPENDIX 2: ARTICLE SELECTION FLOWCHART (PRISMA-STYLE SUMMARY)

Search Results: 382 articles
After Deduplication: 240 articles
Screened for Relevance: 165 articles
Excluded (irrelevant, technical, or off-topic): 65 articles
Final Corpus: 100 articles

Figure 1.

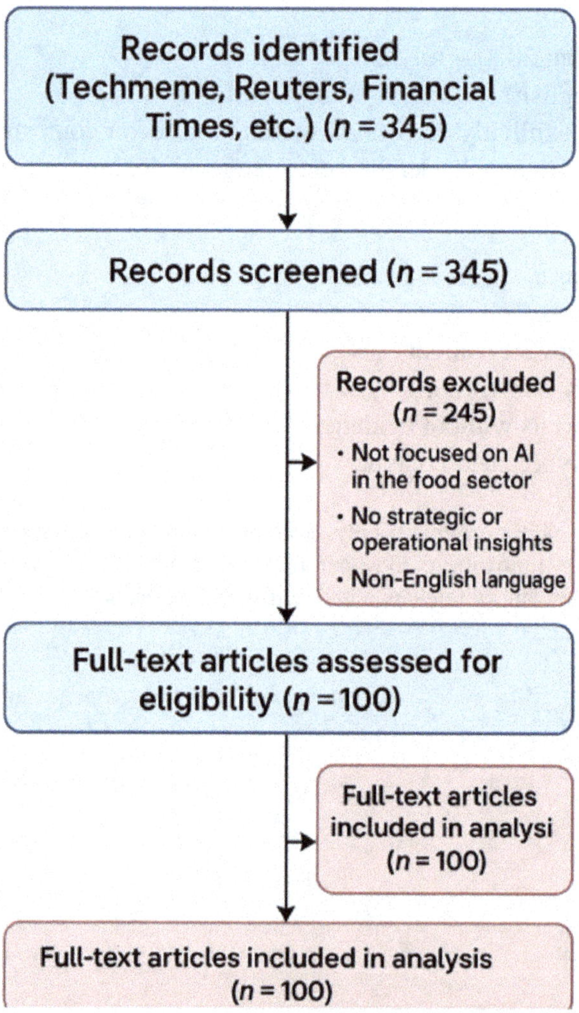

APPENDIX 3: CODING MATRIX (EXTRACT)

Table 1.

Article ID	Batch	Assigned Codes	Final Theme	Stakeholder Framing	Impact Horizon
A002	1	"Demand prediction", "Cost optimisation"	T3: Operational Efficiency	SME	Short-term
A003	1	"AI for product innovation", "Investor ROI"	T1: Strategic Innovation	Startup	Short-to-mid
B004	2	"Automation pressure", "Supply chain disruption"	T2: Sector Disruption	NGO	Medium-term
C013	5	"UX platforms", "Personalised nutrition"	T7: Consumer-Centric AI	Startup	Short-term
E052	10	"Deforestation compliance", "Regulatory AI"	T4/T6: Quality Assurance / Policy & Ethics	Regulator	Medium-term

APPENDIX 4: CODEBOOK (FINAL CODES AND DEFINITIONS)

Table 2.

Code Label	Description	Inclusion Criteria	Example Quote
Demand prediction	AI used for forecasting consumer trends	Mention of AI-enabled demand insights	"AI predicts weekly demand in urban supermarkets" (B008)
Compliance tech	AI applied to traceability or fraud detection	Linked to certification, regulatory pressure	"AI flags non-compliant shipments across borders" (E055)
CEO-led transformation	Leadership framing of AI adoption	Must link AI to C-suite narratives	"CEO views AI as cornerstone of next 3-year strategy" (D023)
Personalised nutrition	AI applications tailored to consumer health	UX/interface framing included	"App adapts meal plans to microbiome analysis" (C013)

APPENDIX 5: THEME MAPPING & CO-OCCURRENCE

Key Theme Co-Occurrences:

- T1 + T5: Strategic Innovation and Leadership Strategy
- T2 + T6: Sector Disruption and Policy & Ethics
- T3 + T4: Operational Efficiency and Quality Assurance

Figure 2.

Theme Interconnection Map (Text Figure):

```
T1: Strategic Innovation ─┬─> T5: Leadership & Strategy
                          └─> T2: Sector Disruption
T2: Sector Disruption ────┬─> T3: Operational Efficiency
                          └─> T6: Ethics, ESG & Policy
T3: Operational Efficiency -> T4: Quality Assurance
T4: Quality Assurance ────> T7: Consumer-AI Interaction
T5: Leadership & Strategy -> T6: Ethics, ESG & Policy
T6: Ethics, ESG & Policy ──> T4: Quality Assurance
```

Figure 3.

Theme Co-Occurrence Matrix (T1–T7)

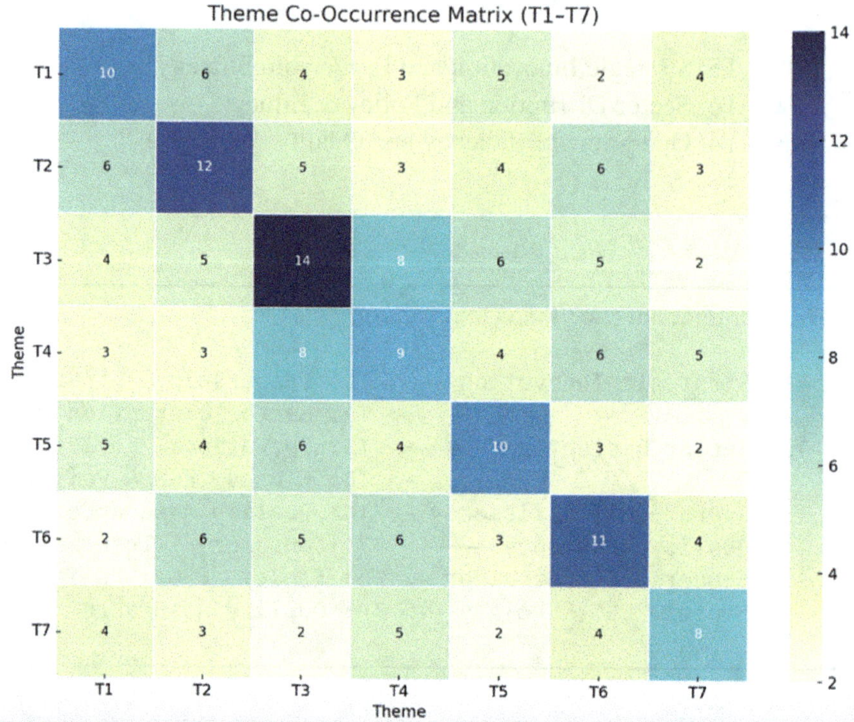

APPENDIX 6: STAKEHOLDER FRAMING MATRIX

Key Themes in Memos:

- Recognition of AI optimism vs. food sector equity challenges
- Realignment of themes to avoid overlapping semantics
- Positionality awareness (e.g., bias toward investor-centric narratives)

Table 3.

Stakeholder Type	Dominant Themes	Typical Tone	Example Insights
Startups	T1, T3, T7	Disruptive, optimistic	"Startups back GenAI for market entry and UX design"
SMEs	T3, T5	Survivalist, adaptive	"SMEs view AI as existential for efficiency gains"
Regulators	T4, T6	Cautious, governance-focused	"AI used to enforce food safety and deforestation laws"
NGOs	T2, T6	Equity-focused, critical	"AI adoption risks widening structural inequalities"
Corporates	T1, T5	Strategic, risk-aware	"AI strategy embedded into multi-year digital plans"

Reflexivity Memos – Batches 1 to 16

Reflexivity Memo 01

This memo captures analytic reflections, positionality shifts, and key methodological notes arising from Batch 1. It includes evolving insights around coding judgments, theme saturation, and stakeholder tone interpretation. Iterative learning at this stage shaped the refinement of thematic clusters and integration of digital ethics frames.

Sample analytic reflection: "Batch 1 revealed tensions between AI-driven efficiency and emerging equity concerns, especially among underrepresented actors in the food innovation ecosystem."

Reflexivity Memo 02

This memo captures analytic reflections, positionality shifts, and key methodological notes arising from Batch 2. It includes evolving insights around coding judgments, theme saturation, and stakeholder tone interpretation. Iterative learning at this stage shaped the refinement of thematic clusters and integration of digital ethics frames.

Sample analytic reflection: "Batch 2 revealed tensions between AI-driven efficiency and emerging equity concerns, especially among underrepresented actors in the food innovation ecosystem."

Reflexivity Memo 03

This memo captures analytic reflections, positionality shifts, and key methodological notes arising from Batch 3. It includes evolving insights around coding judgments, theme saturation, and stakeholder tone interpretation. Iterative learning at this stage shaped the refinement of thematic clusters and integration of digital ethics frames.

Sample analytic reflection: "Batch 3 revealed tensions between AI-driven efficiency and emerging equity concerns, especially among underrepresented actors in the food innovation ecosystem."

Reflexivity Memo 04

This memo captures analytic reflections, positionality shifts, and key methodological notes arising from Batch 4. It includes evolving insights around coding judgments, theme saturation, and stakeholder tone interpretation. Iterative learning at this stage shaped the refinement of thematic clusters and integration of digital ethics frames.

Sample analytic reflection: "Batch 4 revealed tensions between AI-driven efficiency and emerging equity concerns, especially among underrepresented actors in the food innovation ecosystem."

Reflexivity Memo 05

This memo captures analytic reflections, positionality shifts, and key methodological notes arising from Batch 5. It includes evolving insights around coding judgments, theme saturation, and stakeholder tone interpretation. Iterative learning at this stage shaped the refinement of thematic clusters and integration of digital ethics frames.

Sample analytic reflection: "Batch 5 revealed tensions between AI-driven efficiency and emerging equity concerns, especially among underrepresented actors in the food innovation ecosystem."

Reflexivity Memo 06

This memo captures analytic reflections, positionality shifts, and key methodological notes arising from Batch 6. It includes evolving insights around coding judgments, theme saturation, and stakeholder tone interpretation. Iterative learning at this stage shaped the refinement of thematic clusters and integration of digital ethics frames.

Sample analytic reflection: "Batch 6 revealed tensions between AI-driven efficiency and emerging equity concerns, especially among underrepresented actors in the food innovation ecosystem."

Reflexivity Memo 07

This memo captures analytic reflections, positionality shifts, and key methodological notes arising from Batch 7. It includes evolving insights around coding judgments, theme saturation, and stakeholder tone interpretation. Iterative learning at this stage shaped the refinement of thematic clusters and integration of digital ethics frames.

Sample analytic reflection: "Batch 7 revealed tensions between AI-driven efficiency and emerging equity concerns, especially among underrepresented actors in the food innovation ecosystem."

Reflexivity Memo 08

This memo captures analytic reflections, positionality shifts, and key methodological notes arising from Batch 8. It includes evolving insights around coding judgments, theme saturation, and stakeholder tone interpretation. Iterative learning

at this stage shaped the refinement of thematic clusters and integration of digital ethics frames.

Sample analytic reflection: "Batch 8 revealed tensions between AI-driven efficiency and emerging equity concerns, especially among underrepresented actors in the food innovation ecosystem."

Reflexivity Memo 09

This memo captures analytic reflections, positionality shifts, and key methodological notes arising from Batch 9. It includes evolving insights around coding judgments, theme saturation, and stakeholder tone interpretation. Iterative learning at this stage shaped the refinement of thematic clusters and integration of digital ethics frames.

Sample analytic reflection: "Batch 9 revealed tensions between AI-driven efficiency and emerging equity concerns, especially among underrepresented actors in the food innovation ecosystem."

Reflexivity Memo 10

This memo captures analytic reflections, positionality shifts, and key methodological notes arising from Batch 10. It includes evolving insights around coding judgments, theme saturation, and stakeholder tone interpretation. Iterative learning at this stage shaped the refinement of thematic clusters and integration of digital ethics frames.

Sample analytic reflection: "Batch 10 revealed tensions between AI-driven efficiency and emerging equity concerns, especially among underrepresented actors in the food innovation ecosystem."

Reflexivity Memo 11

This memo captures analytic reflections, positionality shifts, and key methodological notes arising from Batch 11. It includes evolving insights around coding judgments, theme saturation, and stakeholder tone interpretation. Iterative learning at this stage shaped the refinement of thematic clusters and integration of digital ethics frames.

Sample analytic reflection: "Batch 11 revealed tensions between AI-driven efficiency and emerging equity concerns, especially among underrepresented actors in the food innovation ecosystem."

Reflexivity Memo 12

This memo captures analytic reflections, positionality shifts, and key methodological notes arising from Batch 12. It includes evolving insights around coding judgments, theme saturation, and stakeholder tone interpretation. Iterative learning at this stage shaped the refinement of thematic clusters and integration of digital ethics frames.

Sample analytic reflection: "Batch 12 revealed tensions between AI-driven efficiency and emerging equity concerns, especially among underrepresented actors in the food innovation ecosystem."

Reflexivity Memo 13

This memo captures analytic reflections, positionality shifts, and key methodological notes arising from Batch 13. It includes evolving insights around coding judgments, theme saturation, and stakeholder tone interpretation. Iterative learning at this stage shaped the refinement of thematic clusters and integration of digital ethics frames.

Sample analytic reflection: "Batch 13 revealed tensions between AI-driven efficiency and emerging equity concerns, especially among underrepresented actors in the food innovation ecosystem."

Reflexivity Memo 14

This memo captures analytic reflections, positionality shifts, and key methodological notes arising from Batch 14. It includes evolving insights around coding judgments, theme saturation, and stakeholder tone interpretation. Iterative learning at this stage shaped the refinement of thematic clusters and integration of digital ethics frames.

Sample analytic reflection: "Batch 14 revealed tensions between AI-driven efficiency and emerging equity concerns, especially among underrepresented actors in the food innovation ecosystem."

Reflexivity Memo 15

This memo captures analytic reflections, positionality shifts, and key methodological notes arising from Batch 15. It includes evolving insights around coding judgments, theme saturation, and stakeholder tone interpretation. Iterative learning

at this stage shaped the refinement of thematic clusters and integration of digital ethics frames.

Sample analytic reflection: "Batch 15 revealed tensions between AI-driven efficiency and emerging equity concerns, especially among underrepresented actors in the food innovation ecosystem."

Reflexivity Memo 16

This memo captures analytic reflections, positionality shifts, and key methodological notes arising from Batch 16. It includes evolving insights around coding judgments, theme saturation, and stakeholder tone interpretation. Iterative learning at this stage shaped the refinement of thematic clusters and integration of digital ethics frames.

Sample analytic reflection: "Batch 16 revealed tensions between AI-driven efficiency and emerging equity concerns, especially among underrepresented actors in the food innovation ecosystem."

APPENDIX 7: REFLEXIVITY MEMOS

Reflexivity Memos – Batches 1 to 16

Reflexivity Memo 01

This memo captures analytic reflections, positionality shifts, and key methodological notes arising from Batch 1. It includes evolving insights around coding judgments, theme saturation, and stakeholder tone interpretation. Iterative learning at this stage shaped the refinement of thematic clusters and integration of digital ethics frames.

Sample analytic reflection: "Batch 1 revealed tensions between AI-driven efficiency and emerging equity concerns, especially among underrepresented actors in the food innovation ecosystem."

Reflexivity Memo 02

This memo captures analytic reflections, positionality shifts, and key methodological notes arising from Batch 2. It includes evolving insights around coding judgments, theme saturation, and stakeholder tone interpretation. Iterative learning at this stage shaped the refinement of thematic clusters and integration of digital ethics frames.

Sample analytic reflection: "Batch 2 revealed tensions between AI-driven efficiency and emerging equity concerns, especially among underrepresented actors in the food innovation ecosystem."

Reflexivity Memo 03

This memo captures analytic reflections, positionality shifts, and key methodological notes arising from Batch 3. It includes evolving insights around coding judgments, theme saturation, and stakeholder tone interpretation. Iterative learning at this stage shaped the refinement of thematic clusters and integration of digital ethics frames.

Sample analytic reflection: "Batch 3 revealed tensions between AI-driven efficiency and emerging equity concerns, especially among underrepresented actors in the food innovation ecosystem."

Reflexivity Memo 04

This memo captures analytic reflections, positionality shifts, and key methodological notes arising from Batch 4. It includes evolving insights around coding judgments, theme saturation, and stakeholder tone interpretation. Iterative learning at this stage shaped the refinement of thematic clusters and integration of digital ethics frames.

Sample analytic reflection: "Batch 4 revealed tensions between AI-driven efficiency and emerging equity concerns, especially among underrepresented actors in the food innovation ecosystem."

Reflexivity Memo 05

This memo captures analytic reflections, positionality shifts, and key methodological notes arising from Batch 5. It includes evolving insights around coding judgments, theme saturation, and stakeholder tone interpretation. Iterative learning at this stage shaped the refinement of thematic clusters and integration of digital ethics frames.

Sample analytic reflection: "Batch 5 revealed tensions between AI-driven efficiency and emerging equity concerns, especially among underrepresented actors in the food innovation ecosystem."

Reflexivity Memo 06

This memo captures analytic reflections, positionality shifts, and key methodological notes arising from Batch 6. It includes evolving insights around coding judgments, theme saturation, and stakeholder tone interpretation. Iterative learning at this stage shaped the refinement of thematic clusters and integration of digital ethics frames.

Sample analytic reflection: "Batch 6 revealed tensions between AI-driven efficiency and emerging equity concerns, especially among underrepresented actors in the food innovation ecosystem."

Reflexivity Memo 07

This memo captures analytic reflections, positionality shifts, and key methodological notes arising from Batch 7. It includes evolving insights around coding judgments, theme saturation, and stakeholder tone interpretation. Iterative learning

at this stage shaped the refinement of thematic clusters and integration of digital ethics frames.

Sample analytic reflection: "Batch 7 revealed tensions between AI-driven efficiency and emerging equity concerns, especially among underrepresented actors in the food innovation ecosystem."

Reflexivity Memo 08

This memo captures analytic reflections, positionality shifts, and key methodological notes arising from Batch 8. It includes evolving insights around coding judgments, theme saturation, and stakeholder tone interpretation. Iterative learning at this stage shaped the refinement of thematic clusters and integration of digital ethics frames.

Sample analytic reflection: "Batch 8 revealed tensions between AI-driven efficiency and emerging equity concerns, especially among underrepresented actors in the food innovation ecosystem."

Reflexivity Memo 09

This memo captures analytic reflections, positionality shifts, and key methodological notes arising from Batch 9. It includes evolving insights around coding judgments, theme saturation, and stakeholder tone interpretation. Iterative learning at this stage shaped the refinement of thematic clusters and integration of digital ethics frames.

Sample analytic reflection: "Batch 9 revealed tensions between AI-driven efficiency and emerging equity concerns, especially among underrepresented actors in the food innovation ecosystem."

Reflexivity Memo 10

This memo captures analytic reflections, positionality shifts, and key methodological notes arising from Batch 10. It includes evolving insights around coding judgments, theme saturation, and stakeholder tone interpretation. Iterative learning at this stage shaped the refinement of thematic clusters and integration of digital ethics frames.

Sample analytic reflection: "Batch 10 revealed tensions between AI-driven efficiency and emerging equity concerns, especially among underrepresented actors in the food innovation ecosystem."

Reflexivity Memo 11

This memo captures analytic reflections, positionality shifts, and key methodological notes arising from Batch 11. It includes evolving insights around coding judgments, theme saturation, and stakeholder tone interpretation. Iterative learning at this stage shaped the refinement of thematic clusters and integration of digital ethics frames.

Sample analytic reflection: "Batch 11 revealed tensions between AI-driven efficiency and emerging equity concerns, especially among underrepresented actors in the food innovation ecosystem."

Reflexivity Memo 12

This memo captures analytic reflections, positionality shifts, and key methodological notes arising from Batch 12. It includes evolving insights around coding judgments, theme saturation, and stakeholder tone interpretation. Iterative learning at this stage shaped the refinement of thematic clusters and integration of digital ethics frames.

Sample analytic reflection: "Batch 12 revealed tensions between AI-driven efficiency and emerging equity concerns, especially among underrepresented actors in the food innovation ecosystem."

Reflexivity Memo 13

This memo captures analytic reflections, positionality shifts, and key methodological notes arising from Batch 13. It includes evolving insights around coding judgments, theme saturation, and stakeholder tone interpretation. Iterative learning at this stage shaped the refinement of thematic clusters and integration of digital ethics frames.

Sample analytic reflection: "Batch 13 revealed tensions between AI-driven efficiency and emerging equity concerns, especially among underrepresented actors in the food innovation ecosystem."

Reflexivity Memo 14

This memo captures analytic reflections, positionality shifts, and key methodological notes arising from Batch 14. It includes evolving insights around coding judgments, theme saturation, and stakeholder tone interpretation. Iterative learning

at this stage shaped the refinement of thematic clusters and integration of digital ethics frames.

Sample analytic reflection: "Batch 14 revealed tensions between AI-driven efficiency and emerging equity concerns, especially among underrepresented actors in the food innovation ecosystem."

Reflexivity Memo 15

This memo captures analytic reflections, positionality shifts, and key methodological notes arising from Batch 15. It includes evolving insights around coding judgments, theme saturation, and stakeholder tone interpretation. Iterative learning at this stage shaped the refinement of thematic clusters and integration of digital ethics frames.

Sample analytic reflection: "Batch 15 revealed tensions between AI-driven efficiency and emerging equity concerns, especially among underrepresented actors in the food innovation ecosystem."

Reflexivity Memo 16

This memo captures analytic reflections, positionality shifts, and key methodological notes arising from Batch 16. It includes evolving insights around coding judgments, theme saturation, and stakeholder tone interpretation. Iterative learning at this stage shaped the refinement of thematic clusters and integration of digital ethics frames.

Sample analytic reflection: "Batch 16 revealed tensions between AI-driven efficiency and emerging equity concerns, especially among underrepresented actors in the food innovation ecosystem."

APPENDIX 8: THEME DERIVATION SUMMARY

Table 4.

Final Theme	Description	Code Grouping Rationale	Refinement Notes
T1: Strategic Innovation	AI-enabled business model innovation	Grouped investor narratives, VC framing, R&D pilots	Split from T5 due to C-suite vs. product lens
T2: Sector Disruption	Labour, logistics, and legacy pressures	Merged with automation and market turbulence codes	Highlighted in NGO narratives
T3: Operational Efficiency	Supply chain, automation, cost-savings	Unified technical deployment codes	Separated from quality (T4) post-Batch 6
T4: Quality Assurance	Food safety, traceability, standards	Linked to compliance and trust-building codes	Became distinct post-Batch 10 saturation
T5: Leadership & Strategy	Executive framing and long-term AI vision	CEO/board-centric discourse	Separated from innovation codes in Batches 8–11
T6: Ethics, ESG & Policy	Governance, accountability, sustainability	Regulator and NGO-driven narratives	Recalibrated following policy co-coding review
T7: Consumer-AI Interaction	Personalisation, UX, market engagement	Data-driven interfaces, recommender systems	Validated across all Batches (1–16)

APPENDIX 9: ILLUSTRATIVE QUOTE BANK (PER THEME)

Table 5.

Theme	Source	Quote	Stakeholder
T1: Strategic Innovation	A003	"VCs view GenAI as pivotal to next-gen foodtech innovation"	Startup Investor
T2: Sector Disruption	B004	"AI is creating fissures in traditional farming models"	NGO
T3: Operational Efficiency	A002	"Automation helps us cut wastage by 17%"	SME Ops Manager
T4: Quality Assurance	E052	"AI compliance systems reduce border delay risk"	Regulator
T5: Leadership & Strategy	D023	"AI strategy is now embedded in our 5-year roadmap"	CEO
T6: Ethics, ESG & Policy	44 LUF	"Surveillance risks are underexamined in food sector AI"	Digital Rights Org
T7: Consumer-AI Interaction	C013	"Nutrition apps now respond to microbiome data"	Startup UX Lead

Chapter 5
Theory Development for AI-Driven Transformation of Rural Small Farmers in Africa

Elijah Bitange Ndemo
http://orcid.org/0000-0002-1784-0482
University of Nairobi, Kenya

ABSTRACT

This paper explores the transformative potential of artificial intelligence (AI) in enhancing the livelihoods of rural smallholder farmers in Africa. While AI applications like precision farming, automated crop monitoring, and market intelligence offer significant benefits, their adoption is hindered by challenges such as digital illiteracy, poor infrastructure, and financial constraints. Drawing on educational theories, particularly Piaget/Vygotsky's and Knowles' frameworks, the study proposes the Baby Steps Theory of Development (BSTD) to facilitate incremental, hands-on learning for AI adoption. By addressing contextual learning needs and fostering gradual skill-building, this approach aims to empower farmers to effectively integrate AI into sustainable agricultural practices.

INTRODUCTION

Artificial Intelligence (AI) is increasingly transforming agriculture by improving productivity, optimizing resource use, and reducing the labour-intensive nature of farming (Dhanta et al., 2024) . In rural Africa, where smallholder farmers form the backbone of food production, AI presents immense opportunities. However,

DOI: 10.4018/979-8-2600-2378-5.ch005

its adoption is often hindered by issues such as digital illiteracy, and challenges in knowledge transfer (Naidoo, 2024). One of the most significant benefits of AI in rural agriculture is its ability to enhance precision farming. AI-powered tools analyse soil conditions, weather patterns, and crop health, allowing farmers to make data-driven decisions that improve yields (Aijaz et al., 2025). Automated crop monitoring is another crucial application, where AI models detect pests, diseases, and nutrient deficiencies through smartphone applications and drone imaging. Additionally, AI-driven market intelligence platforms help farmers access real-time pricing, demand forecasts, and trading opportunities, enabling them to make better financial decisions. In post-harvest processing, AI-powered sorting and grading systems increase efficiency, ensuring that produce meets market standards(Rane et al., 2024).

Despite these benefits, the adoption of AI in rural Africa remains slow due to several challenges including lack of a generalized framework of learning how to adopt these emerging technologies. Many farmers have limited experience with digital tools, making AI training difficult. Poor internet connectivity and inadequate access to electricity further restrict their ability to use AI-powered applications. Financial constraints also play a significant role, as many small-scale farmers cannot afford the necessary AI tools and devices(Dibbern et al., 2024). Perhaps the most pressing challenge is the difficulty in transferring AI knowledge across different applications. Many farmers assume that an AI system trained for one function—such as sorting and grading crops—can also diagnose plant diseases. This misconception highlights a key limitation in AI literacy among rural farmers, where learning remains highly contextual and bound to specific tasks.

THEORETICAL ASSUMPTIONS

Despite these hurdles, successful AI adoption has been observed in cases where farmers learn through hands-on experience, guided by tangible results. A new theoretical framework based on the fundamental concepts of several educational and developmental philosophies is necessary. Jean Piaget's Theory of Cognitive Development can be similarly applied to the development of AI application skills among the impoverished in Africa. Piaget posits that cognitive growth transpires in phases, with each stage contingent upon the proficiency attained in the preceding one (Vygotsky, 1978). Likewise, the introduction of AI application skills to underprivileged groups might use a systematic approach, enabling learners to progressively acquire and enhance their knowledge. Like the developmental stages of children—sensorimotor, preoperational, concrete operational, and formal operational—individuals can initially engage with fundamental AI concepts, including dataset comprehension, predictive analytics, and resource optimization, before advancing to more intricate applications. This approach prevents learners from becoming overwhelmed, facilitating the gradual development of confidence and competence. Piaget's principles can inform the creation of AI training programs that educate impoverished individuals through a systematic learning process, ultimately empowering them to leverage AI technologies to enhance their livelihoods.

Lev Vygotsky's Social Development Theory emphasizes the significance of social interaction and scaffolding in the learning process, positing that learners can attain advanced levels of comprehension with the assistance of more knowledgeable individuals (Vygotsky, 1978). In this regard, a step-by-step approach – more like when babies learn to take their first steps and for the purpose of this paper, hereby referred to as Baby Steps Theory of Development (BSTD) -with incremental learning, guided by specialists, can empower rural farmers to gradually comprehend and utilize AI effectively. This observation supports the notion that intricate skills and knowledge are optimally acquired through a systematic manner, preventing learner overload and fostering confidence as they advance. Utilizing the BSTD allows for the customization of AI projects to meet the distinct demands and settings of African communities, guaranteeing that the deployment of advanced technology is both accessible and sustainable. Engaging in practical, experiential learning and receiving support continuously, individuals can acquire the skills needed to utilize AI for poverty alleviation, so enhancing the overall growth and empowerment of these communities.

LEARNING CHALLENGES FARMERS FACE WHEN ADOPTING AI TOOLS IN AFRICA

Despite the transformative potential of AI in agriculture, many rural farmers in Africa struggle to integrate these tools into their daily farming practices. The challenges they face are deeply rooted in limited digital literacy, the reward-driven nature of adult learning, and difficulties in transferring knowledge across different AI applications (Gardezi et al., 2024a). These factors shape how farmers perceive and engage with AI technology, often determining whether they embrace or abandon it.

One of the most significant obstacles is the lack of digital literacy. Many rural farmers have had little exposure to digital tools, making AI adoption an entirely new experience (Beyene et al., 2024). Unlike traditional farming knowledge, which is passed down through generations and learned through observation and practice (Melash et al., 2023), AI requires an understanding of data interpretation, machine learning models, and digital interfaces (Aldoseri et al., 2023). Farmers often struggle to navigate AI-powered applications, interpret automated recommendations, or troubleshoot errors (Sparrow et al., 2021). When faced with challenges they do not understand, many become frustrated and discontinue using the technology altogether.

Another major barrier is the way adults learn, which is primarily reward-based. Unlike children, who learn through curiosity and exploration, farmers approach AI tools with a results-driven mindset. They engage with new technologies only when they see a direct and immediate benefit, such as increased crop yields, reduced labour, or better market access. If an AI tool does not quickly demonstrate its value, farmers often abandon it, assuming it is ineffective or unreliable. This short-term perspective on learning can hinder the deeper understanding necessary for sustained AI use.

Even when farmers successfully learn to use an AI tool, their knowledge often remains limited to the specific task they were trained on. This inability to transfer knowledge across different AI applications is a significant challenge. For example, a farmer who learns to use an AI-powered system to sort and grade produce may mistakenly assume that the same system can diagnose plant diseases, not realizing that each AI model is trained separately for different tasks. When the tool fails to meet their expectations in an unintended function, they may lose confidence in AI altogether.

Access to structured and continuous training programs is another challenge (Pujianto, 2024). Many AI training sessions in rural Africa are delivered through NGOs, agribusinesses, or technology companies, often in short, one-time workshops. While these initiatives introduce farmers to AI tools, they rarely provide long-term learning support. Without reinforcement or follow-up training, much of the knowledge gained is lost over time. Additionally, language barriers and varying literacy levels make AI training even more difficult, particularly when tools are designed with complex interfaces or require reading proficiency.

Beyond these cognitive and structural barriers, cultural factors also play a role in AI adoption. Many rural farmers have relied on traditional farming methods for generations, trusting practices passed down through their families and communities (Melash et al., 2023). The introduction of AI requires not just learning new tools but also shifting deeply ingrained ways of thinking. Skepticism toward AI-generated recommendations is common, as farmers often prefer to rely on their own experience or community knowledge. Overcoming this resistance requires more than just training—it demands trust-building, demonstration of tangible benefits, and integration of AI tools into familiar farming routine (Ekperi et al., 2024).

Overcoming ingrained cultural practices in rural farming communities requires a nuanced approach that respects tradition while introducing innovation (Burton et al., 2008). By clearly linking the direct benefits of AI to tangible outcomes, farmers are more likely to perceive these tools as allies rather than adversaries. Through pilot demonstrations, incremental learning, trusted endorsements, and long-term support, AI can be seamlessly integrated into farming practices, transforming agriculture while preserving the cultural fabric of these communities. With patience, persistence, and a focus on measurable benefits, the adoption of AI can lead to improved livelihoods, sustainable farming practices, and greater resilience in the face of global challenges (Burton et al., 2008).

Infrastructure and accessibility barriers further complicate AI adoption (Alabdali et al., 2023). Even when farmers are motivated to learn and willing to experiment with AI, poor internet connectivity, unreliable electricity, and the high cost of AI-enabled devices make consistent use difficult. Unlike urban users who can easily access online tutorials and support, rural farmers often depend on in-person training, which may not be frequent enough to help them master AI tools effectively.

Addressing these challenges requires a more adaptive approach to AI training—one that acknowledges the realities of rural farming life and aligns with adult learning principles. Hands-on, practical learning experiences, peer-to-peer knowledge sharing, and ongoing mentorship can help reinforce AI literacy. Additionally, AI tools should be designed with user-friendly interfaces, available in local languages, and requiring minimal digital literacy. By addressing both the learning and infrastructure barriers, AI adoption in rural Africa can become more effective and sustainable, ensuring that farmers can fully harness the benefits of this technology (Tomaškinová et al., 2024).

THE NEED FOR A THEORETICAL FRAMEWORK TO EXPLAIN AI ADOPTION AND LEARNING PATTERNS AMONG RURAL FARMERS IN AFRICA

Despite the growing presence of AI in agriculture, rural farmers in Africa face persistent challenges in adopting and effectively utilizing these technologies. While AI has the potential to improve productivity and efficiency, its successful adoption is highly dependent on how farmers learn, engage with, and apply these tools in their daily practices. Existing learning models do not fully explain the unique ways in which rural farmers interact with AI, particularly their reliance on step-by-step, results-driven learning and the difficulty they face in transferring knowledge across different AI applications. To bridge this gap, there is a need for a robust theoretical framework that captures the learning behaviors, motivations, and limitations of AI adoption in rural farming communities.

Traditional learning theories, such as Piaget's Constructivist Learning Theory, emphasize active, experience-based learning, where individuals build knowledge by interacting with their environment. While this concept is relevant, it does not fully account for the way adults, particularly farmers, engage with AI tools. Unlike children, who learn through exploration and retain knowledge over time, adults—especially those in agriculture—learn in a highly practical and reward-driven manner (Liquin

& Gopnik, 2022,). Farmers are most likely to adopt AI when they see immediate benefits, such as higher yields or reduced labour costs (Sharpe & Elwood, 2024). If the expected outcomes do not materialize, they often abandon the technology, regardless of its long-term potential.

Another critical gap in understanding AI adoption among rural farmers is the issue of context-specific learning. Many farmers struggle to apply AI knowledge beyond the specific task for which they were trained (Benti et al., 2024). For example, a farmer who learns to use an AI tool for sorting and grading crops may assume that the same system can diagnose plant diseases, failing to recognize that different AI applications require separate training models. This limited transferability of knowledge hinders broader AI adoption and creates unrealistic expectations about how AI functions(Gardezi et al., 2024b). Existing learning theories do not adequately explain this challenge, necessitating a framework that accounts for the fragmented and task-specific nature of AI learning in rural contexts.

Furthermore, rural farmers' access to AI training is often informal and unstructured, delivered through short-term workshops or community-based initiatives. Unlike urban professionals who may have continuous exposure to digital tools, farmers typically do not engage with AI in a sustained learning environment. Their understanding of AI is shaped by trial and error, peer discussions, social networks and experiential learning rather than formal education (Slijper et al., 2022). Without a theoretical model to explain how knowledge is acquired, reinforced, and retained in these settings, it is difficult to design effective AI training programs that ensure long-term adoption.

A comprehensive theoretical framework would help address these gaps by integrating principles of constructivist learning, adult learning theories, and reward-based motivation to explain how rural farmers interact with AI. Such a framework would provide valuable insights for technology developers, policymakers, and agricultural extension programs, ensuring that AI training is tailored to farmers' learning behaviors and limitations. By establishing a structured understanding of AI adoption patterns, this framework would contribute to more effective knowledge transfer, improved AI literacy, and ultimately, higher success rates in integrating AI into rural African agriculture.

LITERATURE REVIEW

Applying Piaget's Constructivist Learning Theory to AI Adoption Among Rural Farmers

Piaget's Constructivist Learning Theory explains how individuals build knowledge through active engagement with their environment. Rather than passively absorbing information, people construct their understanding based on experiences, gradually refining their mental models through a process of trial, error, and adjustment. While Piaget primarily focused on how children learn, the core principles of constructivism can also help explain how rural farmers in Africa adopt and interact with AI tools in agriculture (Roberts, 2006).

At the heart of Piaget's theory is the idea that learning occurs through assimilation and accommodation. When individuals encounter new information, they first try to fit it into their existing understanding (Lourenço, 2014). If the new knowledge contradicts their previous beliefs or experiences, they adjust their thinking to accommodate the new reality. This process is evident in how farmers engage with AI tools(Alexander et al., 2024). For example, a farmer introduced to an AI-powered irrigation system may initially assume it operates like a traditional watering schedule but with added automation. However, when they realize that the AI adjusts irrigation based on real-time weather data and soil moisture, they must modify their previous understanding to fully utilize the tool. This shift mirrors the accommodation process Piaget described, demonstrating that farmers, like children, learn best when they are actively engaged in problem-solving rather than simply receiving instructions.

Piaget emphasized that individuals develop knowledge through direct interaction with their environment rather than passive instruction (Miller, 2022). The same is true for rural farmers adopting AI tools. They learn not by attending formal training sessions alone but by using the tools in their daily farming activities. A farmer using an AI-powered disease detection app, for instance, learns by taking images of crops, receiving AI-generated feedback, and adjusting farming practices accordingly. Over time, repeated interactions help them refine their understanding of how the tool works, reinforcing knowledge through practice.

However, unlike children, who often learn through curiosity and open-ended exploration, adults—especially rural farmers—tend to learn in a highly goal-oriented and reward-driven manner (Kaiser & Burger, 2022). Farmers are primarily motivated to adopt AI when they see clear, immediate benefits, such as increased crop yields, reduced labour, or cost savings. If an AI tool does not quickly demonstrate its value, they are likely to abandon it. This results-based approach to learning differs from Piaget's model, which assumes that knowledge acquisition is driven primarily by internal cognitive development rather than external rewards.

Another challenge in applying constructivism to AI adoption is the difficulty of knowledge transfer. In Piaget's theory, as children gain experience, they develop cognitive structures that allow them to apply knowledge to new situations. However, rural farmers often struggle to extend AI knowledge beyond the specific task they were trained for(Singh et al., 2024). For example, a farmer who successfully learns to use an AI-powered system for sorting and grading crops may assume that the same system can diagnose plant diseases, failing to recognize that each AI application requires separate training models. This context-bound learning limits the broader adoption of AI technologies, as farmers may not fully grasp the adaptability and specificity of AI tools.

Despite these challenges, constructivist principles can still inform more effective AI training strategies. Since farmers learn best through hands-on experience, AI adoption programs should prioritize practical, interactive learning rather than relying solely on theoretical instruction Since farmers learn best through hands-on experience, AI adoption programs should prioritize practical, interactive learning rather than relying solely on theoretical instruction (Nurchim & Nurmalitasari, 2025). Since social learning plays a critical role in knowledge construction, peer-to-peer training and community-based learning networks should be integrated into AI education efforts. Farmers are more likely to trust and adopt AI tools when they see fellow farmers successfully using them.

Ultimately, while Piaget's Constructivist Learning Theory provides a useful framework for understanding AI adoption among rural farmers, it must be adapted to account for the reward-driven and context-specific nature of adult learning. By designing AI training programs that emphasize experiential learning, immediate benefits, and social reinforcement, technology developers and agricultural extension programs can improve AI adoption rates, ensuring that farmers not only learn how to use these tools but also integrate them meaningfully into their farming practices.

Key Differences Between Child and Adult Learning, Particularly the Role of Rewards

Learning is a lifelong process, but the way people learn evolves significantly from childhood to adulthood. While children often learn through curiosity and exploration, adults engage with learning in a more goal-oriented and reward-driven way (Bjorklund, 2022). Understanding these differences is crucial, particularly in the context of adult learning in areas like AI adoption among rural farmers.

Children are intrinsically motivated to learn. Their curiosity drives them to explore and experiment with new concepts, and they enjoy the process of discovery without necessarily seeking immediate rewards. This aligns with Piaget's Constructivist Learning Theory, which emphasizes that children build knowledge by interacting with their environment and learning from their experiences (Darling-Hammond et al., 2020). Mistakes are seen as natural and part of the learning process, which encourages children to keep trying new things. Learning is often playful and exploratory, driven by an inherent desire to understand the world around them.

In contrast, adults typically approach learning with a more extrinsic motivation. According to Knowles' Andragogy Theory, adults are practical and goal-oriented learners. They usually seek knowledge and skills to solve real-life problems or improve their situation(Knowles, 1978). For example, a farmer adopting an AI tool for irrigation might not be interested in the tool itself but in the tangible benefits it can offer—such as increased crop yields or cost savings. If an AI system doesn't quickly provide these rewards or if it doesn't align with the farmer's immediate needs, they are likely to abandon it. Unlike children, adults often need a clear, practical reward for their learning efforts, and this rewards-driven learning is crucial when introducing new technologies like AI.

Children, especially in the early stages of development, are more willing to experiment and make mistakes. They learn by trial and error, revisiting concepts and refining their understanding through repeated interactions. Their learning is less about efficiency and more about the experience itself (Carpendale & Lewis, 2004). Adults, however, have limited time and attention to devote to learning. They tend to seek the quickest and most efficient way to acquire new knowledge, particularly when that knowledge can be immediately applied to their daily lives. For rural farmers, this means that AI tools must provide quick, visible benefits to sustain interest (Taylor & and Hamdy, 2013). If the tool requires complex steps or offers slow results, adults may quickly lose interest, as they are more focused on efficiency and practical application rather than exploration for its own sake.

Another key difference lies in how adults and children transfer knowledge. Piaget's theory suggests that children develop flexible cognitive structures that allow them to apply their learning across various contexts (Cantor et al., 2019). For example, a child who learns basic math through games might later apply those skills to different activities, like building something or understanding patterns in nature. In contrast, adults tend to learn in a more task-specific manner. They often compartmentalize new information and may struggle to transfer knowledge from one area to another (Darling-Hammond et al., 2020). In the context of AI adoption, a farmer who learns to use an AI tool for grading produce might assume the same tool can detect plant diseases. However, the farmer may not realize that different AI systems require separate training and datasets for distinct tasks. The lack of transferability of knowledge can be a barrier to broader AI adoption, as farmers may not understand the full potential of AI tools beyond the specific tasks they were taught.

The role of experience is another area where children and adults differ. Children come into learning with fewer prior experiences and more flexibility, which allows them to adapt easily to new concepts (E. G. Liquin & Gopnik, 2022). For example, they often approach learning with an open mind, eager to absorb new information. Adults, however, bring their past experiences and existing knowledge into the learning process. Kolb's Experiential Learning Theory suggests that adults learn best when they can connect new knowledge to their prior experiences. This can be both a strength and a challenge (Kolb, 1984). For farmers, AI tools that are aligned with their existing agricultural practices are easier to adopt. However, if the AI tool contradicts traditional methods or requires a shift in mindset, farmers may resist the technology, even if it could be more effective. The challenge lies in balancing new information with existing beliefs and practices.

Perhaps one of the most important differences between child and adult learning is the role of rewards. Children tend to learn for the sake of learning, driven by intrinsic motivation. Adults, however, need external incentives to sustain their learning (Henderlong & Lepper, 2002). The reward could be economic, such as higher crop yields or reduced costs, efficiency-driven, like saving time or labour, or social, such as peer recognition or improved status within the community. For rural farmers, AI tools must provide immediate, tangible benefits to encourage continued use. If the technology does not show clear rewards, they may abandon it, deeming it ineffective or unnecessary(Atapattu et al., 2024).

Learning processes among farmers in Africa are influenced by context-specific factors, including socio-economic conditions, cultural practices, and levels of technological literacy. Traditionally, farmers have relied on experiential learning, peer-to-peer sharing, and community-based knowledge networks. However, the integration of AI technologies introduces a new dimension, requiring a shift toward more structured and data-driven learning approaches(Dwivedi et al., 2023).

The key differences between child and adult learning—especially regarding the role of rewards, efficiency, experience, and task-specific knowledge—have significant implications for AI adoption, particularly in rural farming communities. By recognizing these differences and designing AI training programs that are practical, reward-driven, and connected to farmers' experiences, we can ensure that AI adoption is more successful and sustainable. Understanding the adult learning process, particularly the need for immediate benefits, will be critical in helping rural farmers integrate AI tools into their agricultural practices effectively.

RESEARCH OBJECTIVE

The objectives of this study were influenced by the author's class project at MIT's Executive Program on the application of AI. As such, several AI startups in agriculture were recruited to be part of the study. Most of the startups had initiated their work with small-scale farmers in Kiambu and Muran'ga counties, to introduce AI in rural Kenya. Their aims were varied including leveraging AI to analyze data on soil health, weather, and crop performance (monitoring) for small scale farmers to better their decisions in planting, irrigation, and fertilization as a strategy to improve productivity and competitiveness in the market. The projects sought to build a culture of using data to: (1) make predictions (2) produce enough food for consumption and (3) export to increase incomes and ultimately solve the problem of poverty. The author's aims were to observe the adoption process.

The project brought together policymakers, researchers, local leaders and a select group of farmers. This was important to dismiss the narrative that the research initiatives were to introduce Genetically Modified Organisms (GMOs) that policymakers wanted to introduce at the time. Further there were undertones that AI was meant to displace workers. Aware of these politics, the researchers explained the intentions of promoting responsible AI as a strategy of improving the country's food security and its competitiveness and ultimately eliminate poverty. Against this back drop the research teams embarked on capacity building. The training focused on getting the farmers to understand the entire agricultural value chain and areas they needed to apply technology. The training also entailed understanding quality produce and aspects of competitiveness. Regulatory organizations too were part of the training leverage technology to minimize obvious quality omissions and improve the country's competitiveness.

The author was guided by the following question: Do we need new theoretical assumptions in AI adoption in Africa? This intriguing query set the foundation for the research, exploring the nuances of integrating AI technologies within the diverse and complex agricultural landscapes of rural Kenya. The study examined how traditional practices intersect with modern AI tools, offering insights into the unique challenges and opportunities that arise in such contexts.

METHOD

The study relied on non-participant observation where the researchers sought to understand the pilot environment, relationships, and interactions through other parties and their evaluations. Further the study also relied on indirect observation where the researcher relied on observations done by others (other researchers), on various types of documentation, recordings and observations(Ciesielska et al., 2018).

In addition, research material used by the author to present a field case study on AI adoption conducted in Kiambu and Muran'ga counties were used to ensure a holistic approach to research, recruitment of volunteers, and monitoring the pilot implementation of AI projects. A small field team visited existing AI initiatives to gain insights into best practices and strategies for integrating AI into agricultural practices. A review of case studies across the continent validated the outcomes of ongoing projects in Kenya. Many these projects are startups in the AI space focused on solving the problems of the small-scale farmers in Kiambu and Muran'ga. The project aimed to leverage AI for data analysis, productivity improvement, and food security. Stakeholder collaboration was crucial for addressing political concerns and misconceptions about AI and its effectiveness in minimizing the chances of exporting produce contaminated by pests such as False Codling Moth (FCM) which leads to rejection of the produce and loss or revenue. and ensuring exports contaminated by GMOs that is not allowed in the European market.

The farmers in these two counties could only do what they could on their modest farms. But they required more help from the government, especially with training and capacity building that would help them learn about the agricultural value chain and how to apply technology in other areas besides their tiny farms. For instance, they wanted the first applications to use machine learning to sort and grade produce, which would lower the costs of waste and rejected exports. Also, the regulatory body should invest in Robotic Process Automation (RPA) to enable them to tell the difference between people who have used technology and those who haven't. The authority knew that investing in RPA was necessary to improve the country's horticulture exports, but they couldn't since it was too expensive. Farmers learnt how to use AI technologies step by step, and they were motivated by the outcomes they saw. They learned how important it was to sort and grade goods correctly, which made sure they had steady income and higher prices for exporting. But investments made by the government take longer.

Policymakers play a critical role in lobbying for investments in advanced AI-powered sorting and grading equipment, which ensures that only high-quality produce meets international standards. This collaboration underscores the importance of a unified strategy, where each stakeholder contributes toward enhancing the competitiveness of the country's exports and reducing losses for farmers. These collaborative efforts brought together policymakers, researchers, local leaders, and farmers, fostering a crucial dialogue about the integration of AI into agriculture. Such partnerships were instrumental in addressing misconceptions surrounding AI, such as fears of job displacement and concerns about ethical implications like the introduction of GMOs. By harnessing local expertise and promoting transparent communication, these synergies laid the foundation for wider acceptance and practical application of AI innovations in rural farming communities.

Pilot Application of Machine Learning and AI

Some pilot applications were ongoing, especially for small scale farmers in export market using machine learning for sorting and grading which proved useful in terms of cost (waste) minimizations. In the past these farmers incurred enormous costs in transporting their produce for export only for the regulators to reject it. Sometimes even the regulators could make mistakes in their visual inspection only to discover too late after the export where the cost is passed back to the producer. To further assist small farmers, especially those in horticultural produce that is aggregated and then exported to Europe, there were three key areas that needed to benefit from investment in robotics and Robotic Process Automation (RPA) for the regulatory organizations. These are (1) sorting and grading, (2) pest control since the European Union has banned the use of pesticides and (3) RPA for regulators' pre-shipment inspection. These will minimize exports that are likely to be rejected in Europe and the producers will bear the cost. There was therefore a need for policymakers too to be lobbied to invest in high-capacity sorting and grading AI equipment. Government investment at the regulatory level is imperative and the earlier it is done the better.

Key Observations from On-Field Work

Farmers learn AI tools step-by-step and are motivated by tangible results. First on the application of the necessary inputs to optimise production followed by monitoring to make interventions where necessary. For example, if some sections of the farm the leaves are changing colour that is not consistent with the rest of the farm, they contact the agronomist to provide the solutions. Initial compliance was lacklustre. However, the first harvest from experiment areas compared to farms that were not in the program showed great improvement by as much as 30 percent. This validation led to greater interest and more farmers wanting to be in the program.

Step 2 was the grading and sorting. In this step the farmers realized less rejects from the control farms than harvest from other farms. Some were not enthusiastic to remove all the rejects arguing that similar rejects were reducing the premium revenues, and they would rather take the risk. However, they learnt that quality guarantees consistent income which was higher than from their previous export. Further importers were willing to pay more for high quality exports. They also realised that poorly graded and sorted without technology sometimes leads to total losses when rejected at the export destination. They realized that such rejects are too costly since they also absorb the shipping cost. The regulator level RPA was not tested given the fact that regulators had not made investment. However, they were part of the experimental programme. Further investment at regulator level could improve the country's horticultural exports.

Farmers were learning at each step and realized the benefits after comparing the outcomes with previous subjective sorting and grading. Although the learning was like Piaget's Theory of Cognitive Development, it was different in the sense that every step required validation, group learning, experimentation, and handholding, like teaching a child how to climb stairs. The farmers learned AI tools step-by-step and were motivated by tangible results.

Although tangible results motivated farmers to engage with AI, they encountered challenges in transferring their knowledge to other AI applications. This difficulty stemmed from the context-bound nature of their learning. Each application demanded specific training and validation, making it hard for farmers to generalize their skills across different AI tools and systems.

Criteria by which success is determined

The application of AI in rural Kenya started a few years ago with universities attempting to establish if the technology could improve productivity and thereby reduce poverty. Some projects use data analytics and machine learning with drones used to monitor crop health and growth patterns. The question of which strategy is relied upon now is often outside the scope of the poor. The application of AI is changing their hopes in farming. Although they have not fully optimized production, improvements have been made. To effectively deal with the situation, more must be done to leverage AI to wholistically solve the problem of poverty. There is need for local collaborations including production, distribution and other enabling AI developers of local language models (LLMs). Local languages are essential for farmers to understand concepts like optimization that although essential are never translated appropriately.

Therefore, for farmers to thrive they need to rely on cost leadership for food production and the focus strategy on horticultural production strategies. In cost leadership, farmers can be able to compete against countries that subsidize their farmers thereby ensuring their food sovereignty. And in focus strategy they can tailor their produce for the European Union imports that provide the most in export earnings ensuring higher incomes, thereby reducing poverty. Part of the initiative to help farmers leverage AI to transition farmers from a desperate situation to a modern economy where they can compete like everyone else.

AI will greatly contribute to greater productivity, lower waste food waste by optimizing demand and supply. Often farmers do not have full understanding of the demand side of their produce. Parts of the country or even neighbouring countries could be having food shortages when some areas have plenty of food that goes to waste. It is estimated that Africa loses more than 40 per cent of food to waste yet the continent is food insecure. AI could change by optimizing production, minimizing waste and increasing revenue to the farmers. Once this is realized in the area, we intend to deploy it, and scale it up using AI chatbots to act as assistants to scale the processes.

Proposed Theory of Development: Reward-Based Constructivist Learning for AI Adoption

The study proposes a Baby-Step-Theory of Development (BSTD) akin to reward-based constructivist learning for AI adoption integrating constructivist learning with reward-driven adult learning. The adoption of AI tools by farmers in rural Kenya has been significantly reinforced by observable productivity gains. These tools offer precise application of inputs, monitoring, and intervention, resulting in improved crop yields and quality. For instance, initial experiments showed a 30% improvement in harvests compared to control farms, which motivated farmers to engage more deeply with AI technologies. As they witnessed tangible improvements, such as reduced rejects and higher export quality, their confidence in AI grew, reinforcing their commitment to its adoption.

Moreover, the ability of AI to optimize demand and supply, thereby reducing food waste, further underscores its value. Farmers realized that AI could help bridge the gap between areas with surplus produce and those facing shortages, thus maximizing revenue. This realization fosters a positive feedback loop: as productivity and profitability increase, so does the willingness to invest in and adopt AI tools, which in turn drives further gains.

AI training specificity plays a crucial role in the effective utilization of technology by farmers. Each AI application requires tailored training that addresses the unique demands and intricacies of specific tasks like monitoring and validating information with experts. This need for specificity helps ensure that the AI systems are correctly implemented and optimized for agricultural practices, such as sorting, grading, and monitoring crop health. Farmers struggle with generalization because their learning experiences are often closely tied to the context in which the training occurs. When they become proficient in using AI tools for one specific application, the knowledge and skills gained may not easily transfer to different AI systems or tasks. Each new AI tool presents its own set of challenges and requires additional training and validation, which can be time-consuming and complex. For example, monitoring crops entails scanning and forwarding the information to experts for interpretation.

This context-bound learning means that farmers must repeatedly engage in the learning process for each new AI application, making it difficult to apply their existing knowledge to new situations. The specificity required in AI training underscores the importance of continuous education and support for farmers, enabling them to adapt to and fully leverage the potential of diverse AI technologies. As the farmers continue to witness tangible improvements in their productivity and profitability, their appreciation for the role of AI in development will naturally deepen. The more they experience the advantages of precise application, optimization of demand and supply, and reduction in food waste, the more they will value the integration of AI into their agricultural practices. This growing recognition will foster a culture of continuous learning and adaptation, enabling farmers to overcome initial hurdles and fully embrace the diverse benefits that AI technologies offer.

Integrating Piaget's and Knowles's learning theories with technological advancements

Learning is a continuous process that evolves across the lifespan, shaped by developmental stages and influenced by the interplay of cognitive, social, and technological factors. The Baby-Step-Theory of Development (BSTD) proposes a structured progression that connects childhood learning, as conceptualized by Piaget's Theory of Learning, to adult learning, as articulated by Knowles's Theory of Adult Learning. Additionally, BSTD incorporates modern methods of adult learning that leverage technology to enhance adaptability, competency, and engagement. BSTD will be an outcome-based learning with an AI chatbot constantly providing farmers with necessary data.

Figure 1. Integrated Learning Progression Framework

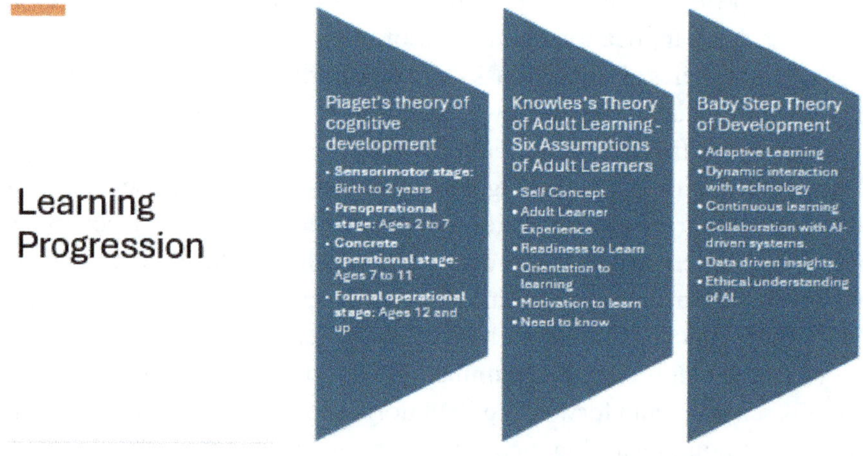

Piaget's key principles include:

- Constructivism: Children actively construct knowledge through interactions with their environment.
- Assimilation and Accommodation: Children integrate new information into existing schemas (assimilation) or modify their schemas to accommodate new experiences (accommodation).
- Stages of Development: Learning occurs in predictable stages that align with cognitive growth.

The foundational concepts of Piaget's theory underscore the importance of tailored instruction that adapts to the learner's developmental stage, promoting curiosity, exploration, and problem-solving.

Knowles's Theory of Adult Learning: The Transition to Self-Directed Learning

Malcolm Knowles introduced key principles of adult learning, or andragogy, which differ significantly from the pedagogical approaches used for children. Adults bring a wealth of prior experiences to the learning process, and they are motivated by practical, goal-oriented applications of knowledge.

Knowles's principles include:

- Self-Direction: Adults prefer to take ownership of their learning and make decisions about their educational journey.
- Experience: Prior knowledge and experiences serve as valuable resources for new learning.
- Relevance: Adults are motivated by learning that has immediate application to their personal or professional lives.
- Problem-Cantered: Learning is most effective when it addresses real-world challenges and provides solutions.

The transition from Piaget's childhood learning principles to Knowles's adult learning framework highlights the shift from externally guided instruction to self-directed exploration, emphasizing autonomy and practical outcomes.

BSTD: BRIDGING THE GAP THROUGH TECHNOLOGY

BSTD builds on Piaget's and Knowles's theories by integrating reward-based constructivist learning with technology. BSTD acknowledges that both children and adults are motivated by observable gains, and it proposes a progression of learning that combines cognitive development with behaviour-driven reinforcement data as delivered by chatbots. Leveraging AI moves learning beyond Piaget and Vygotsky toward more individualized, adaptive, and culturally relevant learning experiences. AI can also collect and analyse large-scale data on learning processes, offering new empirical insights to refine and expand upon these classic theories for contemporary needs.

Reward-Based Constructivist Learning

BSTD employs constructivist learning principles, emphasizing active engagement and problem-solving while introducing a reward system to reinforce desired behaviors. The concept of rewards—whether tangible (e.g., productivity gains) or intangible (e.g., a sense of achievement)—is central to the theory, encouraging learners to persevere and deepen their engagement with educational tools.

Leveraging Technology

Modern methods of adult learning leverage technology to enhance accessibility, personalization, and interactivity. BSTD integrates tools such as:

- Artificial Intelligence (AI): AI-powered platforms provide tailored recommendations, monitor progress, and offer adaptive feedback.
- E-Learning Systems: Online modules and virtual classrooms enable learners to access resources at their convenience.
- Gamification: Incorporating game-like elements into educational systems enhances motivation and retention.
- Mobile Applications: Smartphones and apps facilitate on-the-go learning and real-time decision-making.

These technological advancements align with Knowles's principles of relevance and problem-cantered learning, offering tools that are both practical and impactful.

APPLICATIONS OF BSTD IN AI ADOPTION

BSTD has proven effective in encouraging farmers in rural Kenya to adopt AI technologies for agriculture. Observable productivity gains, such as a 30% improvement in harvests, serve as tangible rewards that reinforce their commitment to learning. The model addresses the specificity of AI training, helping personalizing farmers learning to overcome the context-bound nature of their knowledge and adapt to diverse applications. By integrating AI tools for precise input application, demand optimization, and waste reduction, BSTD facilitates a positive feedback loop of productivity and profitability.

Testing BSTD on AI Adoption in Agriculture: Case Studies and Existing Research on How Farmers Interact with AI in Africa.

The integration of artificial intelligence (AI) into agriculture has the potential to revolutionize farming, particularly for rural farmers in Africa. However, the adoption of AI technologies comes with several challenges, including language, the need for user-friendly tools, proper training, and tangible results. Examining several case studies and existing research provides insights into how farmers interact with AI tools and the various outcomes they experience. However, this is temporal considering the fact that AI can be trained in local languages making adaptation much more simple.

1. AI-Powered Precision Agriculture in Kenya

In Kenya, the Precision Agriculture for Development (PAD) initiative has been working to help farmers increase their yields and reduce costs by using AI-powered tools (Foster et al., 2023). Through PAD, farmers receive personalized recommendations on irrigation, fertilization, and pest control based on local environmental data. These recommendations are delivered via SMS or mobile apps, with machine learning continually refining the advice.

Farmers interact with the AI by using simple tools that provide actionable insights, which is crucial for adoption. Research shows that the simpler the technology and the more accessible it is to farmers, the more likely they are to adopt it. Farmers who received easy-to-understand recommendations, such as SMS messages or voice alerts, found the technology more useful and were more willing to continue using it.

One of the key takeaways from this case study is the reward-driven nature of adult learning. Farmers were more likely to embrace the technology once they saw visible, tangible results—such as increased crop yields or reduced input costs. However, initial skepticism was common, and it took time for farmers to trust the AI's suggestions. As the benefits became apparent, adoption rates increased, showing how rewards can drive continued engagement with AI tools.

2. AI and Crop Disease Detection in India

In Kenya and India, the PlantVillage initiative uses AI to help farmers identify crop diseases by uploading photos of their crops via a mobile app. The AI analyzes the images and provides diagnoses and treatment recommendations, offering an early-warning system for disease outbreaks (Foster et al., 2023).

While many farmers initially showed interest in using the app, they faced challenges in trusting the AI's accuracy. This reluctance was particularly evident when the AI's recommendations did not align with traditional knowledge or when the tool was perceived as unreliable. However, farmers who adopted the tool saw significant improvements in detecting diseases early, which led to better crop health and, ultimately, higher yields.

The success of this project highlights how peer influence and social validation play a significant role in AI adoption. Once early adopters shared their positive experiences, others were more likely to try the tool. This case also shows that farmers were more willing to engage with AI when they felt in control of the process, using their own smartphones to make decisions and receive alerts. The rewards, such as improved crop health, were the driving factors for continued use of the technology.

3. AI for Livestock Monitoring in South Africa

In South Africa, AI-driven systems have been introduced to help farmers monitor the health of their livestock. One example is Cowlar, a smart collar that tracks the behavior and health of cows. The system sends alerts to farmers when abnormalities are detected, such as changes in activity that could signal illness(Lamanna et al., 2025).

Farmers interacted with the AI through mobile apps that provided real-time alerts. While the tool was seen as promising for improving livestock management, there were significant barriers to adoption, particularly around the complexity of the system and internet connectivity issues in rural areas. These barriers made it difficult for some farmers to fully embrace the technology.

Research shows that hands-on training and continuous support were essential for encouraging farmers to use AI tools effectively. Those who received training and had access to ongoing technical support were more likely to trust and adopt AI technology. Additionally, farmers who could see immediate benefits, such as healthier animals and improved productivity, were more inclined to continue using the AI system (Aquilani et al., 2022).

4. AI-Driven Irrigation Systems in Egypt

In Egypt, AI-powered smart irrigation systems have been implemented to help farmers optimize their water usage (Ramadan et al., 2023). These systems use data from weather forecasts, soil moisture levels, and crop types to adjust irrigation schedules automatically, reducing water waste and improving efficiency.

Farmers' interaction with the technology varied, and many initially struggled with the complexity of the systems. Training was needed to help farmers understand how to use the technology effectively, and without this support, some farmers were hesitant to rely on the AI system. Despite these challenges, farmers who received training and were able to see clear benefits, such as reduced water costs and improved crop yields, were more likely to adopt the technology (Lago-Olveira et al., 2023).

This case highlights the importance of making AI tools user-friendly and directly linked to the farmers' goals. When farmers could clearly see how the technology benefited them, they were more likely to engage with it. However, the initial complexity and lack of infrastructure in some areas posed significant hurdles to adoption.

5. AI and Market Access in Nigeria

In Nigeria, AI-powered platforms have been developed to help farmers predict market prices for their crops. These platforms analyse various factors, such as weather patterns and supply-demand data, to predict the best time and place for farmers to sell their produce, potentially increasing their profits (Javaid et al., 2023).

While farmers appreciated the potential benefits of these tools, many were initially skeptical about their accuracy. Some continued to rely on traditional market knowledge and relationships, while others were reluctant to adopt the technology without concrete proof of its effectiveness. However, when early adopters saw real improvements in their profits, they shared their experiences, leading to an increase in adoption.

Research shows that social learning and peer influence were key factors in encouraging broader adoption of AI tools (Hazzan-Bishara et al., 2025). Farmers who were part of networks or communities where AI use was encouraged were more likely to trust and use the technology themselves. Like other case studies, tangible rewards, such as higher profits, were essential for motivating farmers to engage with AI systems.

DISCUSSION AND IMPLICATIONS

The integration of AI tools in agriculture offers significant opportunities to enhance productivity and sustainability. However, the adoption of these technologies by farmers often encounters numerous barriers. The complexity of AI systems, a lack of familiarity with digital technologies, and insufficient infrastructure are major obstacles that hinder widespread acceptance and utilization. In various case studies, the role of training and continuous support emerged as crucial for overcoming these challenges. Farmers who received comprehensive, hands-on training were better equipped to understand and use AI tools effectively. Moreover, ongoing technical support helped to build confidence and trust in the technology, ensuring that farmers could troubleshoot issues and optimize their use of AI systems.

Another key factor influencing adoption is the ability of AI tools to deliver tangible benefits. When farmers witness direct improvements, such as healthier livestock, enhanced crop yields, or reduced water usage, their willingness to adopt new technologies increases. This underscores the importance of designing AI solutions that are user-friendly and aligned with the immediate needs and goals of farmers. Social dynamics also play a significant role in the adoption process. Peer influence and community networks can either facilitate or hinder the acceptance of AI tools. Early adopters who experience positive outcomes can act as influential advocates, encouraging their peers to embrace new technologies. Conversely, skepticism and resistance within a community can slow down the adoption rate.

RECOMMENDATIONS

Improving AI literacy among farmers should consider the key factors identified in the case studies and research. Implementing hands-on training sessions that allow farmers to directly interact with AI tools can significantly enhance their understanding and confidence in using the technology. These sessions should be designed to simplify the technology, making it more accessible and less intimidating for farmers, especially those who may not be familiar with advanced digital tools. Continuous support and technical assistance should be readily available to address any challenges that arise and to reinforce learning. Creating a robust support system can help mitigate the initial resistance and apprehension that many farmers feel towards AI. This support can be provided through various channels, such as helplines, online tutorials, community workshops, and local tech experts.

Moreover, developing AI tools that align closely with farmers' immediate needs and goals is crucial. Tools that offer clear, tangible benefits, such as increased crop yields, reduced costs, or improved market access, will be more appealing and motivating for farmers. Ensuring that these benefits are visible and easily demonstrable can accelerate the adoption process. Social validation and peer learning should also be leveraged to foster a positive perception of AI tools. Encouraging early adopters to share their success stories and experiences can build trust and influence others within their community. Creating networks or communities where AI usage is promoted and supported can further strengthen the adoption rate.

CONCLUSION

The adoption of Artificial Intelligence (AI) in agriculture, while promising, requires thoughtful strategies to address cultural resistance and build sustainable learning progression among farmers. The BSTD offers a powerful framework for guiding this process. BSTD emphasizes incremental progress and manageable learning steps to foster confidence and competence in adapting to new technologies. Applied to agricultural AI adoption, this theory suggests that introducing small, achievable milestones—such as mastering basic functions of AI tools before advancing to more complex operations—can make the transition less intimidating for farmers with limited exposure to digital systems.

Cultural resistance often stems from deeply ingrained traditions and skepticism toward unfamiliar technologies. Addressing these barriers requires both sensitivity and strategic engagement. By leveraging BSTD, training programs can be designed in culturally resonant ways that respect farmers' traditional practices while gradually showcasing the value of AI tools. For example, aligning initial demonstrations of AI's benefits with locally relevant goals, such as improved crop yields or enhanced disease prevention, can provide tangible evidence of its effectiveness. Furthermore, employing local champions or community influencers to advocate for AI adoption can bridge trust gaps, offering peer validation that is often more persuasive than external endorsements.

In addition, adopting a phased approach where farmers see incremental rewards—whether through better market access, reduced resource use, or visible improvements in productivity—can reinforce their commitment to learning. This gradual progression aligns with BSTD's principle of building competence through experience and tangible success, ultimately paving the way for broader and sustained adoption of AI technologies.

To fully realize AI's potential in agriculture, integration efforts must embrace the nuances of cultural dynamics and learning progression. BSTD serves as a foundational strategy to overcome resistance, facilitate understanding, and empower farmers to harness AI tools effectively. By addressing cultural, technical, and social barriers incrementally, AI can unlock transformative benefits for global agricultural systems, enhancing both productivity and sustainability.

Future Studies

Future studies on AI adoption in agriculture should explore several key areas to deepen our understanding of the barriers and facilitators in this domain. Firstly, there is a need to understanding the social dynamics that influence AI adoption is crucial. Future studies can delve into the role of community leaders and influencers in promoting or hindering the uptake of AI technologies. Investigating the cultural and social factors that shape farmers' perceptions and attitudes towards AI can inform more effective strategies for education and engagement.

Secondly, exploring the economic models and funding mechanisms that support AI adoption in agriculture is essential. Identifying sustainable financing options and incentives that can reduce the financial burden on farmers will be key to ensuring broader accessibility and acceptance of AI tools.

Thirdly, research should focus on the development and implementation of AI tools that are customized for specific crops, climates, and farming practices. This could involve collaboration with local farmers to ensure that the solutions are tailored to their unique needs and conditions. Additionally, studying the effects of AI on biodiversity and environmental health can help balance technological advancements with ecological preservation.

Lastly, investigate the long-term impacts of AI tools on agricultural practices and productivity. Examining how AI integration affects the economic stability of farming communities, and the sustainability of agricultural systems can provide valuable insights into the true benefits of these technologies.

In summary, future studies should adopt a multifaceted approach, considering the technical, economic, social, and environmental aspects of AI integration in agriculture. By addressing these areas, researchers can contribute to the design and implementation of AI solutions that not only enhance productivity but also support sustainable and resilient agricultural systems.

REFERENCES

Aijaz, N., Lan, H., Raza, T., Yaqub, M., Iqbal, R., & Pathan, M. S. (2025). Artificial intelligence in agriculture: Advancing crop productivity and sustainability. *Journal of Agriculture and Food Research, 20*, 101762. Advance online publication. DOI: 10.1016/j.jafr.2025.101762

Alabdali, S. A., Pileggi, S. F., & Cetindamar, D. (2023). Influential factors, enablers, and barriers to adopting smart technology in rural regions: A literature review. *Sustainability (Basel), 15*(10), 7908. Advance online publication. DOI: 10.3390/su15107908

Aldoseri, A., Al-Khalifa, K., & Hamouda, A. M. (2023). Re-thinking data strategy and integration for artificial intelligence: Concepts, opportunities, and challenges. *Applied Sciences (Basel, Switzerland), 13*(12), 7082. https://www.mdpi.com/2076-3417/13/12/7082. DOI: 10.3390/app13127082

Alexander, C. S., Yarborough, M., & Smith, A. (2024). Who is responsible for 'responsible AI'?: Navigating challenges to build trust in AI agriculture and food system technology. *Precision Agriculture, 25*(1), 146–185. DOI: 10.1007/s11119-023-10063-3

Aquilani, C., Confessore, A., Bozzi, R., Sirtori, F., & Pugliese, C. (2022). Review: Precision livestock farming technologies in pasture-based livestock systems. *Animal, 16*(1), 100429. Advance online publication. DOI: 10.1016/j.animal.2021.100429 PMID: 34953277

Atapattu, A. J., Perera, L., Nuwarapaksha, T., Udumann, S. S., & Dissanayaka, N. (2024). Challenges in achieving artificial intelligence in agriculture. In Artificial intelligence in agriculture (pp. 7–34). DOI: 10.1007/978-981-97-5878-4_2

Benti, N. E., Chaka, M. D., Semie, A. G., Warkineh, B., & Soromessa, T. (2024). Transforming agriculture with machine learning, deep learning, and IoT: Perspectives from Ethiopia—challenges and opportunities. *Discover Agriculture, 2*(1), 63. Advance online publication. DOI: 10.1007/s44279-024-00066-7

Beyene, E., Bedemo, A., & Gebremeskel, A. (2024). Determinants of digital technology development in sub-Saharan African countries: Evidence from panel data analysis. *Energy Informatics, 7*(1), 21. Advance online publication. DOI: 10.1186/s42162-024-00324-4

Bjorklund, D. (2022). Children's evolved learning abilities and their implications for education. *Educational Psychology Review, 34*(4), 2243–2273. Advance online publication. DOI: 10.1007/s10648-022-09688-z PMID: 35730061

Burton, R., Kuczera, C., & Schwarz, G. (2008). Exploring farmers' cultural resistance to voluntary agri-environmental schemes. *Sociologia Ruralis*, *48*(1), 16–37. DOI: 10.1111/j.1467-9523.2008.00452.x

Cantor, P., Osher, D., Berg, J., Steyer, L., & Rose, T. (2019). Malleability, plasticity, and individuality: How children learn and develop in context. *Applied Developmental Science*, *23*(4), 307–330. DOI: 10.1080/10888691.2017.1398649

Carpendale, J., & Lewis, C. (2004). Constructing an understanding of mind: The development of children's social understanding within social interaction. *The Behavioral and Brain Sciences*, *27*(1), 79–96. DOI: 10.1017/S0140525X04000032 PMID: 15481944

Ciesielska, M., Wolanik Boström, K., & Öhlander, M. (2018). Observation methods. In Qualitative methodologies in organization studies (pp. 33–52). DOI: 10.1007/978-3-319-65442-3_2

Darling-Hammond, L., Flook, L., Cook-Harvey, C., Barron, B., & Osher, D. (2020). Implications for educational practice of the science of learning and development. *Applied Developmental Science*, *24*(2), 97–140. DOI: 10.1080/10888691.2018.1537791

Dhanta, R., Mwale, M., & Dhanta, R. (2024). Transforming agriculture with modern AI: Harnessing artificial intelligence to revolutionize farming. In Handbook of research on AI and machine learning applications in smart agriculture (pp. 350–360). DOI: 10.4018/979-8-3693-1902-4.ch021

Dibbern, T., Romani, L. A. S., & Massruhá, S. M. F. S. (2024). Main drivers and barriers to the adoption of digital agriculture technologies. *Smart Agricultural Technology*, *8*, 100459. Advance online publication. DOI: 10.1016/j.atech.2024.100459

Dwivedi, Y. K., Sharma, A., Rana, N. P., Giannakis, M., Goel, P., & Dutot, V. (2023). Evolution of artificial intelligence research in Technological Forecasting and Social Change: Research topics, trends, and future directions. *Technological Forecasting and Social Change*, *192*, 122579. Advance online publication. DOI: 10.1016/j.techfore.2023.122579

Ekperi, E., Owigho, O., & Ovwigho, B. (2024). Socio-cultural effects of artificial intelligence in agriculture by farmers, extension agents and lecturers in Delta State, Nigeria. *FUDMA Journal of Agriculture and Agricultural Technology*, *10*(1), 10–18.

Foster, L., Szilagyi, K., Wairegi, A., Oguamanam, C., & de Beer, J. (2023). Smart farming and artificial intelligence in East Africa: Addressing indigeneity, plants, and gender. *Smart Agricultural Technology*, *3*, 100132. Advance online publication. DOI: 10.1016/j.atech.2022.100132

Gardezi, M., Joshi, B., Rizzo, D. M., Ryan, M., Prutzer, E., Brugler, S., & Dadkhah, A. (2024a). Artificial intelligence in farming: Challenges and opportunities for building trust. *Agronomy Journal, 116*(3), 1217–1228. DOI: 10.1002/agj2.21353

Gardezi, M., Joshi, B., Rizzo, D. M., Ryan, M., Prutzer, E., Brugler, S., & Dadkhah, A. (2024b). Artificial intelligence in farming: Challenges and opportunities for building trust. *Agronomy Journal, 116*(3), 1217–1228. DOI: 10.1002/agj2.21353

Hazzan-Bishara, A., Kol, O., & Levy, S. (2025). The factors affecting teachers' adoption of AI technologies: A unified model of external and internal determinants. *Education and Information Technologies, 30*(11), 15043–15069. Advance online publication. DOI: 10.1007/s10639-025-13393-z

Henderlong, J., & Lepper, M. R. (2002). The effects of praise on children's intrinsic motivation: A review and synthesis. *Psychological Bulletin, 128*(5), 774–795. DOI: 10.1037/0033-2909.128.5.774 PMID: 12206194

Javaid, M., Haleem, A., Khan, I. H., & Suman, R. (2023). Understanding the potential applications of artificial intelligence in agriculture sector. *Advanced Agrochem, 2*(1), 15–30. DOI: 10.1016/j.aac.2022.10.001

Kaiser, A., & Burger, P. (2022). Understanding diversity in farmers' routinized crop protection practices. *Journal of Rural Studies, 89*, 149–160. DOI: 10.1016/j.jrurstud.2021.12.002

Knowles, M. S. (1978). Andragogy: Adult learning theory in perspective. *Community College Review, 5*(3), 9–20. DOI: 10.1177/009155217800500302

Kolb, D. A. (1984). *Experiential learning: Experience as the source of learning and development.* Prentice-Hall.

Lago-Olveira, S., El-Areed, S. R. M., Moreira, M. T., & González-García, S. (2023). Improving environmental sustainability of agriculture in Egypt through a life-cycle perspective. *The Science of the Total Environment, 890*, 164335. Advance online publication. DOI: 10.1016/j.scitotenv.2023.164335 PMID: 37209738

Lamanna, M., Bovo, M., & Cavallini, D. (2025). Wearable collar technologies for dairy cows: A systematized review of the current applications and future innovations in precision livestock farming. *Animals (Basel), 15*(3), 458. Advance online publication. DOI: 10.3390/ani15030458 PMID: 39943229

Liquin, E. G., & Gopnik, A. (2022). Children are more exploratory and learn more than adults in an approach-avoid task. *Cognition, 218*, 104940. Advance online publication. DOI: 10.1016/j.cognition.2021.104940 PMID: 34715584

Melash, A. A., Bogale, A. A., Migbaru, A. T., Chakilu, G. G., Percze, A., Ábrahám, É. B., & Mengistu, D. K. (2023). Indigenous agricultural knowledge: A neglected human based resource for sustainable crop protection and production. *Heliyon, 9*(1), e12978. Advance online publication. DOI: 10.1016/j.heliyon.2023.e12978 PMID: 36711305

Miller, P. H. (2022). Developmental theories: Past, present, and future. *Developmental Review, 66*, 101049. Advance online publication. DOI: 10.1016/j.dr.2022.101049

Naidoo, G. (2024). The potential of artificial intelligence in South African rural development. *African Journal of Science, Technology, Innovation and Development, 17*(2), 207–218.

Nurchim, N., & Nurmalitasari, N. (2025). Empowering Asian students through artificial intelligence: A workshop on predicting plant growth to support smart farming. *International Journal of Community Service, 5*(1), 30–36. DOI: 10.51601/ijcs.v5i1.839

Pujianto, S. (2024). Effectiveness of employee training and development in increasing productivity and innovation in multinational companies. *Global International Journal of Innovative Research, 2*(9), 2245–2256. DOI: 10.59613/global.v2i9.325

Ramadan, A., Dewedar, O., El-Shafie, A., & Abdelbaset, M. (2023). Improvement of irrigation water management using simulation models and artificial intelligence under dry environment conditions in Egypt: A review. *Egyptian Journal of Chemistry, 0*(0), 0. Advance online publication. DOI: 10.21608/ejchem.2023.151289.6552

Rane, J., Kaya, Ö., Mallick, S., & Rane, N. (2024). Smart farming using artificial intelligence, machine learning, deep learning, and ChatGPT: Applications, opportunities, challenges, and future directions. In AI and machine learning for smart farming (pp. 218–272). DOI: 10.70593/978-81-981271-7-4_6

Roberts, T. G. (2006). A philosophical examination of experiential learning theory for agricultural educators. *Journal of Agricultural Education, 47*(1), 17–29. DOI: 10.5032/jae.2006.01017

Sharpe, S. L., & Elwood, S. A. (2024). E-learning design for older adults in the United States. *Social Sciences (Basel, Switzerland), 13*(10), 522. Advance online publication. DOI: 10.3390/socsci13100522

Singh, N., Wang'ombe, J., Okanga, N., Zelenska, T., Repishti, J. K. J., Mishra, S., Manokaran, R., Singh, V., Rafiq, M., Gandhi, R., & Nambi, A. (2024). Farmer.Chat: Scaling AI-powered agricultural services for smallholder farmers. arXiv. https://doi.org//arXiv.2409.08916DOI: 10.48550

Slijper, T., Urquhart, J., Poortvliet, P. M., Soriano, B., & Meuwissen, M. P. M. (2022). Exploring how social capital and learning are related to the resilience of Dutch arable farmers. *Agricultural Systems*, *198*, 103385. Advance online publication. DOI: 10.1016/j.agsy.2022.103385

Sparrow, R., Howard, M., & Degeling, C. (2021). Managing the risks of artificial intelligence in agriculture. *NJAS: Impact in Agricultural and Life Sciences*, *93*(1), 172–196. DOI: 10.1080/27685241.2021.2008777

Taylor, D. C. M., & Hamdy, H. (2013). Adult learning theories: Implications for learning and teaching in medical education: AMEE Guide No. 83. *Medical Teacher*, *35*(11), e1561–e1572. DOI: 10.3109/0142159X.2013.828153 PMID: 24004029

Tomaškinová, J., West, L., Chircop, T., Judge, J., Vries, S., Marinescu-Muster, R., Hiddink, F., Delicata, F., Pierucci, M., Montalto, M., Camilleri, S., Hermanova, S., Středová, H., Mrkvicova, E., & Stastna, M. (2024). Shaping the future: Navigating the AI era with an innovative training model. In EDULEARN24 proceedings (p. 6303). DOI: 10.21125/edulearn.2024.1486

Chapter 6
A Study on AI-Based Feng Shui Application for Real Estate in Vietnam:
The Conceptual Idea

Thi Thanh Tam Dang
http://orcid.org/0009-0006-4818-8155
Golden Gate University, Vietnam

ABSTRACT

This chapter explores a conceptual study on integrating Feng Shui principles into an AI-based application for Vietnam's real estate. Feng Shui is crucial in property selection and design. A 2018 survey by the Vietnamese Royal Institute of Feng Shui found that 98% of buyers consider Feng Shui, yet few receive satisfactory guidance. This gap presents an opportunity to modernize traditional Feng Shui using AI technologies like machine learning and data analytics. The study examines how an AI-based Feng Shui application can enhance buyers' preferences by aligning with cultural values, improving interaction, increasing satisfaction, and providing tailored recommendations. Grounded in socio-technical and cultural capital theories, this qualitative research uses interviews and surveys with buyers and sellers. Thematic analysis will identify patterns in buyer perceptions and evaluate effectiveness. By merging traditional Feng Shui with AI, this study offers insights into innovation, market differentiation, and improved decision-making, contributing to buyer trust and sustainable growth.

DOI: 10.4018/979-8-2600-2378-5.ch006

INTRODUCTION

Feng Shui, which literally means wind and water, is an ancient Eastern doctrine involving the arrangement of buildings, objects, and space in an environment to achieve living harmony and balance. In real estate, it is believed that a property or land with good Feng Shui would promote good health, wealth, prosperity, and luck to its owners (Cho, 2020b). Originally, Feng Shui has root in early Chinese Taoist philosophy and remains popularly adopted today, influencing spaces across China, Asia, and even in Western cultures along with the current globalization trends of socio-cultural integration. In the West, Wu (Wu, 2019) emphasizes that Feng Shui practice has been used widely for years by architects, interior designers, and landscape architects as a design guideline for pursuing harmony in living environment. Overall, Feng Shui is highlighted as one of the main determinants of making decisions for real estate purchases, as stated by Gibler and Nelson (Gibler & Nelson, 1998).

Vietnam is not an exception among the use of Feng Shui in real estate. In Vietnamese context, people's habits in lifestyles and real estate purchases are highly impacted by Chinese culture. Taking a step back to historical background, the periods of Chinese conquests and Nguyen dynasty reveal shifts on cultural influences. Lately, despite efforts of the French colonialists disrupting the Vietnam's cultural links with China, Vietnamese mindsets continued to reflect the foundational Chinese philosophical and religious doctrines of Confucianism (appeared in structured layouts with central axes and symmetry), Taoism (Feng Shui), and Buddhism (sense of continuity between life and the afterlife) (Vladimirovna & Chu, 2019). Real estate, a part of living space and environment, is considered spiritual goods market in Vietnam (Nghiem-Phu, 2022), which is influenced by these mindsets.

Cultural and Historical Context: Feng Shui, a traditional Chinese practice, has been so deeply embedded in Vietnamese culture for centuries that there is high rate of Vietnamese people who continue to adhere to Feng Shui principles when making decisions about property design, construction, and purchase. According to a 2018 Vietnamese Royal Institute of Feng Shui survey, 2,516 of 2,568 real estate buyers showed concern about Feng Shui in their purchasing decisions. This data suggests that Feng Shui holds significant value in the real estate market, with approximately 98% of buyers influenced by it (Truong, 2020).

Technological Context: Emerging AI technologies, such as machine learning, data analytics, and predictive modelling, have the potential to transform traditional practices by providing more precise, efficient, and scalable solutions. Integrating AI with Feng Shui might potentially modernize the practice, making it more accessible and appealing to contemporary buyers, while preserving its cultural significance.

Business Context: The real estate market in Vietnam is characterized by rapid urban development and a growing middle class with increasing disposable income.

This economic growth has spurred demand for residential, commercial, and industrial properties. However, the market is also highly competitive, with numerous local and international players vying for market share. In this competitive landscape, offering technologically unique and culturally resonant services, such as AI-based Feng Shui, might possibly provide an advantage for enhancing buyers' preferences.

The survey by the Feng Shui Institute, as mentioned above highlights a clear demand for reliable Feng Shui guidance in real estate, with almost all (98%) buyers showing interest in its influence on their property or land. In another survey result, only 30% of these buyers were consulted in this field, while few received satisfactory advice (1%) (T.R.I., 2024). Combining the traditional practice of visiting Feng Shui masters with multiple online channels searches is complicated and time consuming for buyers. This gap suggests several critical insights and opportunities for research on AI-based Feng Shui solutions in real estate to gain high personal impact of Feng Shui for shifting in buyers' preferences and satisfaction.

A description of the proposed AI-based Feng Shui Application including the possible workflow, the functional diagram, and the mock-up screens is demonstrated in Article 2.2, Appendix 2 of this chapter.

The Significance of the AI-based Feng Shui Application involves several areas: cultural relevance and appeal, technological innovation, market differentiation, improvement of buyers' decision-making, and business impact.

1. **Cultural Relevance and Appeal:** As Feng Shui holds significant cultural importance in Vietnam; by incorporating AI-based Feng Shui, real estate enterprises could appeal to the cultural beliefs and preferences of Vietnamese buyers, thereby increasing buyer satisfaction and trust.
2. **Technological Innovation:** Leveraging AI to modernize Feng Shui practices represents a fusion of tradition and technology. Such innovation could streamline and enhance the accuracy of Feng Shui assessments, making them more accessible and reliable for both real estate enterprises and buyers.
3. **Market Differentiation:** Real estate companies that adopt AI-based Feng Shui could differentiate themselves from competitors by offering unique, culturally resonant services, that could attract a niche market segment that values traditional Feng Shui principles in property selection and design.
4. **Improvement of Buyers' Decision Making:** AI-based Feng Shui tools could provide data-driven insights, predictive analytics, and personalized property recommendations, enabling more informed decision-making in property purchasing. This could lead to better alignment of properties with market demand and buyers' preferences, which lead to greater loyalty and word-of-mouth referrals.
5. **Business Impact:** By attracting more buyers who value Feng Shui, real estate enterprises could potentially increase sales and profitability. The business im-

pact extends to the broader real estate market, contributing to its growth and dynamism.

The high percentage of Feng Shui concern undergirds the potential for considering an AI-based Feng Shui application to appeal to a broad audience, especially in Vietnam, where cultural and spiritual practices play a substantial role in purchasing behaviors. It is for this reason, that this study explores the potential for integrating AI-based Feng Shui into the practices of real estate enterprises in Vietnam, to enhance buyers' preferences. For the mentioned purpose, the approach will be intended to limit basic and conceptual ideas of the application and its impact on the real estate industry. The scope of study involves three key areas: technological integration (development of the conceptual idea for an advanced application that integrates AI with traditional Feng Shui principles); socio-cultural integration (real estate buyers' analysis on perception and acceptance, including buyers' preference factors, cultural resonance of AI-based Feng Shui); and business impact (assess how AI-based Feng Shui contributes to the competitive advantage and business performance).

The study focuses on exploring the main question: Considering the interaction of sociotechnical systems and cultural capital, how does the adaptation of an AI-based Feng Shui application in real estate enterprises influence buyers' preferences in Vietnam? To collect insights and feedback for answering this question, interviews and comparative experiments with post-experimental surveys are used. The data is then analyzed using techniques of thematic analysis to ensure all perspectives and insights are well captured for valuable outputs. The theory of sociotechnical-cultural integration will be employed in the study to ensure that the research is grounded in strong theoretical framework while also exploring relevant supporting perspectives to provide comprehensive analysis.

The potential assumptions for this study include: (1) the integration of modern AI and traditional Feng Shui principles might enhance buyer's preferences, leading to enhanced buyers' decision making; (2) the integration of modern AI and traditional Feng Shui principles might enhance buyer's preferences, leading to business effectiveness; and (3) the integration of modern AI and traditional Feng Shui principles might enhance buyer's preferences, leading to competitive advantage of the real estate enterprises.

This exploratory study of the AI-based application in real estate using Feng Shui principles underscores the potential of AI-based Feng Shui to transform real estate practices in Vietnam. By marrying cultural traditions with cutting-edge technology, real estate enterprises could gain buyers' preference and satisfaction, foster sales and profitability for enterprises, and contribute to the sustainable growth of the real estate market.

LITERATURE REVIEW

The context of Feng Shui principles covers the entire range of living environment that impacts on the decision of the land and property selections among large populations in the world (MY Mak, 1998). Especially, when the living standard becomes higher in this modern life, people tend to expect a living place that is not only a "house" for their basic needs, but also a "qualified environment" for bringing good health, comfort, relaxing and enjoyableness. Good Feng Shui is believed to relate to these expectations (Cho, 2020a). For this reason, the integration of Feng Shui principles in nowadays' modern technology such as AI-based application among real estate industry is necessary to help reach end-users' satisfactions when purchasing their land or property.

The literature review themes explored in this study include the concept of Feng Shui, Feng Shui principles, and impacts of Feng Shui on real estate, the concept of AI technologies and its influence in real estate, the approach of AI-based Feng Shui application in real estate, and buyers' preferences in terms of three aspects that relate to the theoretical framework, which are cultural values, property technology (Prop-Tech), and social relevance.

Feng Shui, Feng Shui Principles, and its Impacts on Real Estate: Previous researchers have well mentioned the concept of Feng Shui since early times. The idea of Feng Shui is to create a more comfortable living experience for people by examining how their home is positioned in an environment and how objects are arranged within it (Cho, 2020b). In addition, in the western world, Feng Shui is often translated as Geomancy (S. Lee, 1986). The key evidence that makes the above concept of Feng Shui well persuadable with a large population of users is the consistence of "six scientific theories and principles that have endured more than two thousand years of tests and experiments" (Sang, 2013). These principles are (1) The Yin/Yang Principles, (2) The Five Elements, (3) The Eight Trigrams, (4) The Directions, (5) The Solar System, and (6) The Environment. Feng Shui has significant impacts on the real estate area in terms of pricing and interior design. The pricing impact of Feng Shui practice involves the buyers' willingness to pay higher price for a good Feng Shui living environment (Lin & Chen, 2012). Furthermore, a study of Ahmadnia et al (Ahmadnia et al., 2012) showed that the proper architectural or interior design of the property is essential to the feeling of comfort and safety in the home or workplace.

AI Technologies in Real Estate: Recently, there have been researchers discovering useful findings in AI-based approach in real estate. Conway & Architecture (Conway & Architecture, 2018) attempted to study how AI is applied, and the actionable opportunities for the AI-based application in real estate. Lorenz (2021) also studied rental pricing of residential market using AI-based methods in their interpretability in real estate. The most recent paper of Baur et al (Baur et al., 2023) had brief concept on Ai-based models for valuating automatically based on property descriptions. From the above citations, the adapting of AI has positive impacts on the real estate industry including identifying investment opportunities and valuation.

AI-based Application using Feng Shui in Real Estate: AI-based application in real estate based on Feng Shui principles has not been investigated. As proxy literature review, there are several papers not direct but relevant to the topic. For instance, a patent by Wang et al. (Wang et al., 2014) presents a Fengshui evaluation method and system for automatic identification of key house features (gate orientation, bedroom direction, etc.) for residential house types, allowing rapid and scientific calculations based on a house image. Another related study by Leung and Chan (Leung & Chan, 2024) involves adapting ChatGPT to convert I Ching in modern management. The findings show that ChatGPT enhances accessibility, engagement, and interest in learning the I Ching, bridging ancient wisdom with modern management practices. Both patent and study above show positive insights for the potential to apply similar traditional practice like Feng Shui into modern technology like AI, as being explored in this research.

Another application launched in Vietnam in 2024 named Xhero (Phong Thuy Dai Nam, 2024) is a proxy reference. The application is about offering general Feng Shui consulting services and training courses.

Theoretical Framework: Sociotechnical Systems Theory and Cultural Capital Theory:

The combination of Sociotechnical Systems Theory (STS) and Cultural Capital Theory to provide a framework for the effective integration of technology within a cultural context. It posits that the successful implementation of technology, such as AI, requires a harmonious balance between the technical (application) and social (buyers, culture) components while also recognizing the value of cultural assets (Feng Shui). By considering both technological and cultural factors, enterprises could achieve a more comprehensive and socially relevant application of new technologies. The study is grounded in these two theories that are well-described by theory owners Trist (Trist, 1981) & Bourdieu (Bourdieu, 1986) in their papers.

Buyers' Preferences: For the purposes of this study, the literature review for the theme "Buyers' preferences" involves three aspects that relate to the theoretical framework mentioned above, which are: cultural values, property technology (PropTech), and social relevance:

1. **Buyers' Preference in Alignment with Cultural Values in Vietnam:** Nguyen & Vu emphasized the essence of Feng Shui in the context of Vietnamese cul-

ture and real estate, that Feng Shui plays a significant role in shaping not just the physical aspects of a home, but also the spiritual and cultural dimensions of dwelling (Nguyen & Vu, 2023). This founds the importance to examine how AI-based Feng Shui application might resonate with Vietnamese buyers, potentially influencing their preferences and decisions in real estate.

2. **Buyers' Preference Relating to the Ease of Interaction and Satisfaction with the Prop-Tech in Vietnam:** According to Le, despite the growth of digital technologies, adoption in Vietnam remains slow among users. In the Vietnamese real estate industry, transactions still heavily rely on paperwork, physical documents, and in-person or phone meetings, unlike in most developed countries where these processes are handled digitally. There is limited information on the use of Prop-tech in Vietnam to aid in real estate acquisition and portfolio management (Le, 2022). To resolve this, Cao & Nguyen (Cao & Nguyen, 2021) presented a method for building a real estate chatbot automatically to support customers in Vietnamese. The chatbot is trained with data set collected on Facebook groups and from the famous real estate website in Vietnam.
3. **Buyers' Preferences Based on More Tailored Recommendations from the AI-based Application:** Typically, websites gather user ratings for recommendations. However, with Ai-based tool approach, the recommendations incorporate historical data and the surrounding environment of a real estate location, making it easier for users to determine which place would be more suitable for them (Barua et al., 2019).

To conclude the above literature review, previous researchers have studied different aspects of Feng Shui, its impact on the real estate industry in terms of price valuation and property design. However, the AI-based application in real estate based on Feng Shui principles has not been investigated. This new application is emerging for being studied to provide a tool that is more convenient, quickly recommendable, and user friendly that will be able to add value to the business area, industry of real estate and the transactions for real estate investors. The application, if feasible, will potentially help not only real estate investors to identify the proper land or property quickly and efficiently, but also for sales force and brokers to recommend suitable products for their customers, hence, to increase sales influence.

STUDY METHODS

The design of qualitative methods is used for conducting this study. A qualitative method is to collect information from participants for exploring their insights on feasibility and potential impacts of the AI-based application using Feng Shui principles to the real estate sector.

In the next sections, there will be detailed discussions on the study's theoretical framework, questions and hypotheses with a conceptual model, showing possible themes and their operational definitions, sample population calculation and recruitment, processes of data collection and analysis, background of the researcher and data sources. In addition, external and internal validity and potential limitations of the study will also be mentioned before closing this section.

Theoretical Framework

Theory of sociotechnical-cultural integration is employed in the study to ensure that the study is grounded in a strong theoretical framework while also exploring relevant supporting perspectives to provide comprehensive analysis. The theory underscores the importance of aligning the technical aspects of an AI application with the social and cultural environment in which it will operate. In the context of real estate in Vietnam, where Feng Shui plays a critical cultural role, this alignment is crucial for ensuring that the AI application is both effective and culturally appropriate. The integration of AI with Feng Shui, a form of cultural capital, enhances its acceptance and appeal, contributing to the competitive advantage of real estate enterprises.

Question and Hypotheses

The design of qualitative methods is used to answer the study question: Considering the interaction of sociotechnical systems and cultural capital, how does the adaptation of an AI-based Feng Shui application in real estate enterprises influence buyers' preferences in Vietnam? Accordingly, there are four possible hypotheses for the research questions, which are grounded by the above theoretical framework of sociotechnical-cultural Integration to ensure the test of hypotheses will be logical and reliable.

1. **Hypothesis One:** If the application of AI-based Feng Shui in Vietnam could better align with cultural value, its adaption would enhance buyers' preferences.
2. **Hypothesis Two:** If the application of AI-based Feng Shui in Vietnam could increase the ease of interaction with the technology, its adaption would enhance buyers' preferences.
3. **Hypothesis Three:** If the application of AI-based Feng Shui in Vietnam could enhance the satisfaction with the technology, its adaption would enhance buyers' preferences.

4. **Hypothesis Four:** If the application of AI-based Feng Shui in Vietnam could provide more tailored recommendations, its adaption would enhance buyers' preferences.

Based on these hypotheses, a concept model for the study is described in Figure 1 enclosed in this chapter.

Figure 1. Conceptual model

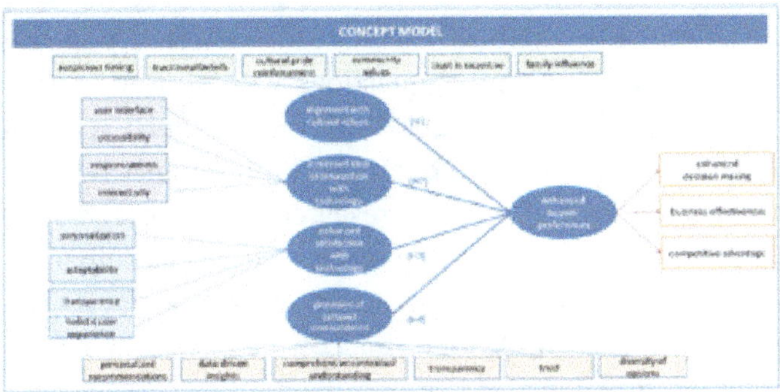

The sub-variables stated in the concept model are ideas that the researcher predicts to get from responses, that actual data might be different. The researcher expects to perceive the valuable findings by collecting as many new ideas as possible from respondents for this behavioral research.

Population Sampling

The targeted population sampling for this study is working adults with relevant experience in selling and buying real estate in Vietnam. For recruitment, the use of true-random and convenient sampling shall be applied, focusing on recruiting thirty participants who could access mobile applications and computers for experiments, from whom ten participants to be randomly invited for online interviews. The recruitment sources are salespeople or brokers working at real-estate agencies and buyers with successful purchasing transactions on real estate trade floors or agents. A testing of interviews and surveys is also conducted with three additional participants to ensure that all instruments work well for completing data collection.

Recruited participants' demographics and responses are committed solely for the purposes of the research. No incentive or compensation will be given to ensure the unbiased data is collected.

Data Collection

While qualitative methods approach in this research is suitable for collecting insights and feedback into "social, emotional and experimental phenomena" (Hs & Minhat, 2015), the study will aim to have two phases with data collected from ten participants:

1. **Phase One:** online interviews; collect insights on the new AI-based Feng Shui application. Other purposes of Phase 1's interviews are for the researcher to describe and identify the variables that relate to research hypotheses. To conduct online interviews, semi-structured questions will be used. Accordingly, there will be a set of 20 to 25 interview questions for each participant. Among these, approximately 50% are structured questions built initially before the interview that relate directly to clarifying the research question and hypotheses. The remaining questions are unstructured and based on participant's responses to explore new ideas. The maximum duration of each interview is 120 minutes and will be audio-recorded. A solid draft of the questionnaire with interview guides is enclosed in Appendix 1 of this chapter.

Why Interviews?

According to Georgescu and Anastaiu (Georgescu & Anastaiu, 2022), interviews are a fundamental tool in this qualitative research grounded by socio-technical and cultural capital, providing an in-depth exploration of participants' perspectives, experiences, and social contexts (Froschauer & Lueger, 2020). Interviews enable researchers to gather nuanced data, capturing the complexities of human behaviors, social dynamics, and cultural influences (Mehrad et al., 2024). Furthermore, interviews allow for the investigation of social meanings and emic (insider) perspectives, fostering a deeper understanding of how individuals navigate and interpret their cultural and technological environments (Lee, 2024). Additionally, interviews contribute to a rigorous and iterative research process that minimizes bias and enhances objectivity (Dursun, 2023). Overall, whether in culturally sensitive research or studies of socio-technological systems, interviews remain a powerful tool for uncovering the intricacies of human behaviors and the interplay between culture, technology, and society (Georgescu & Anastasiu, 2022; Froschauer & Lueger, 2024).

2. Phase Two: comparative experiments; the participants will be provided the currently available application in Vietnam: "Xhero" launched on September 22, 2024, (Le, 2024). After that, a comparative study between the two applications of "AI-based Feng Shui" and "Xhero" will be applied for the same participants from Phase 1 to share feedback through an online survey. The survey questionnaire of

20 questions will be applied online using Qualtrics tool where questions are built based on the variables collected from Phase 1's insights to ensure structured and unique outcomes. The purpose of Phase 2 is to validate the variables and hypotheses on buyer's preferences on the new AI-based Feng Shui application and discover the research findings and results in a robust evaluation. A solid draft of participant guide for comparative experiments, detailed description of the application, and questionnaire for post-experiment surveys are enclosed in Appendix 2 of this chapter.

Why Experiments?

Experiments play a crucial role when being integrated with interviews into research. According to Steils (Steils, 2021), qualitative experiments enhance hypothesis validation in socio-technical and cultural contexts by uncovering processes and behaviors that participants might not articulate. This exploratory approach allows for nuanced comparisons of different case studies, revealing deeper insights into user interactions that enhance the robustness of evaluations. Additionally, according to Yu and Wang (Yu & Wang, 2022), taking the dissemination of a similar traditional practice of Bai Yuchan's thoughts as an example, experiments that integrate cultural terminology into AI models could enhance the relevance and accuracy of the applications, making them more effective in their cultural contexts. Furthermore, Naber (Naber, 2015) emphasizes that the use of qualitative experiments is well suited for examining the process of improving strategies for searching and finding solutions for problems, allowing for a comprehensive analysis of social processes, particularly in innovation research. Finally, understanding user experiences through experiments could inform the design of AI applications that resonate with cultural values, ensuring they meet user expectations and enhance cultural practices (Yuan et al., 2024).

Data Processing and Analysis

For the purposes of this qualitative study, thematic analysis (TA) will be used to analyze the collected data. This structured approach ensures rigorous analysis of interviews and experimental findings in the theorical framework of socio-technical and cultural capital. Its flexibility and adaptability make TA a preferred choice across various disciplines as it uncovers patterns and themes within data, providing insights that enhance rich understanding of participants' experiences and perspectives (Sandhiya & Bhuvaneswari, 2024). Additionally, TA is considered a systematic approach that facilitates rich data interpretation of participants' beliefs and experiences (Saunders, 2023), especially by the adaption of six-phase process proposed by Braun and Clarke (2012) to coding and theme generation, which is essential for extracting meaningful insights from qualitative data (Wæraas, 2022).

Transcripts from interviews and surveys are the primary data sources. Transcript-based is an appropriate strategy for qualitative research in terms of collecting comprehensive insights and open feedback for proper outcomes. The thematic analysis will also identify patterns and themes related to the integration of Feng Shui principles and AI technologies. The steps of analysis shall be 1) prepare and organize data, including transcribing interviews, typing up field notes, and survey responses; 2) review and explore data; 3) develop a data coding system using inductive coding; 4) assign code to data; and 5) identify and recurrent themes.

In addition, inductive analysis shall be applied as well to group relevant codes for creating themes for data analysis and comparison. Illustrations of sampled Interview Transcripts form and Contact Sheets (with theme grouping) are enclosed in the Appendices 2 and 3 of this paper.

Tools and Visual Representations

1. **Zoom for Interviews:** The interviews of participants will be conducted through Zoom meeting tool with automated transcription and audio recording.
2. **Qualtrics CoreXM for Surveys in Comparative Experiments:** shall be used as a tool for data collection and analysis of this qualitative research. According to Content Square (ContentSquare, 2024), Qualtrics is one of the powerful tools designed to collect and analyze insights for better decision-making and removing uncertainty from product development. The platform streamlines processes by enabling researchers to gather qualitative data and act on insights through integrations. Qualtrics is selected as it provides robust analytics capabilities, which provide real-time insights and help optimize surveys for effective data collection.
3. **Matrix structure:** Data collected will be displayed in Matrix format with classifications of participants in rows, and themes in vertical columns. The Matrix structure is helpful for the researcher to present data through rows and columns analyzing two or more dimensions of the topic of interest. Matrix format is chosen because it enables the researcher to collect and arrange data for easy viewing and organizing data by case or chronology (May et al., 2023), to perform detailed analysis of data at the proposition level, and to conduct cross-case analysis (Zhang et al., 2014). The process is based on the participants' responses. The researcher starts with creating codes and grouping similar codes into themes relevantly. Afterwards, the coded data is inserted manually to develop the matrix on the respondent basis. Then, cells in the matrix are used to create a summary of source material. It is done by reference to the information presented in the

relationships of rows and columns. Additionally, the researcher walked through the interview and survey transcripts and went through the rows individually to add information to each cell. The last step of the data presenting through matrix structure was summarizing the material. In the matrix created, empty cells were left for themes that do not relate to the participants' responses.

SOLUTIONS AND RECOMMENDATIONS

External and Internal Validity

If applicable, this new idea of AI-based application will be an internal advancement of the new integration with the traditional Feng Shui in the real estate sector. The application will provide a new AI-based tool for salespeople who have higher confidence while approaching buyers of Feng Shui concern. for more reliable and persuadable offers. Externally, the application shall possibly be an advanced tool that will potentially increase buyers' preferences and fast decision-making. It also assists real estate enterprises to explore higher sales transactions efficiently.

Ethical dilemmas and challenges

Ethical dilemmas and challenges faced in the study might include:

1. **Informed consent:** The study needs ensuring that participants fully understand the study's purpose, procedures, and potential risks and benefits, and obtaining their informed consent to participate.
2. **Conflict of benefits:** With Feng Shui stakeholders, there might be a misunderstanding on the potential disruption of traditional approach that affects Feng Shui stakeholders' profession. The researcher needs to explain clearly that the application is benefitable for them if they could adapt such a tool for their career.
3. **Cultural sensitivity:** The researcher needs to be mindful of cultural beliefs and practices related to Feng Shui and respecting participants' cultural backgrounds when discussing and interpreting findings. The application will be designed to accommodate multiple Vietnamese Feng Shui traditions such as regional geomantic patterns and family-based ancestral beliefs, ensuring that recommendations are both technically valid and culturally meaningful.
4. **Undermining of rich cultural significance:** Despite the positive views on the adapting of Feng Shui into AI-based application, several practitioners mentioned certain challenges. For instance, Zhang (Zhang, 2018) concerned that "Feng Shui, as a locally meaningful heritage, should be cherished and utilized for

contemporary heritage conservation. Over-simplification or automation could undermine its rich cultural significance". For the purpose of this research, focusing on the usage of AI-based application for real estate in Vietnam, the Feng Shui principles will be integrated based on Vietnamese culture and beliefs.

5. **Risks of algorithmic bias:** Although Feng Shui is a traditional and culturally embedded knowledge system, its digital translation could be prone to algorithmic bias, especially if training datasets are regionally limited or skewed toward dominant cultural norms. To mitigate this, it is proposed to diversify the training dataset to include various regional and cultural interpretations of Feng Shui across Vietnam, to conduct regular fairness audits of model performance across demographics; and to implement fairness-aware machine learning practices to monitor and adjust model outputs.

Therefore, development of AI-based application needs to be broader and deeply investigated to ensure that it covers these aspects of concerns. The researcher would need to ensure to address these ethical dilemmas and challenges proactively, adhering to ethical guidelines and seeking appropriate institutional approvals, to ensure the study's integrity and the well-being of the participants involved.

Limitation of the study

The idea of developing an AI-based application in real estate using Feng Shui principles is new to the real estate area. While the conceptual idea of the AI-based application integrates widely accepted Feng Shui principles, it is important to recognize that Feng Shui practices vary significantly across regions and lineages in Vietnam. These differences might include unique geomantic interpretations, symbolic priorities, and emphasis on ancestral traditions. As a result, the AI-based system might not fully reflect the intuitive and personalized insights typically offered by experienced Feng Shui practitioners. Moreover, due to the challenges in pursuing a big dataset that needs deeper investigation into technical aspects, this study area is limited at the initial stage of behavior research only, that means exploring consumer insights to understand how preferable the application will potentially be. Development research will be further studied after this research, to develop and set-up the complete application for commercialization purposes.

FUTURE RESEARCH DIRECTIONS

This study basically provides a theoretical basis on a conceptual idea of the AI-based application using Feng Shui principles in real estate that potentially serves as a foundation for multiple technological and scholarly advancements for future research. Based on the study outcomes, the researcher will be able to develop algorithms and models suitable for the users' demands and preferences. Future research might include:

- Algorithm and model development: Building tailored machine learning models that balance architectural constraints, user preferences, and Feng Shui principles.
- Cross-cultural adaptation: Expanding the model to other culturally sensitive housing markets in Southeast Asia, such as Thailand or Malaysia.
- Integration with immersive technology: Applying Augmented Reality (AR) and Virtual Reality (VR) to enable interactive Feng Shui visualization in property listings.
- Voice and NLP Interfaces: Developing a voice-controlled Feng Shui assistant to support older or non-digital-native users.
- Continuous learning models: Using reinforcement learning to refine recommendations based on user feedback and engagement patterns.
- Ethical AI Practices: Investigating user trust and acceptance of AI-guided decisions in real estate through participatory and transparent design approaches.

CONCLUSION

Overall, a qualitative study will be employed to conduct this research. A qualitative method is to collect information from participants through interviews and experimental surveys to explore their insights on feasibility and potential impacts of the AI-based application using Feng Shui principles to the real estate sector. The AI-based Feng Shui application is designed and implemented with a careful balance of technical capabilities and cultural sensitivity. By modernizing Feng Shui practices through AI, the application not only leverages advanced technology but also preserves and enhances its cultural significance in real estate transactions. Such integration helps real estate enterprises in Vietnam appeal to buyers who value cultural practices, thereby increasing their market position and competitive advantage. In addition, the sociotechnical and cultural dimensions work together to create a solution that is technologically advanced, socially relevant, and culturally respectful. Overall, this approach would provide a solid foundation for analyzing how AI-based Feng Shui could enhance buyers' preferences in the real estate sector in Vietnam while keeping the research focused and manageable.

REFERENCES

Ahmadnia, H., Gholizadeh, M., Bavafa, M., & Rahbarianyazd, R. (2012). *Art of Feng Shui and its Relationship with Modern Interior Design*. academia.edu

Barua, U., Sabir Hossain, M., & Shamsul Arefin, M. (2019). Real Estate Recommendation Using Historical Data and Surrounding Environments. *International Journal of Information Engineering and Electronic Business*, *11*(5), 33–39. DOI: 10.5815/ijieeb.2019.05.05

Baur, K., Rosenfelder, M., & Lutz, B. (2023). Automated real estate valuation with machine learning models using property descriptions. *Expert Systems with Applications*, *213*, 119147. DOI: 10.1016/j.eswa.2022.119147

Bourdieu, P. (1986). The Forms of Capital. In Richardson, J. G. (Ed.), *Handbook of Theory and Research for the Sociology of Education* (pp. 241–258). Greenwood Press.

Braun, V., & Clarke, V. (2012). Thematic Analysis. In H. Cooper, P. M. Camic, D. L. Long, A. T. Panter, D. Rindskopf, & K. J. Sher (Eds), APA handbook of research methods in psychology, Vol. 2: Research designs: Quantitative, qualitative, neuropsychological, and biological. *Washington, DC: American Psychological Association.*, 57–71. DOI: 10.1037/13620-004

Cao, T. D., & Nguyen, Q. H. (2021). An Approach for Building Effective Real Estate Chatbots in Vietnamese. In *Soft Computing for Biomedical Applications and Related Topics* (Vol. 899, pp. 221–229). Springer., DOI: 10.1007/978-3-030-49536-7_19

Cho, A. (2020a). *How to Recognize Good Feng Shui in a Home*. Thespruce.Com. https://www.thespruce.com/signs-of-a-good-feng-shui-house-1274498

Cho, A. (2020b). *How to Use Feng Shui for Wealth and Prosperity in Your Home*. Http://Www.Thespruce.Com. https://www.thespruce.com/feng-shui-money-tips-and-wealth-1275336

ContentSquare. (2024). *11 Best Qualitative Data Analysis Tools and Software*. Contentsquare.Com. https://contentsquare.com/guides/qualitative-data-analysis/tools/

Conway, J., & Architecture, B. A. (2018). *Artificial Intelligence and Machine Learning: Current Applications in Real Estate*.

Dursun, B. (2023). A Qualitative Research Technique: Interview. *Journal Of Interdisciplinary Educational Research*, *7*(14), 100–113. DOI: 10.57135/jier

Froschauer, U., & Lueger, M. (2020). Das qualitative Interview zur Analyse sozialer Systeme. *Utb.*, S. 18. https://books.google.de/books?hl=de&lr=&id=NqfNDwAAQBAJ&oi=fnd&pg=PP1&dq=qualitative+interviews&ots=2MYVGTJn2z&sig=QGSvpuxu5MMHE38DzyKXx1nRLME#v=onepage&q=Trans&f=false

Georgescu, S. D., & Anastaiu, I. E. (2022, February 18). The Interview as a Qualitative Research Instrument. *Proceedings of the 15th International Management Conference "Managing People and Organizations in a Global Crisis."* DOI: 10.24818/IMC/2021/05.14

Gibler, K., & Nelson, S. (1998). Consumer Behavior Applications to Real Estate. *Academia.Edu.*

Hs, M., & Minhat, H. S. (2015). An Overview on the methods of interviews in qualitative research. In *International Journal of Public Health and Clinical Sciences* (Vol. 2, Issue 1).

Le, B. (2022). *The behavioral intention to adopt proptech services in Vietnam real estate market.* Swiss School of Management.

Le. (2024). *Ra mắt Ứng dụng Xhero, bước tiến trong ứng dụng công nghệ hiện đại | Nhịp sống số | Báo Văn Hóa Online.* Baovanhoa.Vn. https://baovanhoa.vn/nhip-song-so/ra-mat-ung-dung-xhero-buoc-tien-trong-ung-dung-cong-nghe-hien-dai-106004.html

Lee, S. (1986). *FS definitions and concepts.* Cornell University ProQuest Dissertations Publishing.

Lee, Y. S. (2024). Qualitative and mixed methods. *Translational Orthopedics, 229–232*, 229–232. Advance online publication. DOI: 10.1016/B978-0-323-85663-8.00010-6

Leung, C. H., & Chan, W. T. Y. (2024). Bridging ancient wisdom and modern management: Enhancing I Ching learning with ChatGPT. *Asian Journal of Contemporary Education, 8*(1), 19–32. DOI: 10.55493/5052.v8i1.5028

Lin, C.-C., & Chen, C.-L. (2012). *An Estimation of the Impact of Feng-Shui on Housing Prices in Taiwan: A Quantile Regression Application* (Vol. 15, Issue 3).

Lorenz, F. (2021). Dissertation_Felix_Lorenz_Pflichtexemplar. *Library of University of Regensburg.*

Mak, M. Y. (1998). *ANZASCA-1998-Mak_FengShui.* Science and Design.

May, K., Taylor, J. S., & Binding, C. (2023). Stratigraphic Analysis and The Matrix: Connecting and reusing digital records and archives of archaeological investigations. *Internet Archaeology, 61.* Advance online publication. DOI: 10.11141/ia.61.2

Mehrad, A., Nguyen, K., & Dihii, J. B. (2024). Social Cognition Approach: Using Observation and Interview Methods via Qualitative Research. *Journal of Education for Sustainable Innovation, 2*(1), 11–17. DOI: 10.56916/jesi.v2i1.718

Naber, A. (2015). Qualitative experiment as a participating method in innovation research. In *Historical Social Research* (Vol. 40, Issue 3, pp. 233–257). GESIS - Leibniz Institute for the Social Sciences. DOI: 10.12759/hsr.40.2015.3.233-257

Nghiem-Phu, B. (2022). Consumer behaviour towards purchasing Feng Shui goods: An empirical study from Vietnam. *Journal of Asian Finance, 9*(1), 83–0092. DOI: 10.13106/jafeb.2022.vol9.no1.0083

Nguyen, T., & Vu, N. (2023). Spiritual factors in Vietnamese consumers' purchasing decisions. *International Journal of Business and Management, 06*(10), 15–27. DOI: 10.5281/zenodo.8407732

Phong-Thuy-Dai-Nam. (2024). *Hướng dẫn sử dụng App Phong Thủy Xhero Đại Nam.* Phongthuydainam.Vn. https://phongthuydainam.vn/news/675134b439cf021bcf2fe4b7

Sandhiya, V., & Bhuvaneswari, M. (2024). Qualitative Research Analysis. In *In book* (pp. 289–310). Design and Validation of Research Tools and Methodologies., DOI: 10.4018/979-8-3693-1135-6.ch014

Sang, L. (2013). *The Principles of Feng Shui - Larry Sang - Google Books.* Books. Google.Com. https://books.google.com.vn/books?hl=vi&lr=&id=bCiLW7_DdI8C&oi=fnd&pg=PR12&dq=the+principles+of+feng+shui&ots=fUu-PgWIut&sig=c_ec_Jwt4rmLV3ynBFoZycrCwWQ&redir_esc=y#v=onepage&q=the%20principles%20of%20feng%20shui&f=false

Saunders, M. N. (2023). Research Methods for Business Students. *Pearson, 9.* https://www.researchgate.net/publication/367780349

Steils, N. (2021). Qualitative Experiments for Social Sciences. *New Trends in Qualitative Research, 6,* 24–31. DOI: 10.36367/ntqr.6.2021.24-31

T.R.I. (2024). *Hiểu biết phong thuỷ - Lợi thế lớn của môi giới bất động sản.* Viện Nghiên Cứu Bất Động Sản (Real Estate Research and Training Institute). https://tri.edu.vn/hieu-biet-phong-thuy/

Trist, E. (1981). The evolution of sociotechnical systems: A conceptual framework and an action research program. In Van De Ven, A. H., & Joyce, W. F. (Eds.), *Perspectives on Organization Design and Behavior* (pp. 19–75).

Truong, T. (2020). Phú Mỹ Gold City mang vượng khí đón Phú quý | Báo Dân trí. *Dan Tri Newspaper*. https://dantri.com.vn/kinh-doanh/phu-my-gold-city-mang-vuong-khi-don-phu-quy-20201118084003622.htm

Vladimirovna, G., & Chu, H. (2019). The gardens and architecture of the imperial burial complexes of Vietnam as a result of the influence of Chinese culture on the countries of Southeast Asia. *Amazonia Investiga*, *8*(21), 491–499.

Wæraas, A. (2022). Thematic analysis: Making values emerge from texts. In *Researching Values: Methodological Approaches for Understanding Values Work in Organisations and Leadership* (pp. 153–170). Springer International Publishing. DOI: 10.1007/978-3-030-90769-3_9

Wang, Y., Huang, X., & Ouyang, S. (2014). Automatic Fengshui evaluation method and evaluation system used for residential house type. In *Worldwide Patent Application #CN103679618A*. Current Assignee Hangzhou Group's Nuclear Information Technology Co Ltd.

Wu, S.-J. (2019). *Feng Shui: A comparison of the original concept and its current Westernized version*. proquest.com

Yu, Y., & Wang, H. (2022). The Application of Artificial Intelligence in the Dissemination of Local Culture in Hainan Province: Taking the Dissemination of Bai Yuchan's Thoughts as an Example. *Advances in Multimedia*, *2022*, 1–10. Advance online publication. DOI: 10.1155/2022/7255853

Yuan, X., Hu, J., & Zhang, Q. (2024). *A Comparative Analysis of Cultural Alignment in Large Language Models in Bilingual Contexts*. https://doi.org/https://doi.org/10.31219/osf.io/6hpcf DOI: 10.31219/osf.io/6hpcf

Zhang, J., Schomer, P. D., Marc, Zhang, Lei, Wu, D., & Boyle, J. (2014). A Metric Matrix Establishment for Cases Studies on the Effectiveness of the Key Environmental Protection Policies for Transportation Pollution Control. *Inter-Noise 2014*.

Zhang, S. (2018). The heritage practices in a Chinese historic neighbourhood: The manifestation of traditional Feng Shui in Langzhong, China. *International Journal of Heritage Studies*, *24*(5), 531–546. DOI: 10.1080/13527258.2017.1390778

ADDITIONAL READING

Ardittis, S. (1994). East-West Migration: An Overview of Trends and Issues. *The Politics of East-West Migration*, 3–46. DOI: 10.1007/978-1-349-23352-6_1

Baldominos, A., Blanco, I., Moreno, A. J., Iturrarte, R., Bernárdez, Ó., & Afonso, C. (2018). Identifying real estate opportunities using machine learning. *Applied Sciences (Basel, Switzerland)*, *8*(11), 2321. Advance online publication. DOI: 10.3390/app8112321

Bhattacherjee, A., Davis, C. J., Connolly, A. J., & Hikmet, N. (2018). User response to mandatory IT use: A coping theory perspective. *European Journal of Information Systems*, *27*(4), 395–414. DOI: 10.1057/s41303-017-0047-0

Bourassa, S., & Peng, V. (1999). *Hedonic Prices and House Numbers: The Influence of Feng Shui* (Vol. 2, Issue 1).

Hassan, M., Tedong, P., & Fadzilah, A. (2023). Feng Shui Superstitious belief: Does it influence young generations in housing purchase intention? *Journal of Malaysian of Planners*, *21*(5), 341–356. DOI: 10.21837/pm.v21i29.1375

Hennink, M., & Kaiser, B. N. (2022). Sample sizes for saturation in qualitative research: A systematic review of empirical tests. *Social Science & Medicine*, *292*, 114523. Advance online publication. DOI: 10.1016/j.socscimed.2021.114523 PMID: 34785096

Kamiri, J., & Mariga, G. (2021). Research Methods in Machine Learning: A Content Analysis. *International Journal of Computer and Information Technology (2279-0764)*, *10*(2). DOI: 10.24203/ijcit.v10i2.79

Manickam, P., Mariappan, S. A., Murugesan, S. M., Hansda, S., Kaushik, A., Shinde, R., & Thipperudraswamy, S. P. (2022). Artificial Intelligence (AI) and Internet of Medical Things (IoMT) Assisted Biomedical Systems for Intelligent Healthcare. In *Biosensors* (Vol. 12, Issue 8). MDPI. DOI: 10.3390/bios12080562

Pandey, K. (2021). The Basic Principles of Feng Shui: *A beginner's guide to the ancient discipline of harmony*. Https://Www.Timesnownews.Com/. https://www.timesnownews.com/spiritual/article/the-basic-principles-of-feng-shui-a-beginners-guide-to-the-ancient-discipline-of-harmony/713070

Thandapani, S., Mahaboob, M. I., Iwendi, C., Selvaraj, D., Dumka, A., Rashid, M., & Mohan, S. (2023). IoMT with Deep CNN: AI-Based Intelligent Support System for Pandemic Diseases. *Electronics (Basel)*, *12*(2), 424. Advance online publication. DOI: 10.3390/electronics12020424

KEY TERMS AND DEFINITIONS

Accessibility: involves cross-platform availability that is compatible with mobile devices, tablets, and desktops, and offline functionality that allows access to core features without requiring constant internet connectivity.

Buyers' Preferences: For the purposes of this study, buyers' preferences are relating to the grounded theoretical framework only, which are "the alignment of the application with cultural value", "the ease of interaction and satisfaction with the technology that the application increases", and "the more tailored recommendations that the application provides".

Cultural Pride: reinforces pride in maintaining and adapting traditional values in modern living.

Data-driven Insights: is a method of behavior data analysis and predictive model by leveraging user interaction data to predict preferences and improve the relevance of suggestions, using AI to anticipate buyer needs and recommend properties aligned with their long-term goals, and incorporating local cultural nuances into recommendations to ensure alignment with traditional values.

Diversity of Options: The application provides multiple property options with explanations of how they align with user preferences and offers insights into how small adjustments i.e. renovations or rearrangements could improve a property's Feng Shui.

Holistic User Experience: includes helpdesk and chatbot support for users to instant access to assistance for technical or content-related queries, sharing tools of land or property layouts or Feng Shui recommendations with family members or agents, user feedback loops with mechanisms for users to report issues or suggest improvements, and AI adaptive learning of user preferences over time to improve recommendations.

Personalization and Adaptability: The definition in this context means the inclusion of customizable settings which allow users to tailor the interface based on preferences (language, themes, or level of detail), dynamic recommendations (adjusting recommendations in real-time based on user input or feedback), and role-based access (features for different user types such as buyers or sellers) to ensure relevance.

Responsiveness and Interactivity: This variable emphasizes quick response time (fast-loading pages and real-time calculations to avoid delays), inclusion of interactive features like drag-and-drop room layouts, interactive maps, and Feng Shui calculators.

The Application of AI-based Feng Shui: means the AI-based application in real estate using Feng Shui principles. For the purposes of this study, the application is limited to an initial and conceptual approach of the application only.

Traditional Beliefs: involve in Feng Shui principles (how well the AI-based application integrates key Feng Shui rules like land and property orientation, layout, and balance of elements), local customs (incorporation of original Feng Shui variations or practices prevalent in Vietnam), and symbolism (use of Feng Shui symbols, colors, and motifs).

Transparency and Trust: In this context, the application emphasizes recommendation rationale (explaining why specific recommendations were made to build user confidence in the AI's suggestions) and trust reinforcement (highlighting data sources, algorithms, and validations that back the tailored suggestions).

User Interface: means the AI-based application able to provide intuitive navigation with simplified menus and clear instructions for ease of use, visual aids i.e. the use of interactive visuals like 3D property models or real-time Feng Shui overlays to enhance comprehension, and cultural relevance in design with aesthetic elements like culturally appropriate colors and symbols to create a familiar environment.

APPENDIX

Questionnaire for Online Interviews

What is your age?

What is your profession?

How long have you been working on this designation?

Question 1: How do you perceive the influence of an AI-based Feng Shui application on your preferences for real estate properties?

Question 2: Unstructured question #1

Question 3: In what ways should an AI-based Feng Shui application reflect traditional beliefs and practices to make it more acceptable and trustworthy for you?

Question 4: Unstructured question #2

Question 5: What features or assurances would increase your trust in an AI-based Feng Shui application compared to advice from a traditional Feng Shui expert?

Question 6: Unstructured question #3

Question 7: How important is it for you to understand how an AI-based Feng Shui application generates its recommendations, and what level of detail would you expect to see in its explanations (for transparency reliability)?

Question 8: Unstructured question #4

Question 9: What features or design elements would make an AI-based Feng Shui application more visually appealing and easier for you to use?

Question 10: Unstructured question #5

Question 11: How valuable is it for you to customize the recommendations or settings in an AI-based application, and what specific features would you prioritize for personalization?

Question 12: Unstructured question #6

Question 13: How crucial is it for an application to respond quickly and provide interactive features, such as real-time adjustments or simulations, to enhance your experience?

Question 14: Unstructured question #7

Question 15: What aspects of an AI-based Feng Shui application would make the overall experience satisfying and memorable for you, and how would you describe the ideal user journey?

Question 16: Unstructured question #8

Question 17: How do you feel about receiving property suggestions tailored to your personal preferences and lifestyle, and what specific features would make these recommendations feel more relevant to you?

Question 18: Unstructured question #9

Question 19: In what ways do you think data and AI insights could enhance the accuracy of Feng Shui recommendations, and how would you like this information to be presented to you?

Question 20: Unstructured question #10

Question 21: What kind of information or explanations would you need to feel confident in the recommendations provided by an AI-based Feng Shui application?

Question 22: Unstructured question #11

Question 23: How would you prefer an application to present alternative property options, and what factors should be emphasized to help you make the best decision?

Question 24: How do you feel about balancing cultural traditions with advanced technologies in real estate decision-making?

Question 25: Overall, in your opinion, how could integrating AI-based Feng Shui applications improve overall satisfaction for buyers in Vietnam's real estate market?

C*onfirmation of the debrief of what has been discussed.*

Do you have any other comments or opinions?

Note: Since this is a semi-structured interview, there are up to 12 unstructured questions between the above series to clarify or supplement the meaning of the provided information or to pursue new ideas. The sequence of unstructured questions might be flexible on position in the questionnaire, as long as it will not exceed the time consumption of the interview.

Description of the Proposed AI-Based Feng Shui Application, and Questionnaire For Post-Experiment Survey

Description of the Proposed AI-Based Feng Shui Application

1. **Possible workflow:** The possible workflow diagram is stated in Figure 2 enclosed in this chapter.

Figure 2. Possible workflow

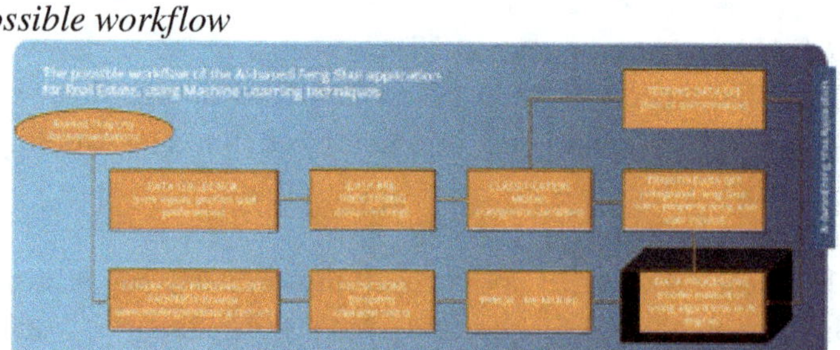

Workflow steps include: (1) Data collection: The first step of the workflow involves collection of user inputs including user's profiles and preferences. It is to gather necessary information to personalize property recommendations; (2) Data pre-processing: Once data is collected, a cleaning process is undergone to ensure quality and readiness for the next steps in the workflow. The process of data cleaning includes handling missing values, normalizing data, and other preparatory tasks; (3) Classification model: Using data of user profile and preference as basis, this step is to categorize the user inputs into relevant categories, ready for data testing and process; (4) Testing and trained data set: The application's performance is tested by using a testing dataset to validate how well it could make predictions and recommendations. The trained dataset consists of integrated Feng Shui rules, property data, and user input. Trained dataset is sourced by Phong Thuy Dai Nam (XheroZone Corporation), founded by the Feng Shui Master Nguyen Trong Manh, Deputy Director - Institute for Research and Application of Human Potential (Vietnam) cum Chairman of the Scientific Council – Global Feng Shui Science Institute. Master Manh is a leading expert in the field of pure, authentic, and systematic Feng Shui in Vietnam with his over 15-year experience in Feng Shui. These rules are used to guide the classification and adaptation of user inputs within the AI engine. Property data is trained data from the best cases in history for providing future predictions. The data is sourced from Dat Xanh Group, a property developer and agent, founded in 2003, one of the top 10 real estate enterprises in Southeast Asia. The data will be strictly committed with confidential use for the purpose of machine learning only. User inputs, including user profile and preferences; Data processing: The data, after being collected and pre-processed, is then processed using algorithms in AI engine. The process incorporates trained datasets of three components of Feng Shui rules, property data, and user inputs for further adaptation. The model is also evaluated to ensure that it produces recommendations that are accurate, reliable and aligned with both user preferences and Feng Shui principles; (5) Predictive model and predictions: A predictive model is created with predictions made about the properties' characteristics to ensure property alignment with user's preferences as well as Feng Shui requirements; (6) Generating personalized property: Based on user profile and preferences and their Feng Shui, personalized property listings are generated, being sourced from the website www.batdongsandanang.com.vn, which is a real estate platform providing real-time properties available for sales; (7) Ranked property recommendations: This is the last step of the workflow. The application provides the user with a list of properties that are ranked according to how well they align with the user's personal preferences and the Feng Shui principles. This final list is the result of the entire workflow, offering the user a curated selection of real estate options that match their unique needs.

Algorithmic transparency: The AI engine will incorporate explainable AI (XAI) frameworks such as SHAP (SHapley Additive exPlanations) and decision trees to increase the interpretability of its property recommendations. These mechanisms will allow users to view the logic behind each suggestion, including which Feng Shui attributes and personal preferences contributed to the outcome. An in-app explanation interface such as the functional diagram as below will also support transparency, building user trust in the system.

2. The functional diagram:

The functional diagram of the proposed AI-based Feng Shui application consists of four layers: user interface layer, input layer, processing layer, and output layer.

Layer 1: User interface: This layer serves as the initial touchpoint for the end user. In this layer, essential personal data of user is gathered through the User Profile component. Information such as the user's birth year, month, date, time, lifestyle preferences, and property goals are collected to personalize the Feng Shui recommendations. By prioritizing user experience, this layer ensures accessibility, ease of interaction, and alignment with user expectations.

Layer 2: Input layer: The input layer of the functional diagram synthesizes three critical data sources: property data, Feng Shui rules, and user input. Property data includes real estate characteristics such as location, orientation, floor plans, exterior and interior materials, etc. Feng Shui rules: This source encapsulates rich data of traditional spatial and metaphysical principles, including directionality, elemental harmony, and flow of energy (Qi), etc. User inputs: The data consists of user-defined criteria, such as birth and gender information, preferred location, budget constraints, and personal auspicious directions.

Layer 3: Processing layer: This layer is the technological core of the application. It comprises two principal components: AI engine and cultural adaptation. The AI engine utilizes machine learning and decision-support algorithms to evaluate property data against Feng Shui principles and user profile and preferences. A distinct and culturally sensitive component to be collected during the research interviews and experiments to ensure the interpretation of Feng Shui rules is consistent with Vietnamese traditions, values, and user expectations. This mechanism enhances the cultural legitimacy, user trust and reliability in the recommendations provided.

Layer 4: Output layer: The output layer translates the processed data into actionable insights for the user. This layer delivers three main outputs: Ranked property recommendations, Feng Shui compatibility score, and visual feedback. Ranked Property Recommendations is the most important output of the application. Properties are prioritized based on their alignment with Feng Shui principles and user criteria. The recommendations will be made based on real-time properties available for sales to users. Feng Shui Compatibility Score is a quantitative metric (percentage) that helps users assess the spiritual and energetic suitability of a property. Graphical interfaces and data visualizations of the recommended properties are presented to enhance comprehension and support decision-making.

The functional diagram is stated in Figure 3 enclosed in this chapter.

Figure 3. Functional diagram

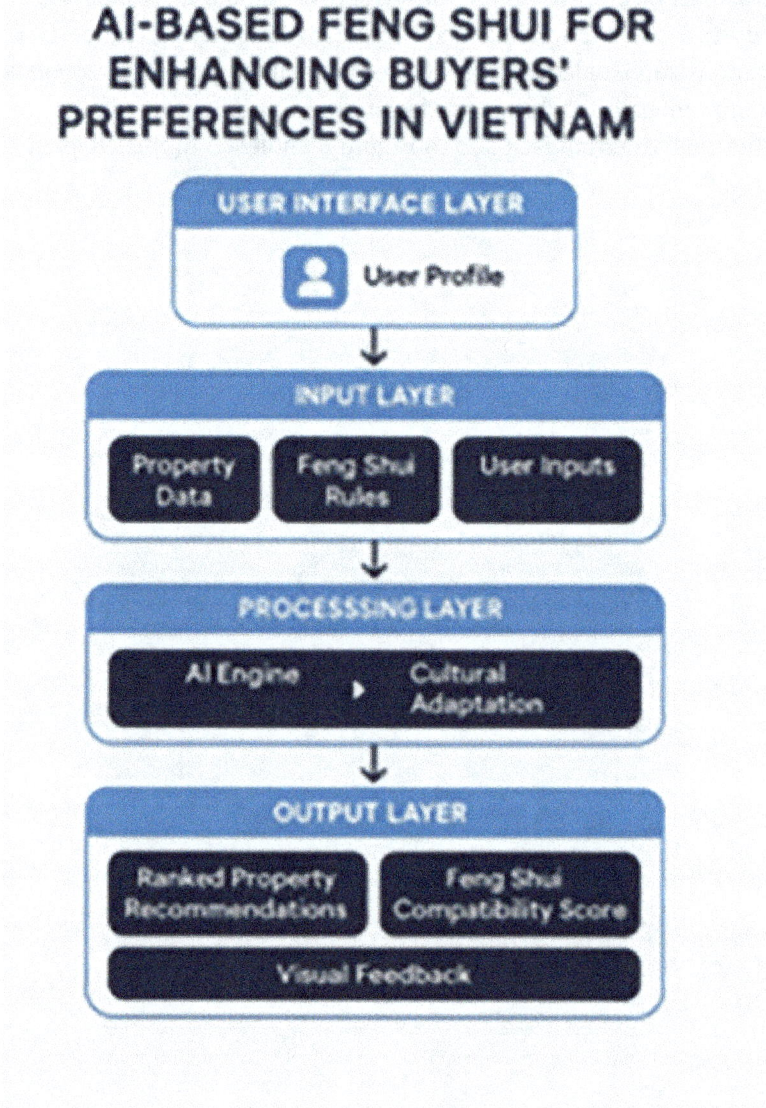

3. The demonstrated mock-up screens of the application:

The demonstration of application includes 07 main screens: (1) Welcome, (2) user profile, (3) Home dashboard, (4) Property listing, (5) Property specification, (6) Feng Shui analyser tool, and (7) Profile and settings.

Welcome & onboarding screen: to introduce users to the application's purpose and gather basic information for customization. This screen consists of the following elements: Logo and tagline: "Find harmony in your next home."; Brief introduction carousel about the benefits of Feng Shui and AI; "Let's Begin" button; Optional buttons to "sign in" or "continue as guest", and Language selection (Vietnamese, English)

Set-up of user profile: to collect user profile and preferences for generating and personalizing property recommendations. The elements of this screen are: Budget range slider; Preferred location; Preferred type of property (apartment or house or villa or land); Personal information relating to Feng Shui (time, date, month, year of birth, gender, Chinese Zodiac, element); Lifestyle preferences (e.g., quiet neighborhood, proximity to schools, etc.); Button "Save & Continue".

Home Dashboard: This is the central hub showing recommendations and status. Elements include Greeting: "Xin chào, [UserName]"; Daily Feng Shui tip or lucky direction; Featured property of the day (with Feng Shui score); "My Recommendations" shortcut; "Check Home Feng Shui" feature; Bottom menu: Home|Explore|Saved|Profile.

Property listing (AI-Personalized): This screen shows AI-selected homes with Feng Shui evaluations. Elements include Photo carousel of the property, Price, location, area, AI Feng Shui Score (0–100); Compatibility with user's birth element; Key Feng Shui highlights (e.g., "Good Qi flow," "East-facing front door"); and "View Details"/ "Save"/ "Share".

Property detail screen: the purpose of this screen is to provide a detailed view of a selected property. Elements: Full photo/video walkthrough; Full Feng Shui analysis: Element compatibility, Orientation and direction, Layout balance (e.g., entrance, bedrooms, kitchen), Energy flow visualization (mini floorplan overlay with energy flow lines), Location map, Nearby amenities; Contact agent button.

Feng Shui Analyzer Tool: This screen allows users to upload blueprints or input addresses for custom Feng Shui analysis. The screen contains these elements: Provision of floorplan or address; Interactive compass overlay; Summary of results (score, highlights, risks); Tips to improve Feng Shui (e.g., "Add a mirror near entrance to redirect energy"); and Option to generate a "Feng Shui Report" PDF file.

Profile & Settings: This screen is about user accounts and preferences, containing elements that allow user to: edit the personal information (birth element, preferences); manage saved properties; download reports; adjust language, theme (dark/light), notification preferences; share feedback and support.

The mock-up screens diagram is stated in Figure 2 enclosed in this chapter.

Figure 4. mock-up screens

Questionnaire for Post-Experiment Survey

Section 1: General Perceptions

Question 1: What were your initial expectations when comparing the Xhero application with the conceptual AI-based Feng Shui application?

Question 2: How do you perceive the potential benefits of an AI-based Feng Shui application compared to Xhero?

Question 3: If the AI-based Feng Shui application were fully developed, how likely would you be to use it instead of Xhero? Why or why not?

Section 2: Cultural Alignment

Question 4: Do you believe an AI-based Feng Shui application could better align with Vietnamese cultural values and real estate decision-making compared to Xhero? Why?

Question 5: How important is it for an AI-based tool to integrate Feng Shui principles into property recommendations?

Question 6: What specific cultural or traditional factors should an AI-based Feng Shui application consider to be effective?

Section 3: Ease of Interaction with Technology

Question 7: Based on the description or demonstration, do you think an AI-based Feng Shui application would be easy to use? Why or why not?

Question 8: What usability features from Xhero should be retained or improved in an AI-based Feng Shui application?

Question 9: What potential challenges or barriers do you foresee in using an AI-based Feng Shui application?

Section 4: Satisfaction with Technology

Question 10: How satisfied were you with the functionality and performance of Xhero? What could be improved?

Question 11: If the AI-based Feng Shui application were developed, what features would make it more satisfying and engaging than Xhero?

Question 12: What role does trust in AI-based recommendations play in your willingness to adopt a new application?

Section 5: Provision of Tailored Recommendations

Question 13: How well does Xhero personalize property recommendations to your needs?

Question 14: Do you think an AI-based Feng Shui application could offer more relevant and personalized recommendations than Xhero? Why or why not?

Question 15: If the AI-based Feng Shui application were to include personalized Feng Shui recommendations, how would this impact your decision-making process?

Question 16: What kind of additional personalization would you like to see in an AI-based Feng Shui application?

Section 6: Future Adoption & Improvements

Question 17: What features from Xhero would you like to see integrated into an AI-based Feng Shui application?

Question 18: What additional AI-driven features do you think would enhance a Feng Shui real estate application?

Question 19: Would you prefer an AI-based Feng Shui app to function as an independent platform or as an add-on to existing real estate apps like Xhero? Why?

Question 20: What are your final thoughts on the potential of an AI-based Feng Shui application for real estate in Vietnam?

Note: This questionnaire is only a demonstration. The content might be adjusted to match the actual responses collected from online interviews

Informed Consent Form

Informed Consent Document

For the research study:

Enhancing Buyers' Preferences by Adapting AI-based Feng Shui Application for Real Estate in Vietnam

Researcher:

Dang Thi Thanh Tam

DBA Candidate at Golden Gate University, San Francisco Contact Email: dthithanhtam@my.ggu.edu

Introduction: The purpose of this form is to provide you as a potential participant with information to help you decide whether or not to participate in this research and to record the consent of those who agree to participate in the study.

Invitation/identification: Dang Thi Thanh Tam, a DBA Candidate at Golden Gate University, invites you to participate in a research study in forms of online interview and/or experimental surveys. It means you are invited to participate in one or both forms.

Purpose: This study aims to explore how an AI-based Feng Shui application for real estate could influence buyers' preferences in Vietnam. Your feedback from one or both forms of online interviews and experimental surveys will help assess how cultural relevance, usability, and personalized recommendations may enhance the effectiveness of real estate platforms.

Description of research study: For the purposes of this study, data will be collected through two forms of online interviews and experimental surveys.

Online interviews: Here is a brief on the nature of the interview:

1) Number of total participants involved in study interviews: 10
2) Interview format: online, one on one
3) Means of recording: audio, and notes by interviewer
4) Duration: approx. 120 minutes
5) Number of questions: 20-25
6) Scope to covered: the interview will be scoped within the criteria below:
    - Alignment with cultural values: How well the AI-based application respects and integrates Vietnamese cultural values and traditions (e.g., Feng Shui).

- Ease of interaction with technology: How intuitive and user-friendly the application is.
- Satisfaction with technology: How adaptable, transparent and holistic the application is in supporting satisfaction in user experience.
- Provision of tailored recommendations: How effectively the application provides personalized property suggestions.

As a participant, you agree and are expected:

1) To spend a maximum duration of 120 minutes for an online one on one interview
2) To the interview recording and it will be in the form of audio and notes.
3) To sit in a quiet venue with no interference with noise.
4) To speak clearly, loud enough and try to minimize repeated contents.
5) The interview would be expected to be relaxing and natural talk to express the best of your honest thoughts.
6) By continuing with this interview, you are providing consent for your responses to be recorded and analyzed for research purposes.

Key consideration: It is understood that there are no right or wrong answers, and your honest thoughts are important. It is clear that your input and expertise would be valuable to this study.

Experiments: The brief of experiments

1) Number of total participants involved in experiments: 30
2) Survey format: online, to fill in survey questionnaire provided in form of link to Qualtrics platform.
3) Number of questions: 20
4) Scope to covered: the interview will be scoped within the criteria below:
   - Alignment with cultural values: How well the AI-based application respects and integrates Vietnamese cultural values and traditions (e.g., Feng Shui).
   - Ease of interaction with technology: How intuitive and user-friendly the application is.
   - Satisfaction with technology: How adaptable, transparent and holistic the application is in supporting satisfaction in user experience.
   - Provision of tailored recommendations: How effectively the application provides personalized property suggestions.

You will go through three main stages:

Stage 1: Familiarization With Xhero

a) You will explore the Xhero platform and its features.
b) Focuses: how the application helps you search for properties, navigate the interface, and receive recommendations.
c) Strengths and weaknesses regarding the above-mentioned criteria will be noted.

Stage 2: Exploration of the AI-Based Feng Shui Application (Conceptual Model)

Since the AI-based Feng Shui application is not fully developed, you will be introduced to its concept through descriptions, mock-ups, and guided demonstrations.

a) Imagine how the features would work in a real-life scenario and consider their potential impact on your decision-making process.
b) Compare its potential usability, personalization, and cultural relevance with Xhero.

Stage 3: Post-Experiment Survey

After exploring both applications, you will answer a qualitative questionnaire about your perceptions. The survey will focus on comparing Xhero's real-world functionality with the potential benefits of an AI-based Feng Shui application. Your feedback will help refine the development of future AI-driven Feng Shui application in Vietnam.

Key considerations:

a) There is no right or wrong answer—please provide honest and detailed feedback based on your perceptions and experiences.
b) Focus on comparing existing functionalities (Xhero) with expected usefulness (AI-based Feng Shui application) rather than direct performance.

Think about how an AI-based Feng Shui application could influence your real estate decisions in Vietnam compared to Xhero.

Risks and benefits:

Risks of participation: Participation in this study involves minimal risks. However, you should be aware of the following:

- Privacy and confidentiality: there is a slight risk of indirect identification and personal experiences as a real estate seller/buyer.
- Emotional or cultural sensitivity: Discussions around Feng Shui might touch on cultural beliefs or personal opinions that you might find sensitive and could be judged.

Benefits of participation:
As buyer participants:

- You might reflect on how cultural beliefs influence your real estate preferences.
- You might be able to benefit from the idea of more personalized, culturally sensitive property recommendations in future applications that better meet your expectations.

As seller participants:

- You might gain insights into how incorporating Feng Shui elements might enhance property appeal.
- This might be an opportunity for you to influence future digital tools that support better marketing and buyer engagement.
- You might be able to benefit from the future application that might help you gain buyers' decision making from its tailored recommendations for your potential buyers.

In general, your contribution to research might be helpful to bridge traditional Vietnamese cultural practices with emerging technologies like AI, that potentially lead to the development of a more user-focused, culturally informed real estate platform for both buyer and seller usages.

Confidentiality: All information obtained in this study is strictly confidential unless required by law. The results of this research may be used in reports, presentations, and publications, but the researchers will ensure anonymizing all data and accessible only to the research team. To protect your privacy and identity, all records of your participation will be given a group number that does not allow anyone (including the project staff) to personally identify you. These records will be stored securely where they will remain for at least 3 years following the completion of this research study or until the records can be safely destroyed.

Compensation: This is voluntary participation. No incentive or compensation will be given. You are free to share insights and perspectives honestly. It helps ensure that data collected will be unbiased.

Questions/contact information: Thank you for your valuable participation. If you have any questions or concerns about this study, please do not hesitate to contact the researcher at Dang Thi Thanh Tam, dthithanhtam@my.ggu.edu or +84 913 944 666.

Disclaimer: Participation in this study is entirely voluntary, and you can quit or withdraw from the study at any time without consequence. In addition, you may refuse to answer any portions of the study without withdrawing from the entire study.

Signature lines: Participant statement: This form explains the nature, demands, benefits, and any risks associated with this research. I have read the informed consent form, have had all my questions answered to my satisfaction, and voluntarily agree to participate in this research study.

Participant signature   Printed name   Date

Researcher statement: I certify that I have orally explained to the above individual the nature and purpose of, as well as the potential benefits and possible risks associated with participation in, this research study; have answered any questions that have been raised; and have received the above digital signature. I have provided (offered) the participant with a copy of this signed consent document.

Researcher signature   Printed name   Date

Chapter 7
Unleashing the Potential of Artificial Intelligence for the Chinese Cultural Heritage of the Forbidden City in China

Poshan Yu
http://orcid.org/0000-0003-1069-3675
Soochow University, China

Jiawen Guan
http://orcid.org/0009-0009-7934-3543
Independent Researcher, China

ABSTRACT

As technology rapidly advances, artificial intelligence (AI) is increasingly used in cultural heritage. Using CiteSpace, this study analyzes three - decade - long domestic and international literature on cultural heritage and AI, uncovering knowledge evolution, paradigm shifts, and research gaps. With the Forbidden City as a case, it examines two decades of AI - related practices and their effects on cultural heritage's production, communication, and identity. AI drives collaborative co - creation in production, enhances cultural - centered communication, and enables participatory cultural identity construction. The research concludes that for successful AI - based cultural heritage transformation, a balance between technological innovation and cultural essence is essential. Also, an inclusive "production - communication - identity" system must be established to meet stakeholder needs and safeguard cultural heritage in the digital age.

DOI: 10.4018/979-8-2600-2378-5.ch007

INTRODUCTION

Cultural heritage serves as a vital link to our past, offering meaningful insights into our identity and development (Aronsson & Price, 2024). It significantly reflects our cultural, historical, and social values while also contributing to sustainable development (Holtorf, 2018). The development of technology, especially in recent decades, has seen the rise of AI. Numerous AI applications have been developed for cultural heritage, such as automated historical document classification and knowledge extraction, automated 3D reconstruction, and identification of cultural heritage assets. These applications could be used to extract or generate information from a digitized cultural heritage object, manage digital heritage databases, or support the development of assisting tools for cultural heritage professionals.(Gîrbacia, 2024)

Artificial intelligence has the potential to revolutionize our engagement with cultural heritage. By analyzing extensive datasets related to cultural artifacts, AI facilitates more accurate restoration and preservation efforts. Utilizing machine learning algorithms, AI can detect patterns in historical documents, artworks, and archaeological discoveries, offering fresh insights into our cultural history. An increasing number of cultural institutions and museums are acknowledging the advantages of AI in enhancing visitor experiences, with many adopting AI-driven virtual tours, interactive exhibits, and tailored content recommendations.

Existing research on AI in cultural heritage has several limitations. Most studies concentrate on case analyses of technological implementation, which results in fragmented findings that hinder a deeper understanding of the internal mechanisms of technology-enabled cultural heritage. In particular, the connections between key aspects such as reproduction, communication, and identity construction remain inadequately explored, lacking a significant theoretical breakthrough. Moreover, the existing literature seldom uses quantitative tools like scientific knowledge mapping to conduct cross-cultural comparisons of research hotspots, which obscures the global research trends and cross-disciplinary characteristics.

This study aims to fill these research gaps. It starts by exploring the fundamental aspects of AI-enabled cultural heritage, using relevant domestic and international policies to understand the evolution of cultural heritage preservation methods. Then, through CiteSpace visual analysis, it examines and contrasts the main research hotspots and trends in the field of AI and cultural heritage over the past three decades. Finally, taking the Forbidden City as a specific case, it explores how AI can empower cultural heritage through the "production - communication - identification" chain while maintaining its core cultural values.

The research questions are as follows:

Q1: How has the policy support for cultural heritage changed in recent years?

Q2: What are the main research hotspots and trends in the field of AI and cultural heritage, both nationally and internationally, identified by analyzing the literature using CiteSpace?

Q3: How can AI systematically contribute to the reproduction, communication, and public identity building of cultural heritage?

This study, for the first time, triplets of policy analysis, bibliometrics and case studies were verified, and a complete evidence chain of "macro policy - mesoscopic research - micro practice" was established.

LITERATURE REVIEW

The sustained and lasting behavior of human beings towards the natural environment is likely to produce civilization on the basis of the technological, scientific and artistic achievements of the human mind scientific and artistic achievements of the human mind (Trivyzadakis, 2023) . According to the European Commission (2014), the civilization of the past consists both of material objects (such as architecture, archaeological sites and pieces of jewelry) and immaterial behaviors (such as traditional crafts, performing arts and social practices). In this way, human beings are linked to their roots, the links between different communities are strengthened, the education of contemporary people is improved, the background for prosperity is created, and the self-awareness of present and future generations is determined. In this framework, these elements are promoted to the category of monuments and become part of the concept of cultural heritage.

The above definition is essential for understanding the underlying logic of AI-enabled cultural heritage, because it reveals the role of cultural heritage as a bridge connecting the past, the present and the future. The material carrier and non-material practice together constitute the memory network of civilization, and the continuity of this network depends on human cognition, protection and inheritance of them. However, the cultural heritage with material as the carrier is facing physical degradation, and the cultural heritage with non-material form is facing challenges such as memory fault. In this context, the digitization of cultural heritage came into being. Its core is to transform the material and intangible cultural heritage into a digital form that can be stored, transmitted and interacted through digital technology (such as 3D scanning, virtual reality and artificial intelligence), so as to break through the limitations of time and space and realize the permanent preservation, dynamic activation and global sharing of cultural heritage (Portalés et al., 2018). Among these digital technologies, artificial intelligence (AI) has emerged as a game-changer, bringing about a new wave of innovation in the cultural heritage sector.

Recent literature studied various aspects of using AI techniques in cultural heritage. Several studies have examined ML techniques for assessing heritage building health conditions(Mishra, 2021), point cloud data acquisition and semantic segmentation algorithms for cultural heritage(Yang et al., 2023), and damage detection and structural health monitoring in masonry structures (Soleymani et al., 2023). Neural networks (NNs) and deep learning (DL) techniques have demonstrated potential in archaeology, specifically in reconstruction, remote sensing, and collection management (Barceló et al., 2022). Emerging techniques like Neural Radiance Fields (NeRFs) have been identified as effective in creating detailed 3D models of cultural heritage, especially those with challenging characteristics for traditional photogrammetry.(Croce et al., 2023) AI and subfield ML/DL techniques have also been used to significantly improve the performance of image-restoration methods (Liu, 2022), extract choreographic sequences from traditional dance (Rallis et al., 2020), improve accessibility of museum and online cultural heritage content (Zhao et al., 2020), and improve the management of archives (Colavizza et al., 2021). Recently, Casillo (2024) explored how AI can enhance visitor interactions with cultural assets, focusing on techniques and tool integration.

However, there are three key shortcomings in the existing research: first, policy analysis is disconnected from academic research, and there is a lack of correlation research; Second, case studies focus on technical details and fail to integrate effectively with theoretical framework. Third, bibliometric multi-static analysis ignores the dynamic influence of policy and technology evolution.

METHODOLOGY

Search Method

Policy Analysis

Policy analysis is a research method that systematically evaluates and interprets government or agency policy documents, aiming to understand the background, objectives, implementation effects and impact of policy formulation. This approach allows for both vertical (policy level) and horizontal (cross-regional) comparisons.

CiteSpace Analysis

CiteSpace, an information visualization software, is utilized to analyze and visualize co-citation networks with the goal of exploring trends and directions within the field (Chen, 2005). This study draws upon data from the CNKI website to provide an overview of domestic research on cultural heritage digitization. Employing bibliometric methods, we selected the CNKI database and the Web of Science Core Collection database. Using keyword screening and visualizations through CiteSpace, we conduct a comparative analysis and summary of the published literature, research hotspots, and developmental pathways in this domain, both domestically and internationally.

Case Study

Case analysis refers to the in-depth analysis of typical cases, observing the manifestation and mechanism of theoretical elements in actual situations, so as to verify or summarize the theoretical framework. On the one hand, it can maintain the integrity and context of the research phenomenon; On the other hand, through the process tracing method, the causal mechanism between variables can be clearly demonstrated.

Policy Analysis

Government and institutional policies are pivotal in guiding and promoting digital heritage preservation, with artificial intelligence emerging as a linchpin in this process. As Throsby (2018) said, effective cultural policies need 'multi-level intervention'. Digital heritage projects, in particular those leveraging AI, are highly dependent on the government to establish the necessary infrastructure and incentive mechanisms. This study compiles the relevant policies of the past decade to explore how they support cultural heritage protection and its digital transformation, with a special focus on the role of AI.

Over the past decade, the policies surrounding global cultural heritage preservation have experienced a notable paradigm shift. There has been a transition from traditional, material-based conservation strategies to digitally enabled and comprehensive protection frameworks, especially AI. The two tables outlining key policies from 2014 to 2024 illustrate that cultural heritage preservation has evolved from reactive salvage interventions to a systematic approach that integrates preventive conservation, technological empowerment, and sustainable utilization. This policy-driven evolution not only represents progress in preservation theory but, more significantly, highlights the transformative power of AI in redefining conservation paradigms (Buragohain et al., 2024). Artificial intelligence has facilitated a more proactive and intelligent approach to the preservation of cultural heritage, leading to more effective resource allocation, enhanced public engagement, and increased global accessibility. Moreover, it has created new opportunities for creative expressions and interpretations of cultural heritage, making it more relevant and engaging for contemporary audiences.

Table 1. Domestic Policy

Year	Policy Name	Authority	Key Content	Potential Impact on Cultural Heritage
2014	*Guidelines on Promoting the Integration of Cultural Creativity and Design Services with Related Industries*	The State Council	Encouraged cultural innovation and heritage-based product development.	Risk of over-commercialization but boosts economic value.
2014	*Guidelines on Strengthening Cultural Relics Protection*	The State Council	Reiterated the principle of "preservation-first, rescue-focused, rational utilization, and enhanced management."	Strengthened legal frameworks for systemic protection.
2016	*Guidelines on Strengthening Cultural Relics Protection*	The State Council	Reiterated the principle of "preservation-first, rescue-focused, rational utilization, and enhanced management."	Strengthened legal frameworks for systemic protection.
2017	*National 13th Five-Year Plan for Cultural Development and Reform*	The State Council	Integrated heritage preservation into national strategy, emphasizing digitization and public services.	Promoted technological adoption and public engagement.
2018	*Opinions on Implementing the Revolutionary Cultural Relics Protection and Utilization Project (2018-2022)*	The State Council	Prioritized conservation and utilization of revolutionary heritage.	Concentrated resources but may divert attention from other heritage types.
2019	*Construction Plan for National Cultural Parks (Great Wall, Grand Canal, Long March)*	The State Council	Establish a unified framework for linear heritage protection and tourism.	Enhanced large-scale heritage conservation but raised tourism pressure concerns.

continued on following page

Table 1. Continued

2021	*Opinions on Strengthening Historical and Cultural Preservation in Urban and Rural Development*	The State Council	Mandated protection of historical landscapes and tiered management systems.	Reduced urban construction damage and encouraged adaptive reuse.
2022	*14th Five-Year Plan for Cultural and Tourism Development*	Ministry of Culture and Tourism	Linked heritage preservation with tourism, emphasizing technology.	Balanced economic benefits with risks of overtourism.
2023	*Implementation Opinions on Reforming Cultural Relics Protection and Utilization*	The State Council	Streamlined administration and encouraged private sector participation.	Improved efficiency but required safeguards against market-driven neglect.

Source: Author

In the Chinese context, the evolution of cultural heritage policies reveals distinct developmental phases. From 2014 to 2016, policy frameworks predominantly focused on emergency conservation, as demonstrated by directives such as the Guidelines on Strengthening Cultural Relics Protection, which emphasized urgent interventions for at-risk heritage. During this period, technological applications were largely limited to basic documentation and archival methods. While AI was still in its early stages within the cultural heritage sector, it began to show promise; for example, initial explorations into AI-powered image recognition were undertaken to categorize cultural relics in a few pilot projects. However, the scope of its application remained narrow, and the technology had not yet matured for large-scale implementation.

Post-2017, the strategic focus expanded toward systemic preservation, with digital applications proliferating into domains such as virtual exhibition and environmental monitoring. AI played an increasingly important role in this digital expansion. In virtual exhibitions, AI-enhanced technologies like intelligent voice-guided tours and personalized exhibition route planning based on user preferences became available. For environmental monitoring of heritage sites, AI-enabled sensors could analyze data in real - time, predicting potential risks to cultural heritage due to environmental changes. This not only improved the efficiency of preservation but also provided more comprehensive protection.

The enactment of the 14th Five-Year Plan for Cultural Relics Protection and Technological Innovation in 2021 marked China's formal foray into an era of technology-driven preservation. This initiative explicitly integrated cutting-edge technologies into conservation systems and pioneered innovative applications. AI emerged as a vital element in this technological transformation. In the realm of conservation, AI-based algorithms were employed to analyze the structural stability of ancient buildings, enabling the prediction of areas that may require reinforcement. Furthermore, AI-powered restoration techniques were developed, allowing machine learning models to draw from a vast database of historical restoration cases to offer more precise restoration recommendations for damaged cultural relics. This policy evolution, with AI at its core in many aspects, has not merely enhanced the scientific precision and efficiency of preservation practices but, more fundamentally, has established a novel "physical - digital" dual-track conservation model that provides both institutional safeguards and technological foundations for sustainable heritage transmission.

Table 2. International Policy

Year (Region)	Policy/Initiative	Key Content	Potential Impact on Cultural Heritage
2014 (EU)	*Conclusions on Cultural Heritage as a Strategic Resource for a Sustainable Europe*	Positioned heritage as central to European identity, fostering cross-border collaboration.	Enhanced regional coordination but faced uneven implementation.
2015 (UN)	*2030 Agenda for Sustainable Development (SDG 11.4)*	Included heritage preservation in global sustainability targets.	Raised awareness but lacked enforcement mechanisms.
2017 (UNESCO)	*Recommendation on Preservation of Digital Heritage*	Advocated digitization to safeguard endangered heritage.	Improved accessibility but highlighted resource gaps in developing nations.
2019 (France)	*Priority Restoration Plan for Cultural Heritage*	Allocated emergency funds for heritage damaged by disasters or neglect.	Accelerated rescue efforts but relied on fiscal budgets.
2021 (Japan)	*Revised Cultural Properties Protection Law*	Expanded coverage to modern industrial heritage and disaster prevention.	Broadened protection scope but increased administrative burdens.
2022 (Egypt)	*National Strategy for Archaeology and Cultural Heritage*	Focused on museum development and international research cooperation.	Improved conservation capacity but intensified restitution debates.
2023 (India)	*Draft National Cultural Heritage Policy*	Centralized resources and combated illicit artifact trade.	Curbed smuggling but faced enforcement challenges.

Source: Author

At the international level, policy development has manifested as a multidimensional, coordinated advancement. The incorporation of cultural heritage preservation into the United Nations 2030 Sustainable Development Agenda established a global governance framework, while the European Union's cultural heritage strategies emphasized transnational collaboration and digital resource sharing. In this context, AI plays a crucial and multifaceted role.

A comparative analysis of domestic and international policy trajectories reveals that the preservation of cultural heritage is undergoing a significant transformation driven by technology. At the forefront of this evolution is AI, which not only enhances technological capabilities but also redefines our engagement with cultural heritage. Through AI-enabled virtual reality (VR) and augmented reality (AR) experiences, individuals can interact with cultural heritage in immersive and interactive ways that transcend geographical and temporal boundaries. This advancement exceeds traditional static displays, revitalizing cultural heritage and making it more accessible and relevant to contemporary audiences.

This change encompasses technical upgrades and a conceptual reorientation. Cultural heritage is increasingly viewed not as a static historical artifact, but as dynamic cultural capital that allows for ongoing contemporary engagement through digital mediation. AI plays a significant role in facilitating this conceptual shift.

CITESPACE ANALYSIS

Research Overview

Time analysis of publication volume

Initially, based on CNKI database, this study systematically collected 1087 Chinese literature between 2004 and 2024 using the search strategy of SU=('人工智能(AI))*('文化遗产 (cultural heritage)' + '遗产保护(heritage protection)' + '遗产(heritage)'). Through the statistical analysis of the annual publication volume, it is found that the research in China presents obvious stage characteristics: the average annual publication volume between 2004 and 2016 remains at a low level, indicating that it is still a theoretical exploration period; the publication volume between 2016 and 2022 increases steadily, indicating that the research enters a growth stage; the growth shows an explosive growth after 2023, and reaches a peak in 2024.

Figure 1. The Number of Publication in China

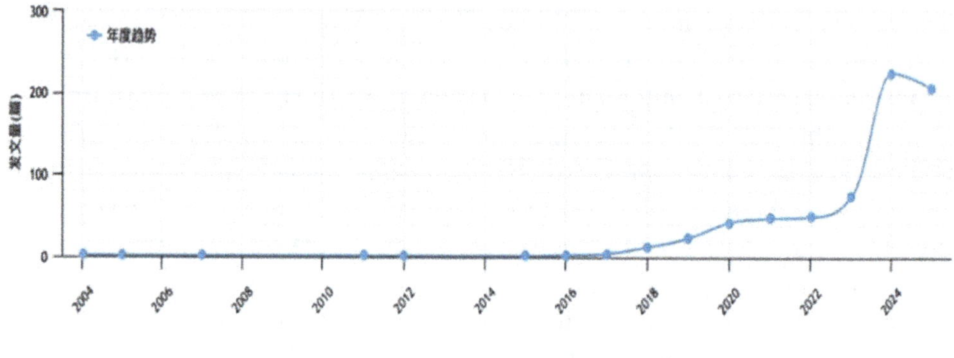

Source: CNKI

Distribution of major disciplines

Figure 2. Distribution of major disciplines

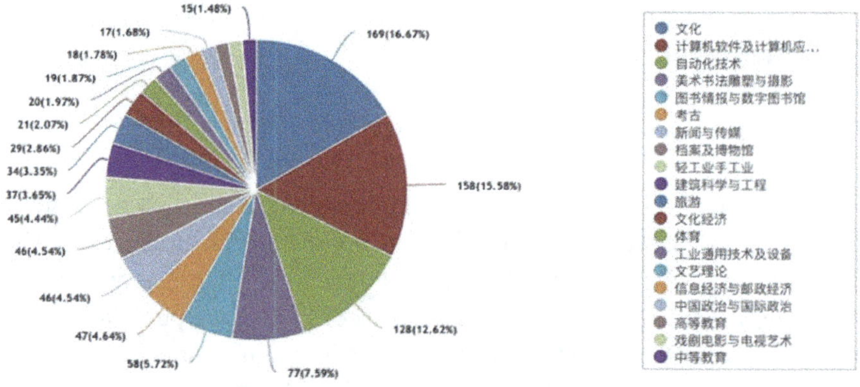

Source: CNKI

Analysis of Research Hotspots

In the medium term, this study adopts CiteSpace software to achieve the visualization of literature information. Firstly, CNKI database and Web of Science core collection database were selected to conduct a pre-search of the research literature on “application of AI on cultural heritage” with multiple combinations of keywords and topics, to test and compare the number of relevant literature under different combinations as well as the validity of the retrieved literature data. The selected search combinations were SU= ('人工智能')*('文化遗产'+ '遗产' + '遗产保护') and TS=(artificial intelligence*(cultural heritage or heritage protection)), which yielded a total of 1087 articles in Chinese and 944 articles in English. In order to improve the quality of the data, further filtering according to the level of journals was carried out, and finally 738 Chinese literature and 843 English literature were obtained, which were used for further research hotspot analysis.

Figure 3. Steps for CiteSpace

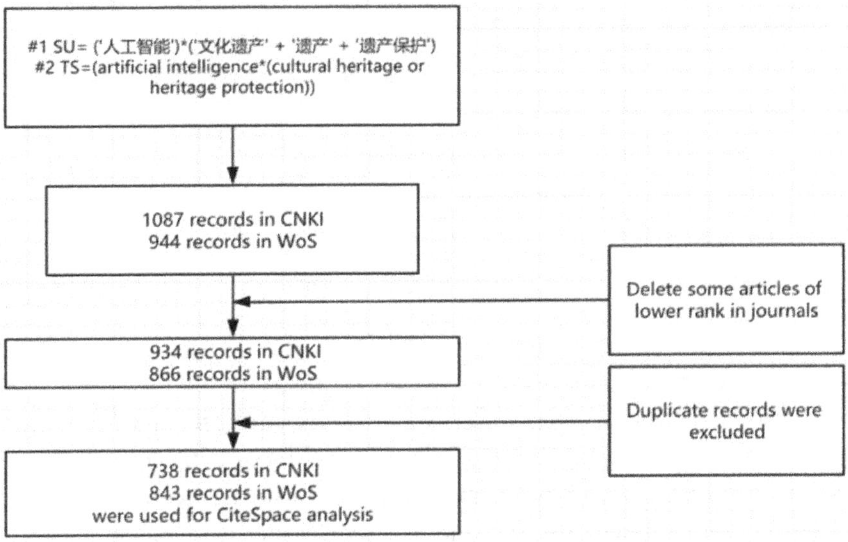

Source: Author

Analysis of domestic research hotspots

In this study, the literature data of CNKI database during 2016-2025 were visualized and analyzed by CiteSpace software, and a keyword co-occurrence network containing 249 nodes and 402 edges was constructed with a network density of 0.013.

Figure 4. Keyword co-occurrence network domestically

Source: Author

From the perspective of keyword frequency distribution, "人工智能(artificial intelligence)" and "文化遗产(cultural heritage)" constitute the most central research themes, both of which appeared in 2019, with mediator centrality as high as 0.88 and 0.1 respectively, occupying a pivotal position in the network structure. The sub-high-frequency keywords include "文化传承(cultural inheritance)", "传统文化(traditional cultural)" and "数字化(digitalization)", whose centrality ranges from 0.06 to 0.15, and which play an important intermediary role in the network.

It is worth noting that some of the high-frequency keywords such as "博物馆(library)" and "传承(inheritance)", although appearing earlier, have relatively low centrality (0.02-0.04), suggesting that these themes have limited connectivity in the network. In contrast, although the frequency of "元宇宙(metauniverse)" is only 16 times, its appearance is the most recent, signaling the direction of future research on AI application.

Table 3. Keywords

Count	Centrality	Year	Keywords
199	0.88	2019	人工智能
34	0.1	2019	文化遗产

continued on following page

Table 3. Continued

30	0.09	2024	文化传承
26	0.04	2020	传统文化
26	0.15	2019	数字化
23	0.07	2021	数字人文
17	0.09	2023	数字技术
16	0.13	2022	元宇宙
13	0.05	2016	传承
11	0.02	2019	创新
10	0.01	2023	乡村振兴
10	0	2020	创新设计
10	0.01	2024	大数据
10	0.02	2021	非遗

Source: Author

To gain a more precise understanding of the current research trends, this paper conducts a clustering analysis based on keyword co-occurrence mapping. The resulting clustering map of domestic research keywords shows a Q value of 0.5727, which exceeds the threshold of 0.4, indicating a significant clustering structure. Additionally, the S value of 0.9134 surpasses 0.7, confirming the credibility of the clustering results. The analysis reveals a total of nine valid clusters.

By combining the keywords with specific literature within these clusters, this study identifies three core research hotspots in the field of application of AI on cultural heritage:

- Digital Transformation and Co-creation in Cultural Heritage with AI

Hotspot 1 integrates clusters 0, 4, and 7, focusing on "artificial intelligence" and "cultural heritage." Key concepts in this area include "AI-empowered cultural heritage," "cultural co-creation," and "digital preservation." Research conducted here explores both fundamental and applied dimensions. On the fundamental side, it investigates how AI algorithms can enhance the analysis and preservation of cultural heritage data. Conversely, the applied aspect emphasizes leveraging AI to encourage public involvement in cultural heritage co-creation, exemplified by initiatives like the Palace Museum's "digital community" for heritage interpretation. Domestic scholars have recognized the gradual erosion of cultural heritage's meaning, the limitations of a singular communication channel, and the intergenerational discrepancies in audience understanding (Xue Ke, 2020). By leveraging AI's capabilities like data analysis and intelligent interaction, practical solutions are proposed. Domestic studies within this hotspot tend to explore the integration of AI with cultural heritage from both theoretical and practical perspectives, aiming to build a comprehensive theoretical and practical model for AI - empowered cultural heritage development.

- Technological Innovation for Inclusive Cultural Heritage Experience

Hotspot 2 encompasses clusters 1, 3, and 9, linked by themes such as "digital inclusion," "accessibility," and "interactive experience." This research highlights three key features: First, it emphasizes inclusive design, illustrated by the Forbidden City's app tailored for the elderly and tactile interactive devices developed for the visually impaired, thereby enhancing access to cultural heritage. Second, it pioneers innovations in interactive experience design, leveraging technologies like augmented reality (AR) and virtual reality (VR) to deliver immersive cultural experiences. Third, it showcases a variety of application scenarios, including cultural heritage-based tourism and digital education. The research aims to dismantle barriers to access cultural heritage, ensuring diverse groups can engage with and enjoy related activities, while simultaneously enriching the overall public cultural experience through technological innovation.

- Digital-Driven Cultural Heritage Industrial Development

Hotspot 3 encompasses clusters 2, 5, and 8, focusing on the "cultural heritage industry," "digital economy," and "innovation design." The key concepts driving this area include "digital cultural products," "cultural industry chains," and "digitization-promoted innovation." This research highlights the development of infrastructure related to cultural heritage and the horizontal and vertical expansion of the cultural heritage industry, spurred by digital technology. It illustrates the transformation of digital resources from mere digitization to large-scale industrial applications, such as the creation of digital cultural relics-themed products and the integration of cultural heritage within the tourism and creative industries. This reflects an innovative shift in the development model of the cultural heritage industry, with digital technology serving as a powerful catalyst for the growth and diversification of the cultural heritage economy.

Figure 5. Clustering map of domestic research

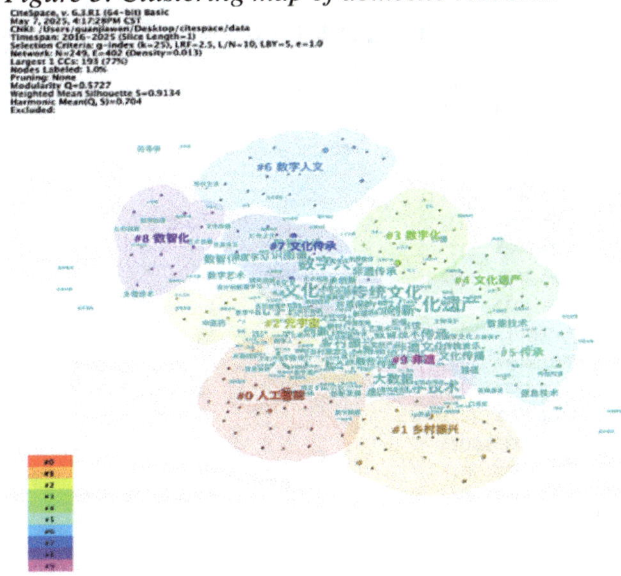

Source: Author

Analysis of domestic research development pathways

Using CiteSpace software to do emergent word analysis and timeline analysis of Chinese literature, we systematically sorted out the evolution path of domestic research in this field. It is found that domestic research presents obvious stage characteristics and technology-driven.

Figure 6. Keyword bursts

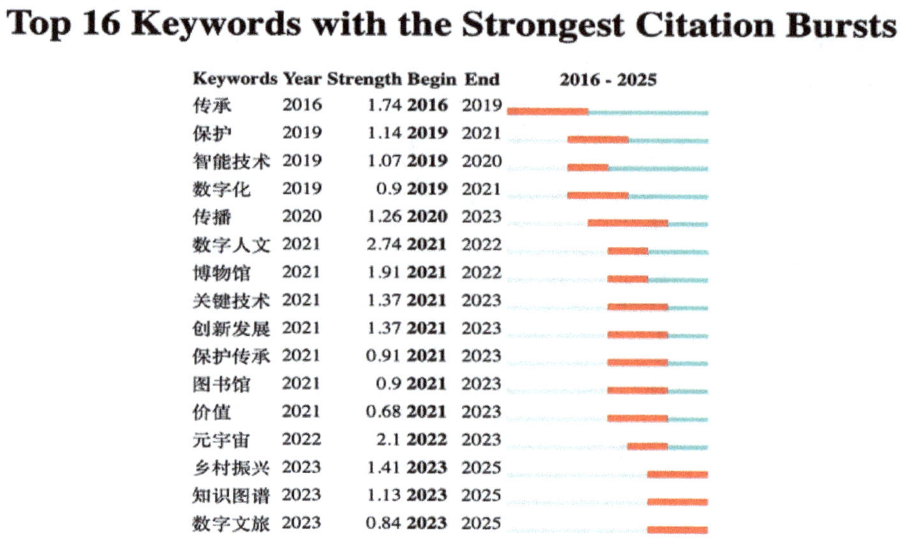

Source: Author

Keyword bursts provide an in-depth analysis of specific keywords over a defined time frame, highlighting trends within a particular field. This study identifies 25 standout keywords that have witnessed the most explosive growth in citations. As illustrated in the figure, "数字人文 (digital humanities)" emerges as the leading keyword, closely followed by "数字化 (digitization)," "元宇宙 (metaverse)," and "文化遗产(cultural heritage)," all boasting a citation strength exceeding one. These keywords represent past research hotspots and offer insights into potential future trends. It is noteworthy that while these explosive keywords arise almost annually, their durations tend to be notably short, typically lasting around one year. In contrast, "信息技术 (information technology)" has maintained its status as a keyword for an impressive nine years, indicative of the long-standing efforts to leverage digital technology in preserving cultural heritage. Keyword bursts provide an in-depth analysis of specific keywords over a defined time frame, highlighting trends within a particular field. This study identifies 16 standout keywords that have witnessed the most explosive growth in citations. As illustrated in the figure, "传承 (Inheritance)" emerges as the leading keyword, closely followed by "保护 (Protect)," "智能技术 (Intelligent technology)," and "数字化(Digitalization)," all boasting a citation strength exceeding one. These keywords represent past research hotspots and offer insights into potential future trends. It is noteworthy that while these explosive keywords arise almost annually, their durations tend to be notably short, typically lasting around one year.

Figure 7. Timeline graph

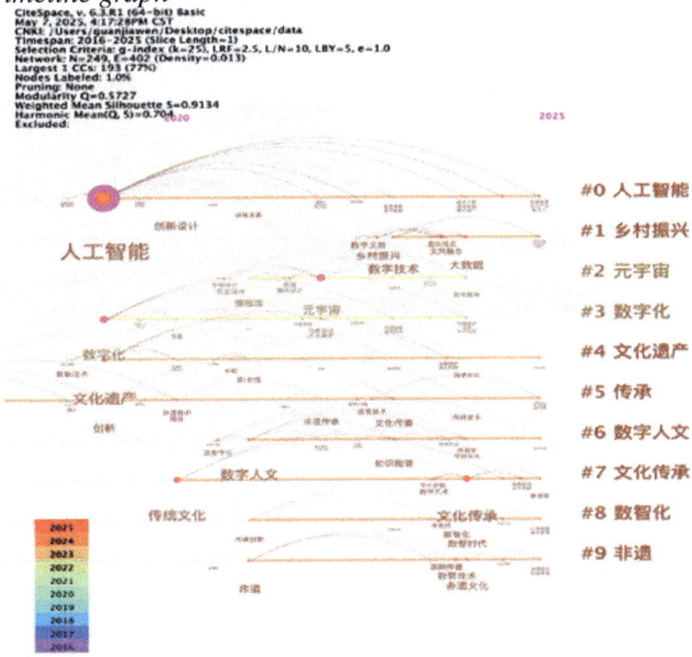

Source: Author

With development time as the X-axis and clustered themes as the Y-axis, themes can be transformed into timeline graphs based on keyword clustering graphs.

The visual analysis of timeline mapping can show the distribution and evolution of keywords in this field in different time regions (Guo Wei, 2022). Based on the timeline diagram, this paper divides the research on the AI - driven transformation path of domestic cultural heritage into three stages.

Figure 8. Threes stages of domestic cultural heritage ages

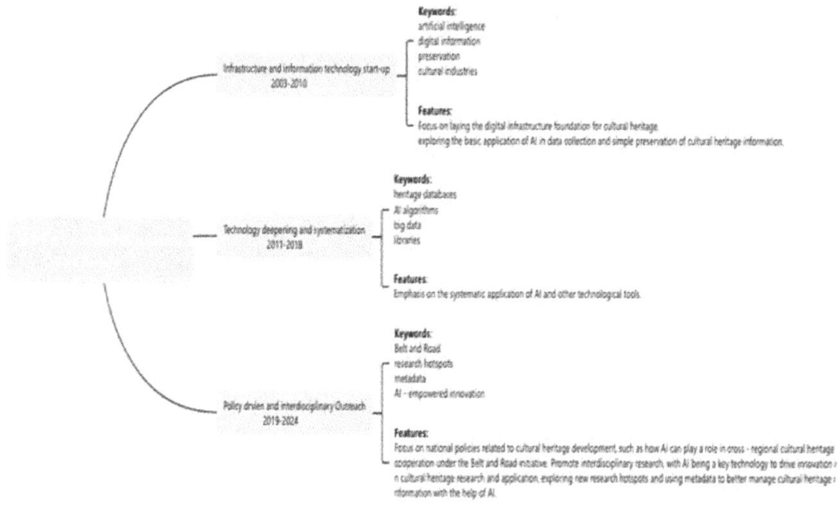

Source: Author

Analysis of foreign research hotspots

The results of the analysis of the keyword co-occurrence network constructed based on CiteSpace software show that the research field of international cultural heritage digitization presents significant knowledge structure characteristics. Statistical indicators show that the research network contains a total of 185 nodes and 456 connected edges.

Figure 9. Keyword co-occurrence network internationally

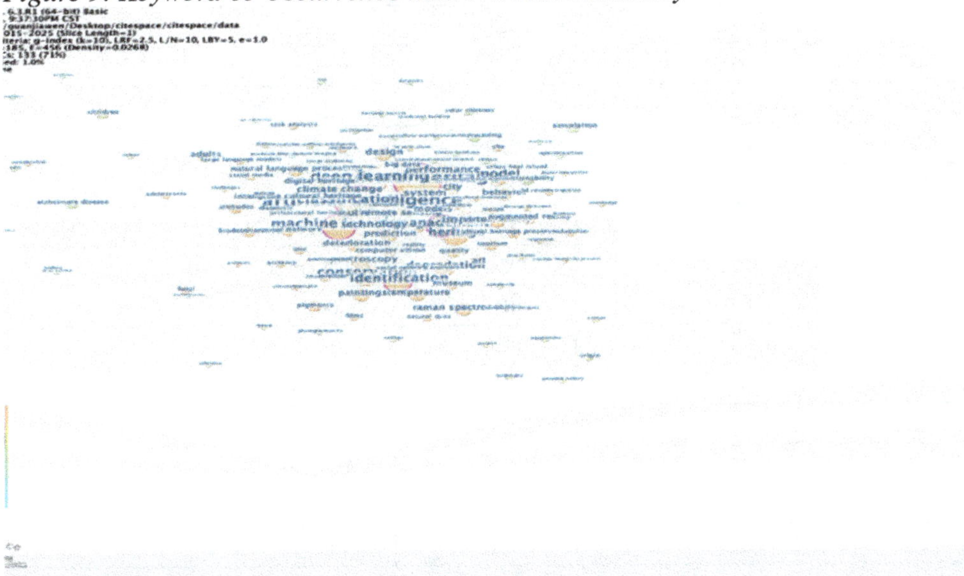

Source: Author

The frequency analysis of keywords shows that "cultural heritage" ranks first with 69 high-frequency occurrences, and its centrality index reaches 0.17, occupying a core hub position in the network structure. The secondary high-frequency keywords include " artificial intelligence ", " deep learning", etc., with a centrality distribution in the range of 0.07-0.2. It is worth noting that about 60% of the keywords have a frequency of less than 15 times, forming a typical long-tail distribution.

Table 4. Keywords internationally

Count	Centrality	Year	Keywords
69	0.17	2017	cultural heritage
67	0.2	2020	artificial intelligence
36	0.07	2020	deep learning
32	0.11	2021	machine learning
23	0.04	2022	classification
23	0.07	2020	conservation
21	0.12	2022	identification
20	0.21	2021	management
19	0.03	2020	model

continued on following page

Table 4. Continued

16	0.02	2023	heritage
14	0.05	2017	impact
13	0.04	2023	degradation
12	0.05	2020	performance
11	0.03	2020	system
11	0.01	2024	technology
10	0.05	2024	buildings
9	0.06	2019	design
9	0.03	2022	city
9	0	2019	climate change
9	0.01	2024	spectroscopy
7	0.02	2020	behavior
7	0	2024	paintings
7	0	2024	virtual reality

Source: Author

To gain a more precise understanding of current research trends, this paper conducts a clustering analysis based on keyword co-occurrence mapping. The resulting clustering diagram of domestic research keywords reveals a significant structure, with Q = 0.5673, which exceeds the threshold of 0.4, and S = 0.8328, surpassing 0.7, thus affirming the credibility of the clustering outcomes. The analysis indicates the presence of nine valid clusters. By combining keywords with relevant literature within these clusters, this study identifies three central research hotspots in the field of digital cultural heritage:

- AI-Driven Cultural Heritage Preservation and Analysis

This hotspot centers around the utilization of artificial intelligence in safeguarding and studying cultural heritage. Clusters 0 and 1, with terms like "artificial intelligence", "deep learning", "raman spectroscopy", and "cultural heritage", illustrate the application of AI in cultural heritage preservation and analysis. In Cluster 0, AI algorithms such as deep learning and convolutional neural networks are used for image recognition in cultural heritage, like remote sensing archaeology and ceramic detection. In Cluster 1, Raman spectroscopy combined with deep learning is applied to pigment analysis and paintings conservation. Trends in this field indicate a shift from the digitization of individual objects to the integrated recording of complex scenes and large-scale heritage. This evolution prioritizes transparent visualization and the long-term accessibility of data (Borissova, 2018). For example, Guadalupe developed a cross-media prototype to provide users with an interactive experience that addresses the low number of visitors to the Afro-Peruvian National Museum and preserves the Afro-Peruvian musical heritage (Guadalupe, 2018).

- AI-Enhanced Immersive Cultural Heritage Experiences

This hotspot emphasizes leveraging AI to create immersive experiences in cultural heritage. Clusters 3 and 8, with keywords like "augmented reality", "virtual reality", "experiential learning", and "ecological forecasting", are key in this area. In Cluster 3, augmented reality and virtual reality technologies are used in educational technology and serious games to provide immersive cultural heritage learning experiences, such as virtual museum tours. Cluster 8 combines ecological and cultural elements, potentially using AI to create immersive virtual environments that blend cultural heritage with natural landscapes for a more holistic experience. Initially, these experiences mainly relied on basic virtual reality technology for simple visual presentations. But currently, with the help of AI, there is a shift towards more personalized and interactive experiences. Current trends indicate that the concept of the metaverse may further drive innovation in blending real and virtual experiences(Sepe et al., 2024). Specifically, the metaverse offers a transformative virtual realm for large-scale interactions that can significantly influence tourism, particularly in the context of cultural heritage (Choi et al., 2023).

- AI-Facilitated Interdisciplinary Research in Cultural Heritage

This hotspot highlights the intersection of AI, digital technology, and various disciplines in cultural heritage research. Clusters 4, 5, and 6, with terms like "natural language processing", "ethnic identity", "archaeology", and "pastoralism", represent this interdisciplinary approach. In Cluster 4, natural language processing and large language models are combined with energy - saving research in the context of cultural heritage buildings, which may involve analyzing historical texts about building design for energy - efficient restoration. Cluster 5 explores the relationship between ethnic identity, language use, and cultural heritage, where AI can be used to analyze language data to understand cultural heritage preservation among different ethnic groups. Cluster 6 uses AI in archaeological research, such as genomic history analysis. Early research in this area was mainly about data collection from different disciplines. Now, with AI, there is a trend towards more in - depth data integration and analysis. AI can integrate data from archaeology, genetics, and linguistics to provide a more comprehensive understanding of cultural heritage. There is a paradigm shift underway: the initial non-embodied paradigm of communication, which prioritized consciousness over materiality, is gradually transforming into an embodied paradigm that highlights the significance of the medium itself (Hilditch, 1995).

Figure 10. Clustering map of international research

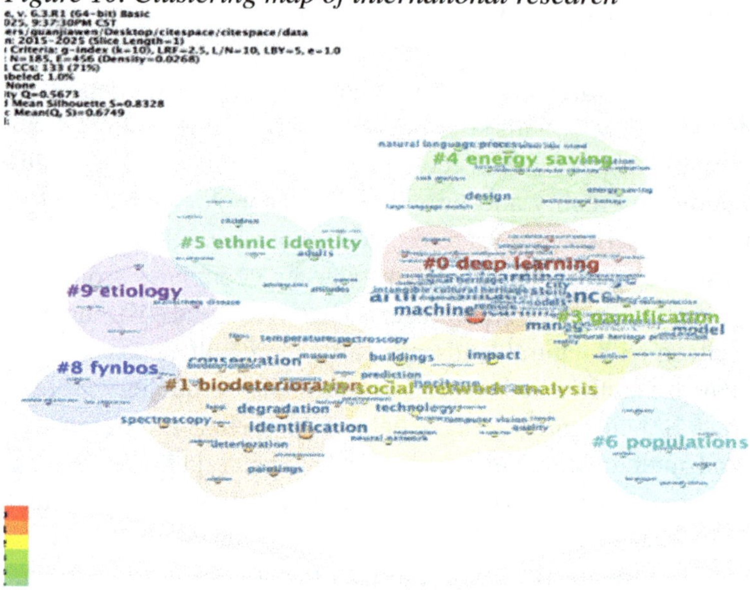

Source: Author

Analysis of foreign research development paths

Figure 11. Timeline graph internationally

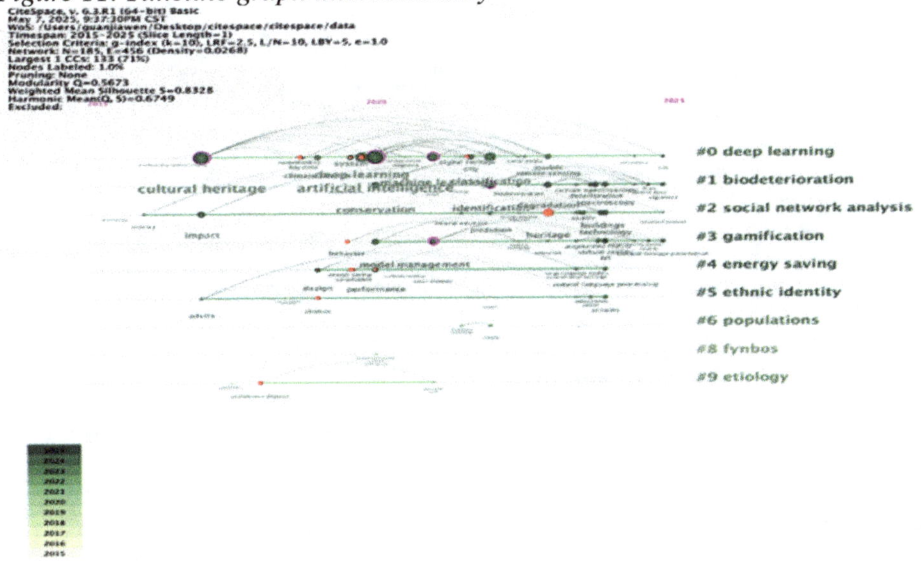

Source: Author

Taking the development time as the X-axis and the clustering theme as the Y-axis, the theme is transformed into a timeline graph. This paper divides the path research of foreign cultural heritage into three stages.

Figure 12. Threes stages of international cultural heritages

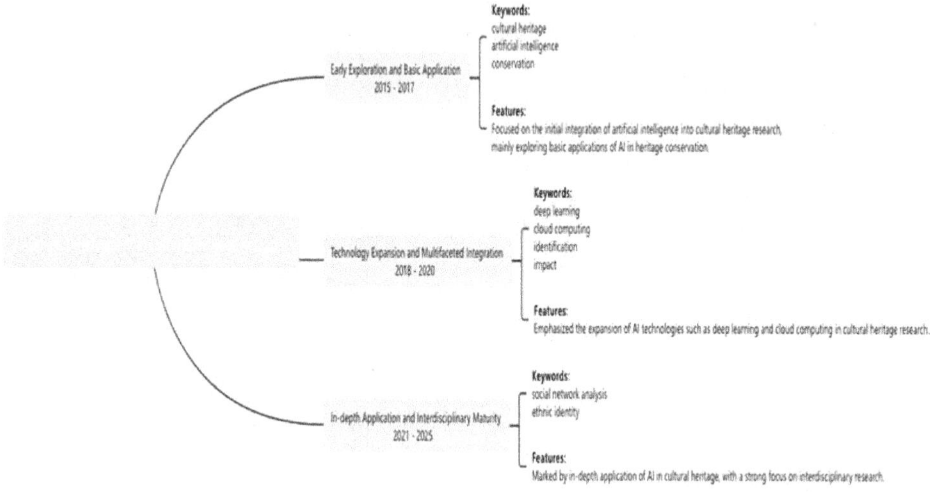

Source: Author

Summary

In summary, research on AI - enabled cultural heritage, both at home and abroad, has established three main paradigms: AI technology application, policy - driven strategies, and AI - enhanced experience design. Nevertheless, the current research has significant limitations. First, the majority of research results focus on a particular area (such as the application of AI in technical production or its dissemination effects), lacking a comprehensive analytical framework. Second, there is a scarcity of empirical evidence on how AI - driven digitization of cultural heritage transforms its social functions in non - Western settings. These shortcomings highlight the pressing need for a systematic theoretical model to explain how AI technologies evolve from mere tools to promoting "living" governance of cultural heritage.

CASE STUDY

Forbidden City: Digitally Enabling Cultural Heritage

Despite the significant advancements in the application of AI within the cultural heritage sector, much of the current research remains primarily concentrated on the effects of AI usage or one-way communication. There exists a noticeable deficiency in systematic examinations that encompass the entire "production-communication-identity" continuum. This gap is particularly pronounced in non-Western contexts, where the pressing theoretical and practical question of how AI can transform the social functions of cultural heritage, promote meaningful public engagement, and contribute to the development of a lasting cultural identity still requires thorough exploration.

In this context, the AI-driven initiatives of Forbidden City in Beijing present an exemplary case study that addresses this research gap. As China's most prominent cultural heritage institution, Forbidden City has undergone significant digital transformation over the past two decades. This transformation spans from the establishment of AI-enabled technological infrastructure to the development of innovative dissemination approaches. It has enabled a paradigm shift from "static conservation" to "living inheritance" through diverse AI-powered interactions. Thus, this study focuses on the Forbidden City, aiming to uncover how AI systematically enhances the reproduction, cross-media communication, and construction of public identity regarding cultural heritage.

Brief description of the research object

As the world's largest and best-preserved complex of ancient wooden structures, the Forbidden City covers an area of 720,000 square meters, with a floor area of about 150,000 square meters, more than seventy palaces of various sizes and more than 9,000 houses. In 1961, it was listed by the State Council as one of the first batch of national key cultural relics protection units, and in 1987, it was included in the List of World Cultural Heritage.

Figure 13. Map of Forbidden City

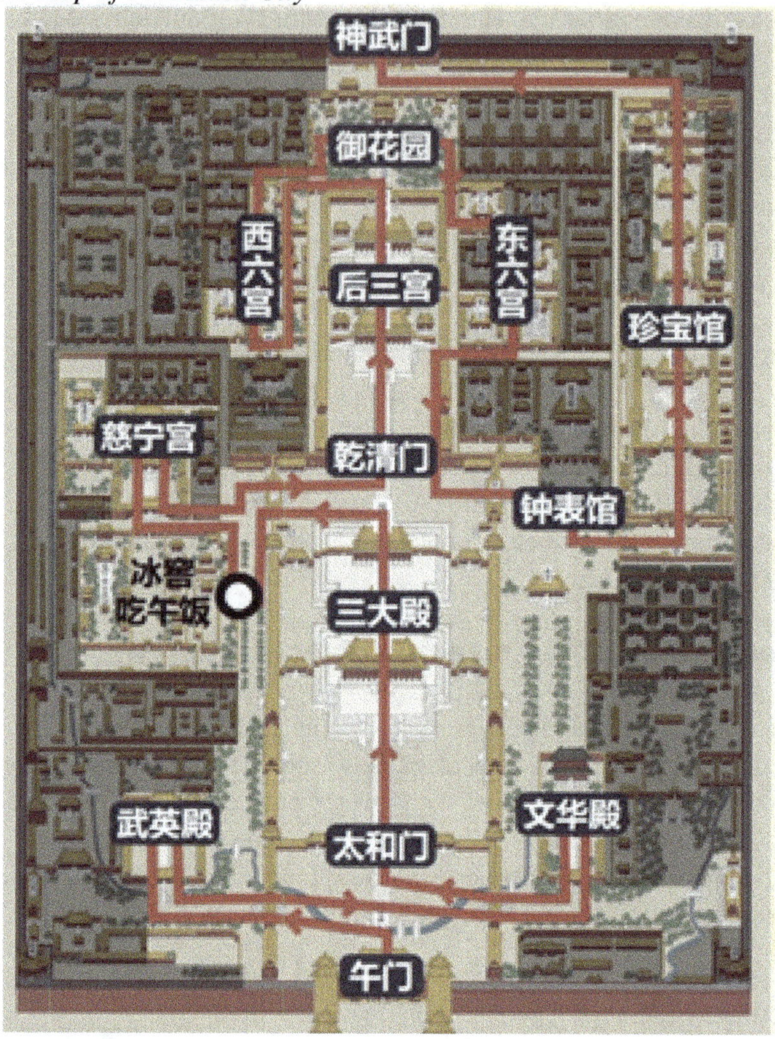

Source: www.dpm.org.cn

Table 5. Main function of each building

Chinese name	English name	Main Function
午门	Wumen	For grand ceremonies like imperial edict announcements and military victory celebrations
太和门	Taihemen	Ming emperors held morning audiences

continued on following page

Table 5. Continued

文华殿	Wenhua Dian	Served as the crown prince's office in the Ming Dynasty and was later used for Confucian classics lectures in the Qing Dynasty
太和殿	Taihe Dian	The most important ceremonial hall, used for coronations, imperial weddings, and grand celebrations.
中和殿	Zhonghe Dian	A resting place for the emperor before ceremonies; also used for reviewing ritual documents.
保和殿	Baohe Dian	Hosted banquets in the Ming Dynasty and became the venue for the imperial examination's final stage (Palace Examination) in the Qing Dynasty.
武英殿	Wuying Dian	Used as the emperor's study and meeting place in the Ming Dynasty; later became a book compilation center.
乾清门	Qianqingmen	The main entrance to the Inner Court, where Qing emperors held daily political meetings.
慈宁宫	Cining Gong	Residence of empress dowagers and imperial concubines; now houses the Palace Museum's Sculpture Gallery.
钟表馆	Clock Exhibition Hall	Displays a collection of exquisite Western and Chinese clocks from the imperial collection.
西六宫	Western Six Palaces	Living quarters for imperial concubines.
乾清宫	Qiaqnqing Gong	The emperor's residence in the Ming and early Qing Dynasties; later used for official affairs.
交泰殿	Jiaotai Dian	The empress received tributes here, and imperial seals were stored in this hall.
坤宁宫	Kunning Gong	The empress's residence in the Ming Dynasty; converted into a Shamanist worship space and the emperor's wedding chamber in the Qing Dynasty.
东六宫	Eastern Six Palaces	Another residential area for imperial concubines; now partially used for exhibitions, including the Treasure Gallery.
珍宝馆	Treasure Gallery	Exhibits rare imperial treasures, including gold artifacts, jade carvings, and precious jewelry.
御花园	Yuhua Yuan	A private garden for the imperial family, featuring pavilions, ancient trees, and decorative rockeries.
神武门	Shenwumen	The northern exit of the Forbidden City, used for imperial guard movements and the selection of concubines; now serves as the museum's exit for visitors.

Source: www.dpm.org.cn

As a significant embodiment of Chinese civilization, the cultural heritage value of the Forbidden City in Beijing can be interpreted through both tangible and intangible dimensions. In terms of tangible cultural heritage, this royal complex exemplifies the exceptional universal value of the "complex" category of heritage as outlined in the Convention for the Protection of the World Cultural and Natural Heritage. On the level of intangible cultural heritage, the Forbidden City serves as a repository of cultural memory, acting as a living legacy. According to the UNESCO Convention for the Safeguarding of the Intangible Cultural Heritage, traditional crafts such as the "official building techniques of the Forbidden City" and "bronze restoration techniques," along with court festivals and customs related to the twenty-four solar terms, as well as academic traditions centered on the study of cultural relics, together form a dynamic system of cultural practices.

Table 6. Process of digital Forbidden City

Application Area	AI Technology/Project	Function & Impact
Digital Preservation	High-resolution imaging & restoration	AI-assisted scanning of 100,000 artifacts for digital archives and restoration support.
	3D modeling (e.g., ceiling structures)	Millimeter-level precision for structural analysis and preventive conservation.
Visitor Experience	VR tours (e.g., virtual Qianlong's study)	Immersive exploration of restricted areas with historical reconstructions.
	AI-guided robots	Personalized tour routes and interactive Q&A about artifacts.
	Interactive games (e.g., voice assistant content)	AI-generated games and cultural stories to enhance engagement.
Heritage Protection	Damage detection system	AI analyzes scans to identify wood deformation, paint cracks, etc.
	Environmental monitoring	Real-time risk alerts for temperature, humidity, and pests.
Cultural Innovation	AI-designed stationery	AI generates modern designs inspired by traditional patterns.
	Interactive drama Dreaming of the Forbidden City	AI-driven multi-ending stories about palace construction history.
	Facial recognition & painting matching (e.g., Along the River During Qingming Festival)	Visitors' selfies matched with historical figures in artworks.

Source: Author

Figure 14. Panoramic Palace Museum Website

Source: https://pano.dpm.org.cn

Production: from technology empowerment to co-creation

In the journey of enhancing cultural heritage with AI, the Palace Museum has undergone a significant transformation in its production approach, evolving from mere digital empowerment to AI-driven collaborative creation involving the community. Initially, the digitalization efforts at the museum primarily focused on the preservation of information. For example, during the database management process, advanced imaging and restoration technologies facilitated AI-assisted scanning of 100,000 cultural relics for digital archiving and restoration support. Additionally, 3D modeling was utilized to achieve millimeter-level precision in structural analysis and preventive maintenance. During this phase, technological advancements were propelled by a single force, led by professional institutions, while the public played a passive role in receiving the outcomes. This approach was characterized by the digital replication and fundamental maintenance of cultural artifacts.

With the advancement of AI technology, the production model has undergone notable changes. In terms of enhancing the tourist experience, innovations such as VR tours, AI tour guide robots, and interactive games employ AI to deliver immersive, personalized, and highly engaging experiences, thereby transforming tourists from passive observers into active participants.

Today, the Palace's digital production is advancing toward a stage characterized by "co-creation and sharing." The introduction of the "Palace Community" platform in 2019 holds considerable significance, allowing users not only to browse digital resources but also to engage in the interpretation of cultural relics and creative design. Statistics show that the platform has accumulated over 230,000 pieces of original user-generated content. President Wang Xudong has expressed that "the future of the Forbidden City will be a shared cultural creation platform for 1.4 billion people." In addition, the Palace Museum collaborates with multiple parties across various fields, using AI as a link to integrate resources, achieving the fusion of technology, culture and creativity, and building an ecosystem of multi-party co-creation. In conclusion, the Palace Museum has achieved a transformation in the production of cultural heritage from static preservation to dynamic creation, and from being solely handled by professional institutions to being co-created by multiple parties in society with the help of AI.

This evolutionary path reveals three deep changes: in the main body of production, from the monopoly of professional organizations to the common participation of society; in the creation of value, from the preservation of information to the co-creation of meaning; and in the logic of technology, from instrumental rationality to the construction of ecology. The practice of the Forbidden City shows that artificial intelligence has not only changed the mode of production of cultural heritage, but also reconstructed the way of production of cultural heritage.

Communication: a return to culture-based value rationality

Freeman Tilden, in his book Interpreting Our Heritage (Tilden, 1967), emphasizes that cultural heritage institutions are not solely responsible for collecting and preserving both tangible and intangible heritage; they also play a vital role in disseminating knowledge in engaging ways. The advent of AI has expanded the channels and scope of communication, transforming cultural heritage communication into a non-linear, scalable, immersive, and collaborative endeavor. This evolution reflects a growing trend towards transmedia narratives that embrace multiculturalism and reach a broader audience (Zhang Bo, 2024). Within this framework, the traditional role of the communicator has been redefined, giving way to a pluralistic and equitable model of multidirectional information dissemination that is becoming increasingly mainstream.

The fundamental principle of AI should consistently align with the preservation of cultural heritage. O'Connor (2007) once reminded digital projects to avoid falling into "technology solutionism" or economic supremacy logic. The application of technology to enhance cultural heritage must not distort its intended purpose but should instead return to its foundational essence (Zhang Bo, 2024). This concept of culture-based communication can be observed in three key areas.

The empowerment of cultural heritage through AI should first and foremost prioritize the unique identity of that heritage. The intrinsic characteristics of cultural heritage must remain intact and not be altered by algorithm-driven symbols or subjected to excessive quantification. For example, the "Daily Forbidden City" app takes a manual approach in selecting cultural relics for its explanations, effectively steering clear of biases that can stem from algorithmic recommendations and ensuring the preservation of the heritage's uniqueness and integrity.

Secondly, AI technology must strive to restore the authenticity of cultural heritage. While the digital realm presents virtual scenes, the data captured can be selectively edited. Therefore, the mimetic representations created through AI technology should aim to uphold the individual authenticity of cultural heritage to the fullest extent possible (Liu & Babar, 2022). The "Digital Duo Bao Ge" project employs millimeter-level 3D scanning to create digital replicas of cultural relics, ensuring that the traces of restoration and aging are preserved. The technical team deliberately avoids "over-restoration" in its virtual representations.

The idea of "digital refugees" underscores the challenges encountered by marginalized and disadvantaged groups within the digital landscape. It is crucial for the communication process to recognize the disparities that exist between these marginalized communities and those who sit at the center of the digital world. Actively adapting to and empowering disadvantaged groups to articulate their needs is essential for fostering inclusivity. This initiative focuses on developing AI application solutions that prioritize humanism and compassionate care, while also transforming the digital public cultural service system to be more inclusive and sensitive to the needs of all age groups. The goal is to establish a more comprehensive communication chain. For example, the Forbidden City has created a specialized version of its app tailored for elderly users, utilizing AI algorithms to analyze their usage habits and needs. This approach streamlines operational processes and incorporates voice-guided navigation, making the app more user-friendly and enhancing the overall experience. Additionally, the Duanmen Digital Museum has introduced an AI-enhanced tactile interactive device, allowing visually impaired visitors to engage with a "touchable Forbidden City." This innovative device enables them to explore the richness of cultural heritage through touch, offering a more intuitive and immersive experience. By 2022, these initiatives have reached over 120,000 visits from special groups, illustrating a significant shift towards inclusivity in technological empowerment (Yisu, 2024).

Identity: a mechanism for participatory cultural construction

The construction of cultural heritage identity can be viewed as a process through which various groups define themselves in relation to cultural heritage, leading to the formation of diverse identities in the context of cultural heritage digitization (Ping, 2025). From the perspective of subject-object unity, Zhang (2022)suggest that the essence of cultural heritage identity encompasses both the subject and the object. In this framework, the object represents the inherent identity characteristics assigned to cultural heritage, while the subject encompasses people's cognition, emotions, evaluations, and behaviors related to it (Zhang Chaozhi, 2022).

For traditional culture to gain wider acceptance and enjoyment among the public, it is essential to establish and maintain a stable cultural identity. Although AI has transformed the format of cultural heritage from offline physical spaces to online virtual environments, it has not altered the intrinsic public value of cultural heritage nor diminished the societal pursuit of cultural identity building, which remains a fundamental function of cultural heritage (De Bari, 2024). This represents not only the fundamental approach to transcending the differences in the cultural expressions of heritage spaces but also serves as the essential foundation for the AI construction of cultural heritage.

In terms of tangible assets, the Forbidden City has redefined the embodiment of cultural heritage through its "Digital Cultural Relics Library" system, aided by AI technology. Utilizing AI-powered image recognition and high-precision scanning techniques, the institution has achieved high-definition digitization of 1.86 million collection items. Furthermore, AI algorithms are employed to generate innovative "digital identity labels" for these cultural relics. These labels not only provide essential information about the artifacts but also analyze and integrate data regarding their historical context, artistic styles, and cultural significance. Consequently, what were once static artifacts are transformed into dynamic, traceable nodes of cultural memory. This thorough digitization not only preserves the material aspects of these items but also establishes a robust and objective foundation for cultural identity.

In addition, the Palace has reconfigured the mechanism for public engagement in identity construction via its "digital community" platform, with AI playing a crucial role. AI-based recommendation engines on this platform are designed to match users with cultural relics that align with their interests, thereby encouraging participation. The introduction of a "heritage interpretation crowdfunding" function is also enhanced by AI. Natural language processing algorithms analyze and categorize the 370,000 user-contributed interpretations of cultural relics, making it easier to manage and display these diverse perspectives.

The experience of the Forbidden City vividly illustrates that when digital technology, especially AI, is intricately integrated into the cultural fabric, it has the remarkable ability to nourish cultural identity and safeguard the core values of culture in the midst of technological transformation.

CONCLUSION

This study employs a literature visualization analysis using CiteSpace to examine the role of artificial intelligence in advancing cultural heritage research domestically and internationally from a theoretical perspective. Findings indicate a dual focus in both contexts on technology-driven initiatives and humanistic orientations. Domestic research emphasizes systematic protection and adaptive utilization of cultural heritage within a policy framework, advocating for the integration of artificial intelligence to transition from static preservation to dynamic inheritance. Conversely, international research prioritizes technology refinement and enhanced user experiences, utilizing artificial intelligence to create personalized cultural heritage encounters.

A notable case is the digital transformation of the Palace Museum, which provides key insights into global cultural heritage protection. Successful digitization requires collaborative innovation across three dimensions: production, dissemination, and identity construction. The production dimension has seen a shift from mere technological replication to collaborative co-creation, as exemplified by the "Forbidden City Community" platform, which fosters public engagement in cultural relic interpretation and design. In terms of communication, the Palace Museum ensures the authenticity of cultural heritage through careful screening of interpretive content and high-precision 3D scanning. It also develops tailored applications to enhance inclusivity. For identity construction, digital tools are utilized to create "digital identity tags" for cultural relics, while AI helps facilitate public participation in defining cultural heritage identities.

This study broadens the theoretical framework of digital research on cultural heritage and proposes a model with practical application value. As technology progresses, artificial intelligence will continue to innovate the protection of cultural heritage while emphasizing the need for sustainable development that prioritizes the preservation of cultural values in the digital age.

REFERENCES

Aronsson, I.-L., & Price, S. (2024). Culture, heritage, memory: Toward a resonant cultural solution for resettlement. *Human Organization, 83*(3), 216–227. DOI: 10.1080/00187259.2024.2307845

Barceló, J., Del Castillo, F., Kayikci, D., & Urbistondo, B. (2022). Neural networks for archaeological classification and typology: An overview. *PCA European Journal of Post-Classical Archaeologies, 12*, 7–32.

Bo, Z. H. F. (2024). Digital technology empowers public cultural services: value presupposition, blocking factors and method adjustment. Study&Exploration(09), 56-62. https://kns.cnki.net/kcms2/article/abstract?v=qKY3Y0De9e7i23-aJAdLH9qxTsP_RFV_uaiLLcGh8UCzXLyS0v8lQLdyp7krbVKjW2r2mlDpR_52vDls6frdGJyGAZIsW0e_IJzG4p-KZzKEd6k5FgzEyiL7XdG4EfRypPjX1AH6ewyNbkv4xYINPHdhuEx4-wHj-8FAsU_26lSlm1nWN5Ri-pCqlZ3f4ZTnfjfLMgiwUpQ=&uniplatform=NZKPT&language=CHS

Borissova, V. (2018). Cultural heritage digitization and related intellectual property issues. *Journal of Cultural Heritage, 34*, 145–150. DOI: 10.1016/j.culher.2018.04.023

Buragohain, D., Meng, Y., Deng, C., Li, Q., & Chaudhary, S. (2024). Digitalizing cultural heritage through metaverse applications: Challenges, opportunities, and strategies. *Heritage Science, 12*(1), 295. Advance online publication. DOI: 10.1186/s40494-024-01403-1

Casillo, M., Colace, F., Gupta, B. B., Lorusso, A., Santaniello, D., & Valentino, C. (2024). The role of AI in improving interaction with cultural heritage: An overview. Handbook of Research on AI and ML for Intelligent Machines and Systems, 107-136.

Chen, C. (2005). CiteSpace II: Detecting and visualizing emerging trends and transient patterns in scientific literature. *Journal of the American Society for Information Science and Technology, 57*(3), 359–377. DOI: 10.1002/asi.20317

Choi, M., Choi, Y., Nosrati, S., Hailu, T. B., & Kim, S. (2023). Psychological dynamics in the metaverse: Evaluating perceived values, attitude, and behavioral intention in metaverse events. *Journal of Travel & Tourism Marketing, 40*(7), 602–618. DOI: 10.1080/10548408.2023.2276435

Colavizza, G., Blanke, T., Jeurgens, C., & Noordegraaf, J. (2021). Archives and AI: An overview of current debates and future perspectives. [JOCCH]. *ACM Journal on Computing and Cultural Heritage, 15*(1), 1–15.

Croce, V., Caroti, G., De Luca, L., Piemonte, A., & Véron, P. (2023). Neural radiance fields (nerf): Review and potential applications to digital cultural heritage. *The International Archives of the Photogrammetry, Remote Sensing and Spatial Information Sciences, 48*, 453–460. DOI: 10.5194/isprs-archives-XLVIII-M-2-2023-453-2023

De Bari, M. (2024). A study for innovation of methodologies and processes in digitisation and digital creation for improving the interaction with the Cultural Heritage. *Procedia Computer Science, 237*, 187–194. DOI: 10.1016/j.procs.2024.05.095

Gîrbacia, F. (2024). An Analysis of Research Trends for Using Artificial Intelligence in Cultural Heritage. *Electronics (Basel), 13*(18), 3738. DOI: 10.3390/electronics13183738

Guadalupe, A. I. S. (2018). User-Centered Design to a Digital Cultural Heritage Installation Based on Afro-Peruvian Musical Legacy. 2018 IEEE Games, Entertainment, *Media Conference (GEM)*

Guo Wei, H. R. (2022). Research Characteristics and Trends of Cultural and Creative Product Design Based on Bibliometrics. *PACKAGING ENGINEERING, 43*(14), 326–334. DOI: 10.19554/j.cnki.1001-3563.2022.14.040

Hilditch, D. J. (1995). *At the heart of the world: Merleau-Ponty and the existential phenomenology of embodied and embedded intelligence in everyday coping*. Washington University in St. Louis.

Holtorf, C. (2018). Embracing change: How cultural resilience is increased through cultural heritage. *World Archaeology, 50*(4), 639–650. DOI: 10.1080/00438243.2018.1510340

Ke, X. L. J. (2020). Digital Communication of Chinese Intangible Cultural Heritage: New Challenges and New Solutions. Cultural Heritage(01), 140-146. https://kns.cnki.net/kcms2/article/abstract?v=qKY3Y0De9e4T346Jsn9f2RN88NfWfhsdlGTfxWnYDbDYsz_DxIGNPB9d6lyriH7bSgTbOcOCUwaY5VrtKOAKZC7VECmpfbYCm7TJSCQrI7qwOsXWGdYJIP2G1q8TwZRhaIUZsNWGUjBa8XUsJGeSmL14aXKsr8cu6-tAH0j7B3cPvqcyecnR5G73kdL_gmSc__GZ3fMXH48=&uniplatform=NZKPT&language=CHS

Liu, J., & Babar, M. (2022). Digitally Protecting and Disseminating the Intangible Cultural Heritage in Information Technology Era. *Mobile Information Systems, 2022*, 1–10. DOI: 10.1155/2022/1115655

Liu, Z. (2022). Literature review on image restoration. *Journal of Physics: Conference Series*.

Mishra, M. (2021). Machine learning techniques for structural health monitoring of heritage buildings: A state-of-the-art review and case studies. *Journal of Cultural Heritage*, *47*, 227–245. DOI: 10.1016/j.culher.2020.09.005

O'Connor, J. (2007). *The cultural and creative industries: a review of the literature.* Arts Council England.

Ping, M. (2025). *Digitalization of cultural heritage and construction of cultural identity*. Jiangsu Social Sciences., DOI: 10.13858/j.cnki.cn32-1312/c.20250319.004

Portalés, C., Rodrigues, J. M. F., Rodrigues Gonçalves, A., Alba, E., & Sebastián, J. (2018). Digital Cultural Heritage. *Multimodal Technologies and Interaction*, *2*(3), 58. Advance online publication. DOI: 10.3390/mti2030058

Rallis, I., Voulodimos, A., Bakalos, N., Protopapadakis, E., Doulamis, N., & Doulamis, A. (2020). Machine learning for intangible cultural heritage: a review of techniques on dance analysis. Visual Computing for Cultural Heritage, 103-119.

Soleymani, A., Jahangir, H., & Nehdi, M. L. (2023). Damage detection and monitoring in heritage masonry structures: Systematic review. *Construction & Building Materials*, *397*, 132402. DOI: 10.1016/j.conbuildmat.2023.132402

Throsby, D. (2018). Art, politics, money: Revisiting Australia's cultural policy. Platform papers(55).

Tilden, F. (1967). *Interpreting our Heritage. Chapel Hill.* North Carolina Press.

Trivyzadakis, N. (2023). A tool for quantitative research on the social awareness for the digitization of cultural heritage: A pilot application. *Journal of Archaeological Science, Reports*, *51*, 104202. Advance online publication. DOI: 10.1016/j.jasrep.2023.104202

Yang, S., Hou, M., & Li, S. (2023). Three-dimensional point cloud semantic segmentation for cultural heritage: A comprehensive review. *Remote Sensing (Basel)*, *15*(3), 548. DOI: 10.3390/rs15030548

Yisu, S. (2024). Evolution iteration and development direction of audio visual communication in the Palace Museum in the "digital intelligence era". (05), 52-55. DOI: 10.13994/j.cnki.stj.2024.05.028

Zhang Chaozhi, Y. J. (2022). Construction and negotiation: Rethinking the relationship between cultural heritage and Tourism. Tourism Tribune/Lvyou Xuekan, 37(11).

Zhao, M., Wu, X., Liao, H.-T., & Liu, Y. (2020). Exploring research fronts and topics of Big Data and Artificial Intelligence application for cultural heritage and museum research. *IOP Conference Series. Materials Science and Engineering*, *806*(1), 012036. DOI: 10.1088/1757-899X/806/1/012036

APPENDIX

Table 7.

	Pinyin	Meaning
文化遗产	Wénhuà yíchǎn	Cultural Heritage
人工智能	Réngōng zhìnéng	Artificial Intelligence (AI)
数字人文	Shùzì rénwén	Digital Humanities
保护	Bǎohù	Preservation / Protection
数字技术	Shùzì jìshù	Digital Technology
传承	Chuánchéng	Inheritance / Transmission
博物馆	Bówùguǎn	Museum
虚拟现实	Xūnǐ xiànshí	Virtual Reality (VR)
图书馆	Túshūguǎn	Library
元宇宙	Yuán yǔzhòu	Metaverse
知识图谱	Zhīshì túpǔ	Knowledge Graph
数字资源	Shùzì zīyuán	Digital Resources
文旅融合	Wénlǚ rónghé	Cultural and Tourism Integration
午门	Wǔmén	Meridian Gate (Forbidden City)
太和门	Tàihémén	Gate of Supreme Harmony
文华殿	Wénhuá diàn	Hall of Literary Brilliance
太和殿	Tàihé diàn	Hall of Supreme Harmony
中和殿	Zhōnghé diàn	Hall of Central Harmony
保和殿	Bǎohé diàn	Hall of Preserving Harmony
武英殿	Wǔyīng diàn	Hall of Martial Valor
乾清门	Qiánqīngmén	Gate of Heavenly Purity
慈宁宫	Cíníng gōng	Palace of Compassion and Tranquility
钟表馆	Zhōngbiǎo guǎn	Clock Exhibition Hall
西六宫	Xī liù gōng	Western Six Palaces
乾清宫	Qiánqīng gōng	Palace of Heavenly Purity
交泰殿	Jiāotài diàn	Hall of Union
坤宁宫	Kūnníng gōng	Palace of Earthly Tranquility
东六宫	Dōng liù gōng	Eastern Six Palaces
珍宝馆	Zhēnbǎo guǎn	Treasure Gallery
御花园	Yùhuā yuán	Imperial Garden
神武门	Shénwǔmén	Gate of Divine Prowess

continued on following page

Table 7. Continued

数字故宫	Shùzì gùgōng	Digital Palace Museum
故宫社区	Gùgōng shèqū	Palace Museum Community (online platform)
数字文物库	Shùzì wénwù kù	Digital Artifact Library
数字标签	Shùzì biāoqiān	Digital Tag / Metadata
数字传播	Shùzì chuánbō	Digital Communication

Source: Author

Chapter 8
Digital Art Museums in the Digital Transformation of Ice and Snow Cultural Tourism:
The Role, Development Trajectory, and Case Study of the Harbin Ice and Snow World Digital Art Museum

Poshan Yu
http://orcid.org/0000-0003-1069-3675
Soochow University, China

Yupian Zhang
http://orcid.org/0009-0001-5580-0684
Independent Researcher, China

ABSTRACT

This paper explores the digital transformation of regional ice and snow cultural tourism, focusing on the role of digital art museums. Using the Harbin Ice and Snow World's Digital Art Museum as a case study, it analyzes policy contexts and applies CiteSpace to visualize 2015–2025 literature, identifying research trends. Findings highlight digital art museums' advantages in technical display, education, and cultural communication: Harbin's museum uses AIGC to enhance tourist experiences

DOI: 10.4018/979-8-2600-2378-5.ch008

and address traditional cultural expression gaps. Grounded in innovation diffusion theory, the study demonstrates digital innovation's mass acceptance and provides theoretical and practical insights for ice and snow tourism's digital shift, urging further exploration of project-specific transformations and innovative models.

INTRODUCTION

A digital art museum is a place that uses digital technologies to showcase artworks. It presents artworks in a digital manner, encompassing forms such as, but not limited to, virtual museums and digital art exhibitions. It can be a digital extension of a physical venue or a completely web-based virtual space. By leveraging technologies like 3D modeling, virtual reality (VR), augmented reality (AR), and artificial intelligence, it provides audiences with a unique artistic experience.

From the perspective of technology and display forms, a digital art museum is a venue that utilizes digital technologies to display artworks. It presents artworks digitally, including but not restricted to virtual museums and digital art exhibitions. It can either be a digital extension of a physical venue or a purely web-based virtual space. Virtual museums make use of various visualization technologies to combine artworks with related image archives, enabling audiences to explore artworks interactively, which reflects the characteristics of digital art museums in terms of technological application and display forms (Meinecke et al., 2022).

Digital art museums also play a crucial role in education. They can provide abundant resources for art education, support teaching activities carried out by students and teachers, and facilitate students' understanding of art and the improvement of their artistic skills. Digital painting courses conducted in a museum setting enhance the social and emotional skills of teenagers through artistic creation, indicating that digital art museums can serve as an educational platform to promote the development of art education (Kastner et al., 2021).

What is most important for the tourism industry is that digital art museums attach great importance to the audience experience and interaction. By providing diverse interactive methods, they enhance the connection between the audience and artworks. In the digital art experience, digital art museums can improve the audience experience by optimizing the way information is presented (Castellotti et al., 2024). Digital art museums can offer audiences the opportunity to participate in art activities remotely, promoting discussion and exploration. Resources like "Look, Think, Do" encourage audiences to discuss and explore around museum objects (Noble, 2021a).

Currently, cultural tourism promotion emphasizes the combination of culture and tourism. Digital art museums, as innovative carriers of cultural dissemination, can break through geographical limitations, spread art and culture, and promote cultural exchange and understanding. At the same time, they can also build communities, enabling audiences to establish connections and share artistic experiences in virtual spaces. During the pandemic, many museums provided services to the public through digital platforms, creating digital learning resources and activity packages, which promoted cultural exchange within the community. This reflects the role of digital art museums in cultural dissemination and community building (Noble, 2021b).

After a thorough review of the development history of digital art museums, their applications in cultural tourism, and the current status of related research, it is found that although digital art museums have demonstrated many advantages in technical display, education, cultural dissemination, etc., and their applications in the cultural tourism field are becoming increasingly widespread, there are still some key issues that need further exploration.

On the one hand, from the perspective of the digital transformation of cultural tourism at a macro level, although the integration of digital technology and the cultural tourism industry has become a research hotspot, there is currently no clear and systematic understanding of the specific models for achieving digital innovation and transformation through digital art museums. Existing research has not delved deeply enough into key aspects such as how digital art museums can be deeply integrated with the cultural tourism industry and in what ways they can promote the innovative development of cultural tourism, and there is a lack of summaries of operable models.

On the other hand, focusing on the specific field of ice and snow cultural tourism, Harbin Ice and Snow World, as an important representative of ice and snow cultural tourism, faces challenges in cultural expression and the improvement of the tourist experience during the process of digital transformation. At the same time, the research on the application of the innovation diffusion theory in the digitalization of ice and snow cultural tourism is still insufficient, and its potential in promoting the dissemination and application of digital innovation achievements in ice and snow cultural tourism has not been fully tapped.

Based on the above-mentioned deficiencies in the current research status, in order to further deepen the understanding of the digital transformation of ice and snow cultural tourism and promote the effective application of digital art museums in this field, this study proposes the following key questions:

1. Through the analysis of empirical literatures by CiteSpace, what are the key models for achieving digital innovation and transformation by applying digital art museums in cultural tourism?

2. In the case study of Harbin Ice and Snow World, how can AIGC (Artificial Intelligence Generated Content) help with the digital transformation of ice and snow cultural tourism?
3. What is the digital revitalization model of ice and snow cultural tourism under the innovation diffusion theory?

This study will conduct a visual analysis through CiteSpace to sort out the relevant literatures on digital art museums from 2015 to 2025. It will also focus on the collaborative relationship between the development and transformation of digital art museums in China and the innovation of cultural tourism in China. Moreover, through the specific case of the Ice and Snow Digital Art Museum in Harbin Ice and Snow World, the study will explore the topics of integrating digital art museums into cultural promotion and driving the development of the tourism industry through digitalization, so as to answer the above questions.

LITERATURE REVIEW

Research Background of Digital Art Museums

In the 1960s, early computerized library systems emerged, mainly used for the automated management of libraries, such as library cataloging and borrowing management, laying the technical foundation for the development of digital venues (Lischer-Katz, 2022). In the same period, art museums began to use computers to assist in organizing, cataloging, and coordinating their collections. During this time, computer art was also introduced in countries like Sweden, and digital art started to take root. For instance, since then, digital art in Sweden has been continuously developing, and digital technologies, methods, and themes have gradually become part of the Swedish art scene. In 1974, the Stedelijk Museum in Amsterdam, the Netherlands, hired the first time-based art curator, Dorine Mignot, becoming a pioneer in this field. During this period, the attention paid by museums to time-based art laid the foundation for the subsequent development of digital art (Braden & Teekens, 2020).

In the early 1990s, digital technology began to be widely applied in the cultural field. In 1990, the Library of Congress in the United States launched the "American Memory" project, initiating the construction of the digital library project in the United States. In 1992, UNESCO launched the "Memory of the World" program, which aimed to rescue and protect precious archival and documentary heritage on a global scale (Fabbri & Pretelli, 2014). During the same period, European museums such as the Louvre Museum in Paris, France, and the British Museum in the United Kingdom also began their digitalization construction.

Entering the 21st century, with the popularization of the Internet and the rapid development of digital technology, digital art museums, as a branch of digital museums, also began to rise rapidly. Most museums have improved their websites, updated their archives, promoted virtual tours, or digitized their collections into increasingly open-access databases. However, although these transformations have changed the appearance of many museums to varying degrees, the persistence of the "physical" still largely determines the form of the so-called digital art museums, which to some extent hinders their potential to fully embrace effective virtual transformations.

In December 2005, the China Digital Science and Technology Museum project jointly constructed by the China Association for Science and Technology, the Ministry of Education, and the Chinese Academy of Sciences was officially launched. It utilized network multimedia technology and virtual reality technology to display scientific knowledge in various forms (Ch'ng et al., 2019). The Shaanxi Digital Museum, which was initially built in 2011 and went online in August 2012, opened characteristic columns such as a virtual reality museum and digital thematic exhibitions (Gao et al., 2024). During this stage, digital venues have been continuously expanding in terms of types and functions, evolving from single display to integrating functions such as popular science, education, and cultural dissemination.

New experience technologies such as holographic presentation, digital twin, ultra-high definition, multi-sensor fusion, AR (Augmented Reality), and VR (Virtual Reality) are developing at a rapid pace, providing more possibilities for the development of digital venues. In 2017, the British Museum collaborated with the VR company Oculus to launch a virtual reality tour (Motta et al., 2019). The Louvre Museum in France combined digital technology with fashion, reinterpreting its collected works through digital means and holding digital light and shadow fashion shows (Evrard & Krebs, 2018). The application scenarios of digital technology in museums have been expanded to three types: on-site, online, and present. Concepts such as digital museums, virtual museums, mobile museums, ubiquitous museums, and smart museums have emerged successively.

Digital venues continue to innovate in terms of technological application and user experience. On May 1, 2016, the "Digital Dunhuang" resource database was launched online, and high-definition digital images of 30 classic caves and virtual roaming experience programs were officially made available on the Internet. This concept was first proposed by Fan Jinshi in the late 1980s. She put forward the concept of "Digital Dunhuang" for the first time, aiming to use computer digital technology to permanently preserve the precious materials of Dunhuang murals and colored sculptures. Since then, the Dunhuang Academy and related units have implemented several scientific research projects, laying the foundation for the breakthrough of key technologies in the digitization of the Mogao Grottoes. In the early 1990s, Fan Jinshi further proposed the idea of Digital Dunhuang, hoping to use high-tech means to establish digital archives for Dunhuang murals, clay sculptures, and caves, providing new possibilities for the protection of cultural heritage (Zhou et al., 2022). On January 6, 2025, the "'Pattern' Carrying the Tao — The Forbidden City Immersive Digital Experience Exhibition" jointly hosted by BOE (Beijing Oriental Electronics Group Co., Ltd.), the Palace Museum, and Tencent opened at the Suzhou Bay Digital Art Museum. At the beginning of the centenary of the establishment of the Palace Museum, BOE used cutting-edge digital technology to digitally connect the "national treasures" of the Palace Museum, promoting the Forbidden City culture to show new vitality in Suzhou Bay (www.suzhou.gov.cn, 2025.January 7.).

Today, the digital content of cultural relics in museums around the world is experiencing explosive growth. The development of digital venues is constantly deepening, and the way audiences learn about cultural relics is evolving towards a comprehensive experience that combines the virtual and the real, with a deep integration of somatosensory interaction.

Application of Digital Art in Cultural Tourism

Digital art can effectively enrich the carriers and forms of dissemination in tourism culture. For example, digital media art can create digital film and television content for cultural tourism. Through vivid images and wonderful plots, it showcases the natural scenery, historical culture, and humanistic customs of tourist destinations. At the same time, it can innovate the digital network and utilize digital network platforms such as travel websites and social media to disseminate tourism culture. The digital network can also enable interaction between tourists and tourist destinations, such as online booking and virtual tours, which improves tourists' participation and experience. By applying digital art to digital museums, digital technologies are used to digitally display cultural resources such as cultural relics and historical sites. Moreover, through technologies like virtual reality and augmented reality, a more realistic and vivid visiting experience can be provided for tourists.

Digital art also plays a role in cultural protection in cultural tourism of specific fields. For instance, in the cultural heritage of cave art sites, advanced photogrammetry, terrestrial laser scanning, structured light scanning, and high-resolution photography technologies can digitally preserve the complex details of cave art. Humidity, temperature, and carbon dioxide sensors can be used to provide a comprehensive digital monitoring system to track the condition of cave art (Kotoula et al., 2018). If the digitization of cave art is properly utilized, it is possible to accurately protect the site and, at the same time, enable tourists to experience this art form in ways that were previously unattainable. The application of advanced digital technologies in the protection of cave art is not only an option but also a necessity, considering the impermanent nature of these paintings in the natural environment (Le Fillâtre et al., 2021).

RESEARCH METHOD & DATA SOURCE

The main research method of this paper is the knowledge graph visualization analysis method, and the CiteSpace (6.4.R1) software is adopted as the analysis tool. As an information visualization software, CiteSpace is used in this paper to analyze the keywords in the literatures, display the co-occurrence frequency and interrelationships among the keywords, so as to excavate the research hotspots and frontier topics within the field, and intuitively present the evolution trends of the research topics. The analysis results are presented in an intuitive graphical way, such as a knowledge graph composed of nodes (representing literatures, authors, keywords, etc.) and connection lines (representing the relationships among them), making the complex literature relationships clear at a glance and facilitating researchers to understand and interpret.

Since this paper discusses the application of the digital art museum in Harbin Ice and Snow World in China based on the keywords "digitization" and "cultural tourism", the data from two major platforms, namely the Web of Science (WoS) and the China National Knowledge Infrastructure (CNKI), are combined and used to make up for the insufficiency of the current literatures in the research on the digital transformation of cultural tourism in the field of ice and snow cultural tourism. On WoS, the keywords "digitization", "culture", "tourism", "digital museum", and "digital art museum" are used to screen the articles one by one. Since the core case discussed is the contribution of the digital art museum in Harbin Ice and Snow World in China to cultural tourism, the relevant data from CNKI are also introduced as the data source for CiteSpace (Figure 1)(Figure 2).

Figure 1. Flowchart of the data collection in CNKI

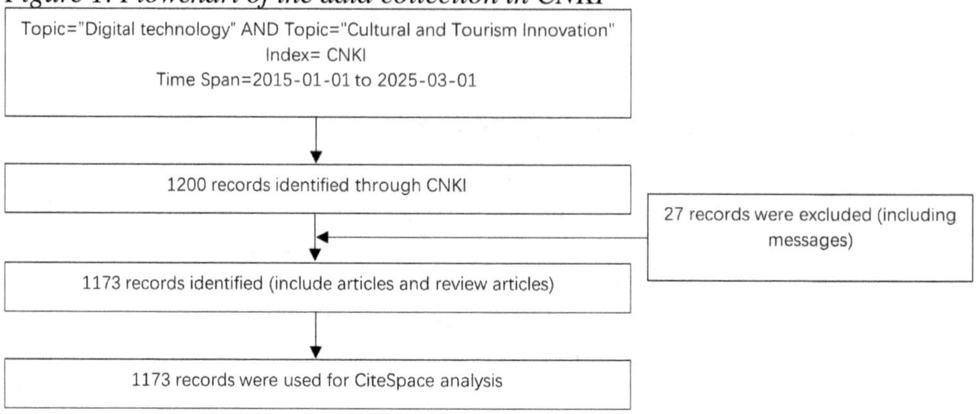

Figure 2. Flowchart of the data collection in WoS

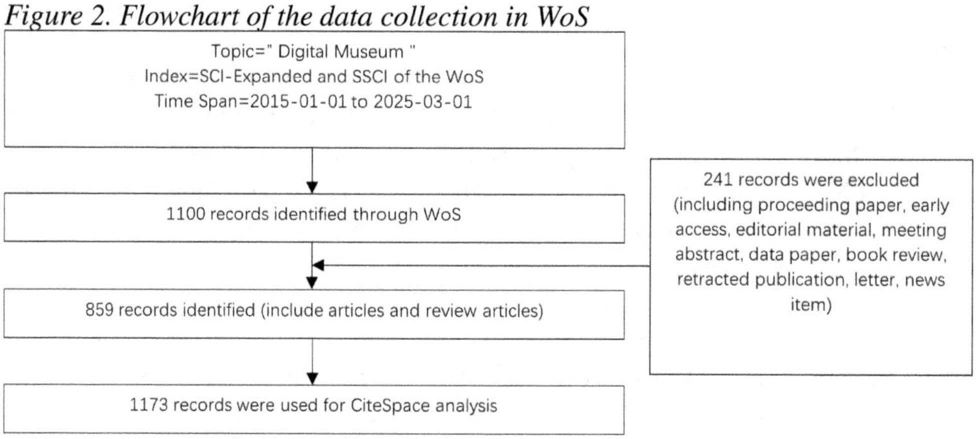

Cluster Analysis of the Keywords "Digitization" and "Cultural Tourism" in CNKI

Figure 3. Cluster Analysis of the Keywords "Digitization" and "Cultural Tourism"

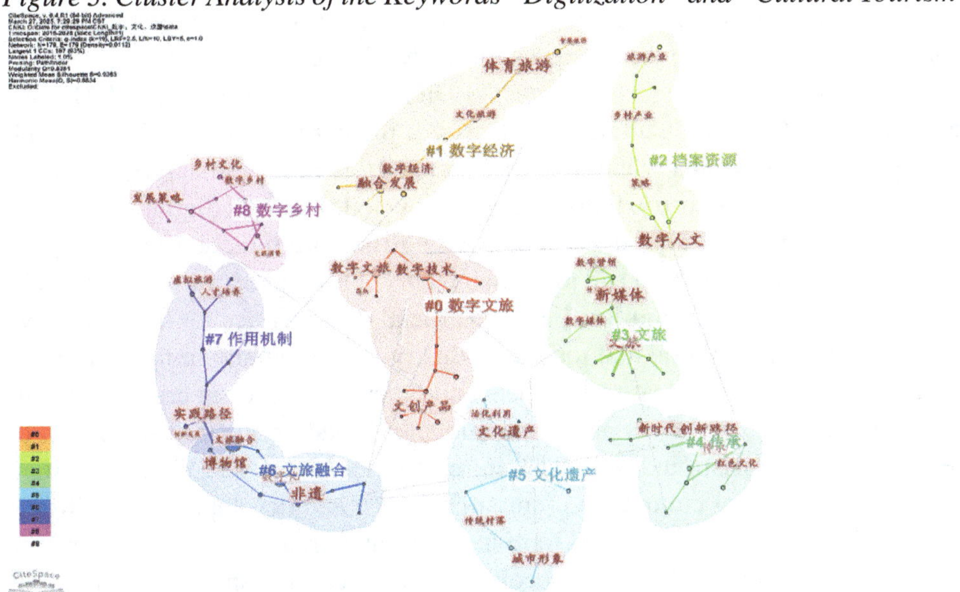

Source: CNKI, 2015 – 2025

The connection lines between keyword nodes represent the co-occurrence relationship, that is, keywords appear simultaneously in the literature. The thickness of the connection lines reflects the intensity of co-occurrence. The thicker the line, the more frequent the co-occurrence, indicating that the corresponding research topics are closely related. The thick connection line between the nodes of "Digital Technology" and "Cultural and Tourism Innovation" indicates that the two frequently appear simultaneously in relevant literatures, and the research often combines these two aspects for discussion (Figure 3).

Keyword clustering is to cluster the same keywords that appear in the co-cited literature. A large number of research literatures gather around a research topic, thus forming a research hotspot. On the CNKI website, an advanced search was conducted with the subject terms of "digitization" and "cultural tourism", and 1,200 articles were selected as the data source. Invalid data such as low-quality journals, conference literatures, literature sources from non-target fields, and literatures without clear authors were screened out. Finally, 1,173 valid data entries were obtained. This paper studies the development of "digitization" and "cultural tourism" over the past 10 years. The starting year is set as 2015, the ending year is set as 2025, and the slice length is set as 1 year, so as to observe the research changes in different time periods. Using CiteSpace software and combining the LLR (Log-Likelihood Ratio) for keyword burst detection to conduct cluster analysis, 9 keyword clustering labels of the hot research fields on digital cultural tourism were obtained. Combining the high-frequency keywords of the clustering and the situation of the literatures, these clusters are classified into three aspects for analysis (Figure.1).

First, the core key category includes #0 Digital Cultural Tourism and #1 Digital Economy. #0 Digital Cultural Tourism is at the core of the knowledge graph and represents a key area where digital technology is deeply integrated with the cultural and tourism industry. It encompasses key nodes such as digital technology and cultural and creative products, symbolizing the core direction of the digital transformation of the cultural and tourism industry. The application of digital technology in the cultural and tourism field is a crucial driving force for the innovative development of the cultural and tourism industry at present and in the future. By developing cultural and creative products and other means, it can create new cultural and tourism experiences and economic value, leading the transformation of the cultural and tourism industry. Therefore, it holds a core and pivotal position in the entire system of cultural and tourism research and development. #1 Digital Economy is closely linked to sports tourism, cultural tourism, etc., emphasizing the integrated development of the digital economy and the cultural and tourism industry. In the era of the digital economy, the integration model and development path are of vital importance for the upgrading of the cultural and tourism industry. Through integration, it can expand the economic model of the cultural and tourism industry, enhance industrial competitiveness, and promote the innovative development of the cultural and tourism industry at the economic level. It is an important economic support direction for the development of the cultural and tourism industry and is of core significance.

Second, the important support category includes #3 Cultural and Tourism, #6 Integration of Culture and Tourism, and #8 Digital Countryside. #3 Cultural and Tourism focuses on the application of digital marketing, new media, etc. in the cultural and tourism industry. In the current era of rapid information dissemination, new media and digital media are important means for the cultural and tourism industry to promote itself externally and attract tourists. Through effective digital marketing, it can enhance the popularity and influence of cultural and tourism products and services, providing strong communication and promotion support for the development of the cultural and tourism industry. It is an important link for the cultural and tourism industry to achieve market expansion and value enhancement. #6 Integration of Culture and Tourism mainly studies the integration of cultural resources such as museums and intangible cultural heritages with the tourism industry. The integration of culture and tourism is an important trend in the development of the cultural and tourism industry. The combination of rich cultural resources such as museums and intangible cultural heritages with the tourism industry can not only tap the tourism value of cultural resources but also inject profound cultural connotations into the tourism industry, promoting the development of both sides. It is an important support direction for the connotative development of the cultural and tourism industry. #8 Digital Countryside explores the application of digital technology in the rural cultural and tourism industry and the development strategies of rural culture. Rural cultural and tourism is an important part of the cultural and tourism industry, and the construction of a digital countryside brings new opportunities for the rural cultural and tourism industry. With the help of digital technology, it is possible to tap the cultural characteristics of the countryside, promote rural industrial revitalization, enrich the formats of the cultural and tourism industry, and provide important support for the diversified development of the cultural and tourism industry.

Third, the characteristic development category includes #2 Archival Resources, #4 Inheritance, #5 Cultural Heritage, and #7 Mechanism of Action.

Cluster Analysis of the Keywords "Digital Museum" in WoS

Figure 4. Cluster Analysis of the Keyword "Digital Museum"

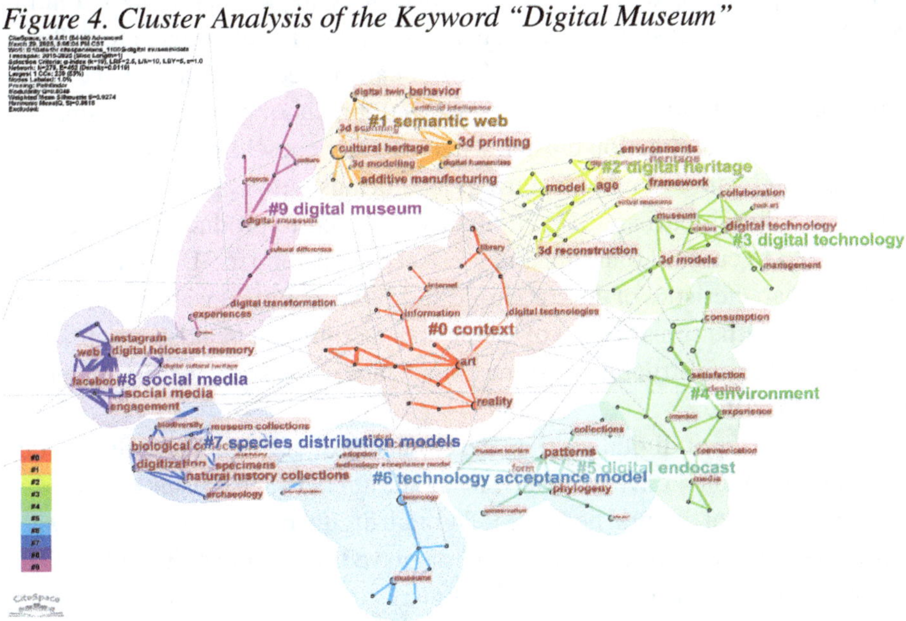

Source: WoS, 2015 – 2025

An advanced search was conducted on the WoS website with the subject term "digital museum", and 1,100 articles were selected as the data source. Invalid data such as low-quality journals, conference literatures, literature sources from non-target fields, and literatures without clear authors were screened out. Finally, 859 valid data entries were obtained. This paper studies the development of "digitization" and "cultural tourism" over the past 10 years. The starting year is set as 2015, the ending year is set as 2025, and the slice length is set as 1 year to observe the research changes in different time periods (Figure 2).

The literatures were exported in the Plain text file format, and the data were transformed and processed to draw the relevant knowledge graph. A cluster analysis was carried out on the keywords, and 10 keyword clustering labels of the hot research fields on digital cultural tourism were obtained (Figure 4). Combining the high-frequency keywords of the clustering and the situation of the literatures, these clusters were classified into three aspects for analysis.

First, the core cluster of digital technology, #3 digital technology, serves as the key support. Technologies such as 3D models and digital protection involved in it are of great significance in the digitization of cultural tourism. 3D models can be used for digital modeling of cultural heritages and tourist attractions, enabling tourists to experience the scenery of the attractions immersively online. For example, high-precision 3D modeling of ancient buildings can be carried out for tourists to visit virtually. Digital protection technology can be used to protect precious cultural heritages. Through digital storage and management, the risk of damage to cultural heritages caused by natural and human factors can be reduced, and at the same time, it provides rich digital resources for the development of cultural tourism. The emerging technology application cluster, #1 semantic web, with technologies such as 3D printing and additive manufacturing, also brings new opportunities for the digitization of cultural tourism. 3D printing can be used to produce characteristic souvenirs related to cultural tourism, and tourists can customize 3D printed handicrafts with the characteristics of the attractions. Additive manufacturing has potential application value in the construction of scenic area facilities and landscape creation, and it can realize the construction of some unique landscape shapes, bringing novel experiences to tourists.

Second, the cluster of digital tourism development of cultural heritage includes #2 digital heritage and #4 environment. The #2 digital heritage cluster emphasizes the digital processing of cultural heritages. Through digital means, cultural heritages are transformed into resources that can be displayed, disseminated, and utilized online. For example, 3D reconstruction and digital storage of historical sites can not only better protect these heritages but also be developed into online tourism products for global tourists to visit, breaking through the limitations of time and space and expanding the influence and dissemination scope of cultural heritages. The #4 environment cluster indicates that in the process of digitization of cultural tourism, attention should be paid to the integration with the environment. The digital display of cultural tourism attractions should fully consider their natural and humanistic environment, and the harmonious symbiotic relationship between the attractions and the surrounding environment should be demonstrated through digital technology. For instance, digital technology can be used to display the scenery of traditional villages and the surrounding mountains, rivers, and pastoral areas, allowing tourists to fully experience the charm of rural cultural tourism and enhancing tourists' awareness of environmental protection at the same time.

Third, the cluster of social media and cultural tourism promotion, #8 social media, indicates the importance of social media in the digital dissemination of cultural tourism. Through social media platforms, cultural tourism attractions and related activities can be widely publicized and promoted. Scenic areas can release exquisite digital pictures, videos, and other contents to attract the attention of potential tourists. At the same time, tourists can also share their cultural tourism experiences on social media, forming word-of-mouth communication and further enhancing the popularity of cultural tourism destinations.

Summary of Keyword Clustering Analyses: "Digitization" and "Cultural Tourism" on CNKI vs "Digital Museum" on WoS

Through the keyword clustering analysis of "Digitization" and "Cultural Tourism" on CNKI, the following conclusions are drawn: Digital cultural tourism and digital economy are the core and key categories in the digital transformation of the cultural and tourism industry. The application of digital technology in the cultural and tourism field is a crucial driving force for innovative development, which can create new experiences and economic values. The integration of the digital economy and the cultural and tourism industry is of great significance for industrial upgrading. Cultural and tourism, the integration of culture and tourism, and digital countryside are important supporting categories, playing significant roles in digital marketing and promotion, the integration of cultural resources and the tourism industry, and the development of the rural cultural and tourism industry respectively.

Through the keyword clustering analysis of "Digital Museum" on WoS, the following conclusions are obtained: Digital technology is the key support for the digitization of cultural tourism. Technologies such as 3D models and digital protection are of great significance for the digitization of cultural heritages and the display of tourist attractions. Emerging technologies such as 3D printing and additive manufacturing also bring new opportunities. In the digital tourism development of cultural heritages, attention should be paid to digital processing and integration with the environment. Social media are of vital importance in the digital dissemination of cultural tourism, which is conducive to publicity and promotion and enhancing the popularity of tourist destinations.

Their common core lies in the following aspects. Firstly, digital technology occupies a central position and plays a key role in the process of the digital transformation of cultural tourism. It is not only an important driving force for the innovative development of the cultural and tourism industry, capable of creating new cultural and tourism experiences and economic values and promoting industrial upgrading, but also of great significance in the digitization of cultural heritages, the display of tourist attractions, and the application of emerging technologies. Secondly, elements such as the integration of culture and tourism, as well as related digital marketing, social media promotion, and digital countryside construction, provide important support for the digital transformation of cultural tourism from different perspectives. They are interrelated and work synergistically to jointly drive the development of the cultural and tourism industry in the digital direction.

Policy:The State's Support for the Digital Transformation of the Cultural and Tourism Industry

Table 1. 2020 – 2024 Cultural and Tourism Digitization Policies

Time	Policy	Issuing Entity
2020/11/18	Opinions of the Ministry of Culture and Tourism on Promoting the High - quality Development of the Digital Cultural Industry	Ministry of Culture and Tourism
2022/5	Opinions on Promoting the Implementation of the National Cultural Digitization Strategy	General Office of the Central Committee of the Communist Party of China, General Office of the State Council
2022/1	Development Plan for the Tourism Sector during the 14th Five - Year Plan Period	The State Council
2022/2/18	Several Policies on Promoting the Recovery and Development of the Difficult Sectors in the Service Industry	National Development and Reform Commission, Ministry of Culture and Tourism and other 14 departments
2023	Notice of the Ministry of Industry and Information Technology and the Ministry of Culture and Tourism on Strengthening the Collaborative and Innovative Development of 5G + Smart Tourism	Ministry of Industry and Information Technology, Ministry of Culture and Tourism
2024/5/13	Action Plan for the Innovative Development of Smart Tourism	General Office of the Ministry of Culture and Tourism

In the process of the cultural and tourism industry moving towards digitization, the state attaches great importance to it and provides comprehensive support through a series of policies. The "Opinions on Promoting the Implementation of the National

Cultural Digitization Strategy" issued by the General Office of the Central Committee of the Communist Party of China and the General Office of the State Council lays a solid foundation for the digitization of culture and tourism from a strategic height. By constructing a national cultural big data system and integrating various cultural resources, it provides rich data support for the digital transformation of the cultural and tourism industry, promotes the efficient utilization and widespread dissemination of cultural resources, and establishes a macro-development pattern. The "Development Plan for the Tourism Sector during the 14th Five-Year Plan Period" issued by the State Council elevates the digitization of tourism to a crucial position in the overall industrial development. It clarifies the development goals and key tasks of tourism digitization during the 14th Five-Year Plan period, points out the path for the digital transformation of the cultural and tourism industry during this period, and guides the cultural and tourism industry to make an orderly layout to avoid blind development.

The "Opinions of the Ministry of Culture and Tourism on Promoting the High-quality Development of the Digital Cultural Industry" introduced by the Ministry of Culture and Tourism focuses on the internal innovation and upgrading of the cultural and tourism industry. It encourages cultural and tourism enterprises to use digital technology to innovate products and services, and cultivate new types of cultural enterprises and business forms. Driven by this policy, innovative products that integrate digital technology are constantly emerging in the cultural and tourism market, such as immersive cultural and tourism experience projects and digital cultural and creative products. These have enriched the supply of cultural and tourism products, optimized the industrial structure, and effectively promoted the digital upgrading of the cultural and tourism industry.

The notice jointly issued by the Ministry of Industry and Information Technology and the Ministry of Culture and Tourism actively promotes the integrated development of 5G and smart tourism. By leveraging the characteristics of high speed, low latency, and large connectivity of 5G technology, the cultural and tourism industry has created many innovative application scenarios, such as 5G panoramic live broadcasts of tourist attractions and 5G+VR immersive tourism experiences. These have significantly enhanced the tourism experience of visitors and increased the attractiveness of the cultural and tourism industry. The "Several Policies on Promoting the Recovery and Development of the Difficult Sectors in the Service Industry" provides a crucial opportunity for the cultural and tourism industry during the process of digital transformation. It helps cultural and tourism enterprises optimize their operation models, improve operation efficiency through digital means, such as promoting online reservation and intelligent tour guide services, and reducing operation costs. At the same time, it encourages enterprises to innovate business models, expand the online cultural and tourism market, and enhance their ability to respond to market changes.

The "Action Plan for the Innovative Development of Smart Tourism" implements the spirit of the 20th National Congress of the Communist Party of China and the requirements of documents such as the "14th Five-Year Plan for the Development of the Digital Economy" and the "Development Plan for the Tourism Sector during the 14th Five-Year Plan Period". Taking digitization, networking, and intelligence as the development directions, it cultivates the product forms of smart tourism. By encouraging enterprises to play a dominant role and promoting collaborative cooperation among industry, university, and research institutions, it continuously enriches smart tourism products and promotes the cultural and tourism industry to move towards a higher level of digitization.

The central government encourages local governments to fully explore the local cultural characteristics, create cultural spaces with regional characteristics, and improve the coverage and timeliness of public cultural services. Local governments, based on their local operation capabilities and market demands, steadily promote the construction of digital infrastructure, deeply tap local cultural and tourism resources, ensure that the digital transformation of culture and tourism can be carried out in accordance with local conditions and in an orderly manner, and achieve the balanced development of the digitization of the national cultural and tourism industry.

THE APPLICATION OF AIGC TECHNOLOGY IN THE CASE OF THE DIGITAL ART MUSEUM OF ICE AND SNOW WORLD

On January 5, 2024, at the "2024 China Ice and Snow Tourism Development Forum" held in Harbin, the Harbin Ice and Snow World successfully challenged the Guinness World Record with a park area of 810,000 square meters and won the title of "the World's Largest Ice and Snow Theme Park" (Harbin Municipal People's Government, January 5, 2024). According to the Organizing Committee of the Ice and Snow World, since December 7, 2023, until the opening of the park, the project team has collected 6,000 cubic meters of ice blocks daily from the ice surface of the Songhua River and transported them to the Ice and Snow World park. In terms of the quantity of ice used and the scale of the park, this year's Harbin Ice and Snow World has reached the maximum values in both indicators compared with previous years. The natural ice blocks taken from the Songhua River have been applied in the construction of large ice and snow buildings such as 14 super ice slides, the 43-meter-high main tower of the "Crown of Ice and Snow", and the "Dance in the

Sky" created by using 3D composite ice technology (CHINA CENTRAL TELEVISION, January 11, 2024).

In the current light and shadow feast, Harbin's unique ice and snow culture is presented in a rather vague way. As a city known as the "Ice City", Harbin has a profound ice and snow cultural heritage, covering many unique cultural elements, ranging from the winter fishing culture on the Songhua River, the ice lantern tradition on Central Street, to ice and snow sports and winter folk customs. However, in this light and shadow show, these cultural elements have not been fully demonstrated. Instead, what is presented are uniform lighting effects, which lack uniqueness and can also be applied in many other ice and snow scenic areas.

This lack of cultural expression directly weakens the characteristics of the project. When tourists recall the Ice and Snow World, it is difficult for them to form a unique cultural connection with the city of Harbin. The lack of characteristics makes it difficult for the Ice and Snow World to stand out among numerous ice and snow tourism projects, which is not conducive to establishing a differential advantage in the fierce market competition and hinders its further development.

The immersive AI digital art museum of "One Ice and Snow, One World" brings hope for solving this problem. The immersive AI digital art museum of "One Ice and Snow, One World" is the first digital art immersive experience project initiated and launched nationwide (Departure of Culture and Tourism of Heilongjiang Province, December 26, 2024). The museum is located on the second floor of the Dream Ice and Snow Hall of the Harbin Ice and Snow World, and it is jointly created by Inspur Intelligent Terminal, Jindong Digital Creativity, and Heilongjiang Fenggu Jindeng Agricultural Company. The museum has a large ice and snow theme space of 2,300 square meters, and it is ingeniously arranged with three major exhibition areas and seven exhibition items, presenting an immersive experience of the new Chinese aesthetic under the integration of ice and snow culture and avant-garde art (Chinanews.com, December 26, 2024).

The Ice and Snow Digital Art Museum has many technical and exhibition highlights. Firstly, digital art is integrated with traditional culture. With the innovative application of AIGC (Artificial Intelligence Generated Content) technology, the "Ninety-Nine Cold-Relieving Chart" is integrated into the ice and snow in the form of digital art, creating the longest digital interactive scroll with an ice and snow theme in the country.

The AIGC is closely related to the core conclusions drawn from the CiteSpace analysis. In terms of enriching the application scenarios of digital technologies, as a cutting-edge digital technology, AIGC can expand the application of digital technologies in cultural tourism. When conducting 3D modeling, it can automatically generate more creative virtual scenes and characters, enhancing the immersive feeling of tourists during online tours. When making cultural and tourism souvenirs, it can quickly generate personalized design schemes according to the characteristics of scenic spots and the needs of tourists, enhancing the uniqueness of 3D printed souvenirs. At the same time, AIGC helps the digital cultural and tourism industry create more novel cultural and creative products, such as generating digital illustrations and virtual short dramas based on historical stories, enriching the content of cultural and tourism experiences, promoting the development of digital cultural and tourism, and creating more economic value. AIGC deepens the digital development of cultural heritage, conducts creative processing on the digitized resources of historical sites, and carries out virtual restoration and re-creation of damaged cultural heritage based on existing data, presenting the stories behind them in a dynamic and interactive form, attracting more tourists' attention, expanding the influence of cultural heritage, providing more attractive content for the integration of culture and tourism, and exploring the deeper tourism value of cultural resources.

AIGC refers to the content generated by using advanced generative artificial intelligence technology. Different from the content created by humans, it can automatically create a large amount of content in a short time. Its generation process generally includes two steps: extracting intent information from human instructions and then generating content based on the extracted intent. The core technologies include basic models (such as Transformer and pre-trained language models), reinforcement learning from human feedback (RLHF), etc. Transformer, with its self-attention mechanism, can effectively capture long-distance dependency relationships when processing sequential data, providing an architectural foundation for many advanced models (such as GPT-3, DALL-E-2, etc.). Pre-trained language models can be divided into two categories according to training tasks: autoregressive language modeling and masked language modeling, and they perform excellently in natural language processing tasks. RLHF optimizes the model through human feedback, making the output of AIGC more in line with human preferences (Xing et al., 2025).

In the section of "The Application of AIGC Technology in the Digital Art Museum of Ice and Snow World", the application of AIGC technology in the Digital Art Museum of Harbin Ice and Snow World demonstrates powerful innovation capabilities and application values. By integrating the "Ninety-Nine Cold-Relieving Chart" into the ice and snow in the form of digital art, creating the longest digital interactive scroll with an ice and snow theme in the country, constructing the iconic 81-meter "Scroll of the Coldest Days of Winter", setting up more than 200 characters and more than 40 interactive games, and integrating the interactive technology created by AIGC with offline physical devices, etc., it has greatly enriched the playing experience of tourists and enhanced the interestingness and attractiveness of the project.

From the perspective of academic research, an in-depth analysis of this application of AIGC technology precisely responds to the key issues raised in this study and is closely related to the core theme of the role and development of the digital art museum in the digital transformation of cultural tourism.

Regarding research question 2, that is, "In the case study of Harbin Ice and Snow World, how can AIGC (Artificial Intelligence Generated Content) help the digital transformation of ice and snow cultural tourism?", this application provides a highly valuable practical example. Traditional ice and snow tourism projects are often limited to simple displays of ice and snow landscapes, with cultural expressions remaining on the surface, lacking depth and innovation, and it is difficult to meet the growing diversified cultural experience needs of tourists. The intervention of AIGC technology, through the digital innovative interpretation of traditional cultural elements, transforms the "Ninety-Nine Cold-Relieving Chart", a traditional folk cultural element, into a digital art work full of a sense of technology and interactivity, injecting profound cultural connotations into ice and snow cultural tourism. At the same time, the rich and diverse interactive content generated by using AIGC technology, such as AI ice sculpture interactive devices, AIGC-UGC human-computer interesting interactive devices, and AI costume-changing projects, has realized the transformation of tourists from passive viewing to active participation, providing a practical path and model for the digital transformation of ice and snow cultural tourism, and powerfully proving the effectiveness and innovation of AIGC technology in solving the problem of the lack of cultural expression in traditional ice and snow tourism projects.

In terms of the core theme of the role and development of the digital art museum in the digital transformation of cultural tourism, the application of AIGC technology further highlights the key role of digital technologies. Echoing the relevant content about the application of digital art in cultural tourism in the previous literature review, AIGC technology, as a cutting-edge field of digital technologies, not only enriches the communication carriers and forms of cultural tourism but also deepens the digital development of cultural heritage. In terms of communication carriers and forms, the digital art works and interactive experience projects generated by AIGC technology, with their unique creativity and interactivity, attract more tourists' attention and participation, expanding the communication scope and influence of cultural tourism. In terms of the digital development of cultural heritage, AIGC technology conducts creative processing on the digitized resources of historical sites, carries out virtual restoration and re-creation of damaged cultural heritage based on existing data, and presents the stories behind them in a dynamic and interactive form, which is consistent with the role of digital art in protecting cultural heritage in the literature review, and further explores the deeper tourism value of cultural resources.

In addition, the application of AIGC technology is closely related to the discussion of the digital innovation of ice and snow cultural tourism under the innovation diffusion theory in the later part. From the perspective of the innovation diffusion theory, the application of AIGC technology in the Ice and Snow Digital Art Museum significantly enhances the relative advantage, compatibility, trialability, and observability of the digital innovation of ice and snow cultural tourism. In terms of relative advantage, AIGC technology brings tourists an unprecedented immersive experience, which is different from traditional ice and snow tourism projects, meets tourists' pursuit of novelty and personalized experiences, and improves the competitiveness of ice and snow cultural tourism products. In terms of compatibility, AIGC technology integrates traditional cultural elements into the display of digital art, which is compatible with the values of the public regarding cultural inheritance and innovation and is easily accepted by tourists of different age groups and cultural backgrounds. In terms of trialability, the digital art museum can leverage the flexibility of AIGC technology to quickly update the display content and interactive links, conduct innovation trials at a low cost and high efficiency, and optimize the project according to tourists' feedback. In terms of observability, the unique digital art scenes and interactive experiences created by AIGC technology have inspired tourists' enthusiasm for sharing on social media. Through the dissemination of user-generated content (UGC), it has attracted the attention of more potential tourists, enhanced the popularity and influence of ice and snow cultural tourism projects, and thus promoted the dissemination and application of the digital innovation of ice and snow cultural tourism in a wider range.

In conclusion, the application of AIGC technology in the Digital Art Museum of Harbin Ice and Snow World is of great significance both in theory and practice. It provides new ideas and directions for the digital transformation of ice and snow cultural tourism and also provides rich empirical evidence for relevant academic research.

THEORETICAL FRAMEWORK: THE APPLICATION OF THE INNOVATION DIFFUSION THEORY IN THE DIGITALIZATION OF ICE AND SNOW CULTURAL AND TOURISM INDUSTRY

In the 1960s, Everett Rogers, an American communication scholar and sociologist, began to conduct a systematic study on the issues of innovation and diffusion. He analyzed 506 issues related to the innovation and diffusion processes in history. In 1962, through in-depth investigations into the adoption and popularization processes of new varieties, new pesticides, new machinery and other new things in rural areas, he and Shoemaker published the book "Diffusion of Innovations" (E.M. Rogers, 2016).

The innovation diffusion theory was proposed by Everett Rogers. It holds that the innovation decision-making process consists of five stages: knowledge, persuasion, decision-making, implementation, and confirmation. The adoption of technology shows an S-shaped curve and is influenced by factors such as the characteristics of innovation (such as relative advantage, compatibility, complexity, trialability, observability), communication channels (mass communication and interpersonal communication), time, and social systems (Valente et al., 2015). The innovation diffusion theory aims to explain why people adopt new ideas and technologies and indicates that the innovative attributes affect users' adoption of innovation (Kim et al., 2019).

The innovative attributes that individuals perceive when adopting innovation include Relative Advantage, Compatibility, Complexity, Trialability, and Observability (Table 2).

In terms of relative advantage, relying on cutting-edge digital technologies, the digital art museum demonstrates significant superiority over traditional art museums in many aspects. In terms of the exhibition space and the number of exhibits, traditional art museums are limited by physical space, resulting in a limited exhibition scale and the number of collections on display. In contrast, the digital art museum, with the help of virtual space, breaks through this limitation and can achieve massive storage and display of artworks. Theoretically, it can gather art treasures from around the world on the same digital platform for the audience to browse and appreciate (Meinecke et al., 2022). In terms of the time dimension, the fixed opening hours of traditional art museums restrict the audience's visits, while the digital art museum can be open around the clock without interruption. The audience can start their art exploration journey anytime and anywhere according to their own schedules. In terms of the display form, the application of digital technology means, such as 3D modeling, VR (Virtual Reality), and AR (Augmented Reality) technologies, has transformed the presentation of artworks from the traditional static display to a dynamic, interactive, and immersive experience (Castellotti et al., 2024). For example, through 3D modeling technology, ancient paintings are digitally reconstructed. With the help of VR devices, the audience can not only view the details of the paintings from a 360-degree perspective but also simulate being in the scene of the era when the works were created through technology, deeply feeling the environmental atmosphere and historical background when the painters created the works. This immersive experience with multi-sensory participation is difficult to achieve in traditional art museums. From the perspective of economic model innovation, the digital art museum has created new profit channels such as paid viewing of online virtual exhibitions, copyright transactions of digital artworks, and online sales of peripheral cultural and creative products, established a new cultural consumption model, realized the diversified realization of artistic value, and promoted the economic growth of the cultural and artistic industry (Zhao et al., 2023).

In terms of compatibility, the digital art museum skillfully integrates modern technology and traditional cultural elements, which is highly consistent with the values and behavior patterns of the general public in contemporary society. With the acceleration of the modern life rhythm, the public's demand for convenient cultural experiences is increasing day by day. The online platform and convenient interactive design of the digital art museum enable the audience to enjoy art resources through the Internet without being restricted by region and time, meeting the needs of this era (Noble, 2021a). Against the background that cultural inheritance and innovation have become the common pursuit of society, the digital art museum uses digital means to display traditional cultural artworks, endowing traditional culture with new vitality. For example, by using digital methods such as high-definition videos and 3D disassembly demonstrations to display ancient folk handicrafts, the audience can have an in-depth understanding of the exquisite processes of traditional techniques. Digitally recording traditional opera performances and adding functions such as interactive bullet screen explanations and character background introductions not only preserves the core charm of traditional culture but also integrates modern technological elements, presenting traditional culture to the public in a form that is easier to understand and accept, echoing the public's value pursuit of cultural inheritance and innovation, and thus being acceptable to audiences of different age groups and cultural backgrounds (Noble, 2021b).

In terms of complexity, during the design and operation process of the digital art museum, with the user experience as the core guiding principle, it is committed to reducing the difficulty for the audience to understand and participate. In the design of interactive devices, it follows the principle of simplicity and intuitiveness, avoiding complex operation procedures and technical jargon. For example, common digital art interactive screens adopt simple operation methods such as touch and swipe, which can realize functions such as work switching, zooming in and out, and special effect display. Even elderly audiences who lack an understanding of digital technology can easily operate them (Kastner et al., 2021). In terms of constructing the guided tour system, it is equipped with intelligent voice navigation and visual maps. When the audience enters the art category or work name they are interested in, the system can quickly provide accurate browsing routes and detailed work introductions. For children, a parent-child interaction area is specially designed. Through means such as interesting games and guidance by cartoon characters, children can feel the charm of art in a relaxed and pleasant atmosphere, ensuring that audiences of different age groups and technical levels can obtain a good art experience in the digital art museum.

Trialability is a unique advantage of the digital art museum with the help of digital technology. The museum can update the exhibition content at a relatively low cost and high efficiency and launch new interactive projects for experimentation. During the operation of virtual exhibitions, according to the trends of art development and the feedback data from the audience, new artworks are added in a timely manner, and the display order and layout are adjusted. By analyzing data such as the time the audience stays on the work page and the click popularity, the display effect is evaluated, and then the exhibition is optimized. For newly developed interactive projects, such as the AI (Artificial Intelligence) assisted art creation experience area, a small-scale open test is carried out first, and various opinions such as the audience's feelings during the operation process and creative feedback are collected. Based on this, targeted improvements are made to the difficulty setting, interestingness, and technical stability of the project, effectively reducing the innovation risk, enabling the digital art museum to continuously maintain a sense of freshness and attractiveness, and accurately matching the changing needs of the audience.

From the perspective of observability, the digital art museum has significantly improved its communication effect by means of diversified communication channels such as social media and online platforms (Kim et al., 2019). After visiting the digital art museum, due to its unique art experience, the audience has a strong willingness to share. They post screenshots of exquisite digital artworks, videos of novel interactive experiences on social platforms, accompanied by vivid text descriptions. These user-generated contents (UGC) spread rapidly on the Internet, attracting the attention of a large number of potential audiences. At the same time, the official of the digital art museum actively operates social media accounts, regularly releases exhibition previews, artist interviews, art popular science content, etc., and holds online interactive activities, such as voting for digital artworks and creative solicitation, further enhancing the popularity and influence of the digital art museum, attracting more people to visit and experience, and effectively promoting the communication and development of the digital art museum in the field of culture and art, making it gradually become an important force in the communication of culture and art.

Table 2. Five Innovation attributes

Relative Advantage	The superiority of the innovation perceived by individuals in economic, social and other aspects compared with the original things.	Compared with other traditional social methods or platforms, MSNS (Mobile Social Networking Platform) has a relative advantage in providing real experiences for the elderly. Through MSNS, the elderly are likely to obtain richer and more unique travel-related experiences, thus attracting them to use this platform (Kim et al., 2019).
Compatibility	The degree of fit between the innovation and the individual's existing values, behavior patterns, and past experiences.	Taking the Flexible Pallet Bench project as an example, the unconventional bench designed with pallets does not conform to users' cognition and usage habits of traditional seats, resulting in a relatively low utilization rate. This phenomenon highlights the difference between the formal design of the project and users' environmental perception habits, that is, there is a lack of compatibility (Kim & Jin, 2024).
Complexity	The degree of difficulty for an individual to understand and apply an innovation.	When studying the ratification of the global tobacco control treaty, the complexity of an innovation (such as new provisions in the treaty) may affect the manifestation of its relative advantage. If the new provisions are too complex and countries face difficulties in understanding and implementing them, it may undermine their original advantages in tobacco control, thereby influencing the willingness of countries to ratify the treaty (Valente et al., 2015).
Trialability	The degree to which an innovation can be tried within a limited scope.	In the article "Interpreting Tactical Urbanism through Innovation–Diffusion Theory: Insights from a Collaborative Design Studio Experience", the role of testability is fully demonstrated in the tactical urbanism project. This project is characterized by its temporariness. Initially, the staff of the welfare center was passive towards the specific project plans. However, when they learned that the facilities would be installed and operated for only one week, they were more likely to accept these plans. This is because the short - term trial made the staff feel that the risks were controllable, and if the trial did not go well, it would not have long - term negative impacts (Kim & Jin, 2024).
Observability	The ease with which the innovation results can be perceived by others.	In the waste separation collection station project within the tactical urbanism project, during the implementation process, students successfully persuaded uncooperative shopkeepers to participate by allowing them to observe other shopkeepers and residents using the collection station. This observable usage situation enabled potential adopters to see the practical effects of the innovation, enhanced their confidence in the innovation, and thus promoted the adoption of the innovation (Kim & Jin, 2024).

As a classic theory in the field of communication, the innovation diffusion theory has constructed a highly explanatory theoretical framework for in-depth exploration of the communication patterns and application mechanisms of various innovations within social systems. The digital art museum, as an innovative practice in the field of culture and art, has reshaped the way culture and art are displayed and disseminated.

The Ice and Snow Digital Art Museum of Harbin Ice and Snow World represents a digital innovation exploration in the field of ice and snow cultural tourism industry. Theoretically, this case falls within the analysis scope of the innovation diffusion theory, providing a real-world sample for theoretical verification and expansion.

In the previous research, the origin, core connotation of the innovation diffusion theory, and the advantages demonstrated by the digital art museum based on this theory at multiple dimensions have been systematically explained. The Harbin Ice and Snow World and its Ice and Snow Digital Art Museum are the materialized manifestations of the innovation diffusion theory in the scenario of the ice and snow cultural tourism industry, possessing typical value for case studies.

When the research perspective shifts from the general characteristics of the digital art museum to the specific case of the Harbin Ice and Snow World and its Ice and Snow Digital Art Museum, the application value of the innovation diffusion theory is more fully demonstrated. In the ice and snow cultural tourism industry, digital innovation achievements also face the practical challenges of being recognized, accepted, and widely disseminated by the target audience. The five key elements proposed by the innovation diffusion theory, namely relative advantage, compatibility, complexity, trialability, and observability, provide a systematic analytical framework for analyzing the Ice and Snow World and its digital art museum. There are internal logical connections among these elements, and they interact with each other, jointly revealing the internal development mechanism and communication path of the digital innovation in the ice and snow cultural tourism.

Firstly, Relative Advantage refers to the superiority of an innovation over the original things in economic, social, and other aspects. As an iconic project of the ice and snow economy, the Harbin Ice and Snow World has unique ice and snow landscapes and rich entertainment activities, giving it a significant relative advantage in the winter tourism market. With its large scale, it attracts a large number of tourists every year, generating considerable tourism revenue and driving the development of related industries such as catering, accommodation, and transportation in the local area. The immersive AI digital art museum of "One Ice and Snow, One World" uses digital technology to bring tourists a brand-new ice and snow experience, which is different from the traditional ice and snow viewing mode. This innovative experience attracts more tourists who pursue novelty and personalized experiences, expands the consumer group of the ice and snow economy, and demonstrates a relative advantage in terms of cultural experience.

Secondly, Compatibility involves the degree of fit between an innovation and an individual's existing values, behavior patterns, and past experiences. Various ice and snow activities held in the Harbin Ice and Snow World, such as the Ice and Snow Festival and ice sculpture competitions, are in line with people's traditional habits of celebrating in winter and appreciating art, and they meet the public's cognition and needs for winter entertainment and cultural activities. The immersive AI digital art museum of "One Ice and Snow, One World" integrates traditional cultural elements (such as the "Ninety-Nine Cold-Relieving Chart") into the digital art display, allowing tourists to experience the charm of modern technology while also feeling the charm of traditional culture. It is compatible with people's values of cultural inheritance and innovation and is easily accepted by the public.

Thirdly, Complexity reflects the degree of difficulty for an individual to understand and apply an innovation. In the Ice and Snow World and the Ice and Snow Digital Art Museum, the operators of the projects try their best to reduce the complexity of tourists' participation. The various amusement facilities and viewing areas in the Ice and Snow World are reasonably laid out with clear signs, enabling tourists to easily find the content they are interested in. The interactive devices in the immersive AI digital art museum of "One Ice and Snow, One World" are simple and easy to understand, and even tourists who are not familiar with digital technology can easily participate in the interactive experience, avoiding participation barriers caused by technical complexity and enabling tourists of different ages and technical levels to enjoy the fun brought by the ice and snow economy.

Fourthly, Trialability reflects the degree to which an innovation can be tried within a limited scope. The Ice and Snow World and the Ice and Snow Digital Art Museum can allow tourists to try by launching short-term themed activities or experience projects. For example, the Ice and Snow World sets up a new ice sculpture theme area within a specific time period, observes the tourists' feedback first, and then decides whether to keep it for a long time. The immersive AI digital art museum of "One Ice and Snow, One World" can also take advantage of the flexibility of digital technology to quickly update some of the display content or interactive links, optimize them according to the tourists' experience feedback, reduce the risk of innovation, and improve the adaptability and attractiveness of the project.

Fifthly, Observability is related to the ease with which the innovation results can be perceived by others. The Harbin Ice and Snow World, with its large scale, prominent geographical location, a large number of media reports, and tourists' sharing on social platforms, makes its ice sculpture art, entertainment activities, and other achievements easily observable by the outside world, attracting more people to come and experience. The immersive AI digital art museum of "One Ice and Snow, One World" conducts online promotion, displays the magnificent digital art scenes and novel interactive experiences in the museum, triggers widespread attention, enables potential tourists to intuitively feel the charm of the project, and thus attracts them to visit, promoting the dissemination and development of the ice and snow economy.

CONCLUSION

This study focuses closely on ice and snow cultural tourism with regional characteristics in cultural tourism, delves deeply into its digital transformation path, and gives priority to researching the crucial role of digital art museums, a branch of digital museums, in this process. Taking the Ice and Snow Digital Art Museum of the Harbin Ice and Snow World as the core case, this study conducts a comprehensive and systematic analysis by means of various research methods, and has achieved a series of results with both theoretical and practical values.

Relying on digital technologies, digital art museums have revolutionized the way artworks are displayed, and have achieved remarkable results in aspects such as technological demonstration, education, audience experience, and cultural dissemination. They break through the limitations of physical space and time. With the help of technologies such as 3D modeling, VR (Virtual Reality), and AR (Augmented Reality), they not only enable the massive storage and round-the-clock display of artworks, but also provide rich resources for art education and enhance the audience's art learning experience. Meanwhile, through diversified interactive designs, digital art museums strengthen the connection between the audience and the artworks, disseminate culture beyond geographical limitations, and build cultural exchange communities. However, existing research has gaps in the specific models of how digital art museums promote the digital transformation of cultural tourism, especially in the field of ice and snow cultural tourism. During the digital transformation process of the Harbin Ice and Snow World, there are also challenges in terms of cultural expression and the enhancement of tourists' experiences. The applied research on the digitalization of ice and snow cultural tourism based on the innovation diffusion theory also needs to be strengthened.

To make up for these research deficiencies, this study uses CiteSpace software to integrate the data from Web of Science and CNKI (China National Knowledge Infrastructure) for the visual analysis of knowledge maps. From the clustering analysis of the keywords “digitization” and “cultural tourism” in CNKI, it can be known that digital cultural tourism and the digital economy are the cores of the digital transformation of the cultural tourism industry. Digital technologies drive the innovation of cultural tourism, creating new experiences and economic values. The integration of the digital economy and cultural tourism promotes industrial upgrading. Cultural tourism, the integration of culture and tourism, and digital countryside, as important supports, respectively play key roles in digital marketing, the integration of cultural resources and tourism, and the development of rural cultural tourism. The clustering analysis of the keyword “digital museum” in WoS (Web of Science) shows that digital technologies are the key to the digitalization of cultural tourism. Technologies such as 3D models and digital protection help with the digitalization of cultural heritage and the display of tourist attractions. Emerging technologies such as 3D printing bring new opportunities. The digital development of cultural heritage needs to focus on integration with the environment, and social media is of great importance in the digital dissemination of cultural tourism. In general, digital technologies are the core driving force for the digital transformation of the cultural tourism industry, and elements such as the integration of culture and tourism work in concert to promote the digital development of the industry.

The state attaches great importance to the digitalization of the cultural tourism industry and has issued a series of policies to support the digital transformation of ice and snow cultural tourism. The "Opinions on Promoting the Implementation of the National Cultural Digitization Strategy" builds a national cultural big data system. The "14th Five-Year Plan for the Development of the Tourism Industry" clarifies the development direction of tourism digitization. The "Opinions of the Ministry of Culture and Tourism on Promoting the High-quality Development of the Digital Culture Industry" encourages cultural tourism enterprises to innovate. The Ministry of Industry and Information Technology and the Ministry of Culture and Tourism jointly promote the integration of 5G and smart tourism. The "Several Policies on Promoting the Recovery and Development of Difficult Industries in the Service Sector" helps cultural tourism enterprises optimize their operations. The "Action Plan for the Innovative Development of Smart Tourism" cultivates smart tourism formats. These policies provide all-round support for the digitalization of ice and snow cultural tourism.

The immersive AI digital art museum of "One Ice and Snow, One World" in the Harbin Ice and Snow World, with the help of AIGC (Artificial Intelligence Generated Content) technology, integrates the "Ninety-Nine Cold-Relieving Chart" into the ice and snow, and creates characteristic exhibition items such as interactive scrolls, solving the problem of the lack of cultural expression in traditional ice and snow tourism and enriching tourists' experiences. The application of AIGC technology not only responds to the issue in this study regarding its assistance in the digital transformation of ice and snow cultural tourism, but also highlights the key role of digital technologies in the digitalization of cultural tourism, enriches the communication carriers, and explores the tourism value of cultural heritage. From the perspective of the innovation diffusion theory, this technology has enhanced the relative advantage, compatibility, trialability, and observability of the digital innovation of ice and snow cultural tourism, strengthened the competitiveness of products, is in line with the public's demand for cultural inheritance, makes it convenient to optimize projects according to tourists' feedback, attracts more potential tourists through social media, and promotes the dissemination of digital innovation in ice and snow cultural tourism.

The innovation diffusion theory provides a theoretical framework for studying the digital innovation and dissemination of ice and snow cultural tourism. The Harbin Ice and Snow World and its Ice and Snow Digital Art Museum, as practical cases of the application of this theory in the ice and snow cultural tourism industry, in terms of relative advantage, attract tourists with unique landscapes and digital technology experiences; in terms of compatibility, the activities and display content conform to the public's cultural habits and values; in terms of complexity, they lower the participation threshold, making it convenient for tourists to have experiences; in terms of trialability, they reduce the innovation risk through short-term activities and project optimization; in terms of observability, they attract tourists with the help of media and online promotion, and promote the development of the ice and snow economy.

Although this study provides theoretical and practical references for the digital transformation of ice and snow cultural tourism, there are still limitations. In terms of research content, the study of the digital transformation models of different ice and snow cultural tourism projects is not in-depth enough; in terms of technological applications, there is insufficient exploration of emerging technologies; in terms of user research, the study of the changes in tourists' behaviors and needs is not meticulous enough. Future research can be deepened in multiple aspects, such as exploring targeted models for the digital transformation of different types of ice and snow cultural tourism projects, studying the applications of emerging technologies such as blockchain and the metaverse, strengthening the research on user behaviors and experiences, exploring the connotations of ice and snow culture, and promoting the sustainable development of ice and snow cultural tourism.

REFERENCES

www.suzhou.gov.cn. (2025, January 7). The Immersive Digital Experience Exhibition of the Palace Museum Opens. https://www.suzhou.gov.cn/szsrmzf/szyw/202501/78b4e73690c241e6b1abd216905052e8.shtml

Braden, L. E. A., & Teekens, T. (2020). Historic networks and commemoration: Connections created through museum exhibitions. *Poetics*, *81*, 101446. Advance online publication. DOI: 10.1016/j.poetic.2020.101446

Castellotti, S., D'Agostino, O., & Del Viva, M. M. (2024). Effectiveness of labels in digital art experience: Psychophysiological and behavioral evidence. *Frontiers in Psychology*, *15*, 1342667. Advance online publication. DOI: 10.3389/fpsyg.2024.1342667 PMID: 39011289

Ch'ng, E., Cai, S. D., Leow, F. T., & Zhang, T. E. (2019). Adoption and use of emerging cultural technologies in China's museums. *Journal of Cultural Heritage*, *37*, 170–180. DOI: 10.1016/j.culher.2018.11.016

CHINA CENTRAL TELEVISION. (2024, January 11). *Harbin ice sculptors: Store ice one year in advance and build the Harbin Ice and Snow World within 15 days.* https://news.cctv.com/2024/01/11/ARTIRhIt4TDjrhYWtXSy7qAJ240111.shtml

Chinanews.com. (2024, December 20). *The immersive AI digital art museum named "One Ice and Snow, One World" has made its debut at the Harbin Ice and Snow World.* https://www.toutiao.com/article/7450392175508062783/?upstream_biz=doubao&source=m_redirect

Departure of Culture and Tourism of Heilongjiang Province. (2024, December 26). *Top - tier Harbin has something new! The "One Ice - Snow, One World" immersive AI digital art museum has made a stunning debut at Harbin Ice - Snow World.* https://wlt.hlj.gov.cn/wlt/c114172/202412/c00_31796073.shtml

Evrard, Y., & Krebs, A. (2018). The authenticity of the museum experience in the digital age: The case of the Louvre. *Journal of Cultural Economics*, *42*(3), 353–363. DOI: 10.1007/s10824-017-9309-x

Fabbri, K., & Pretelli, M. (2014). Heritage buildings and historic microclimate without HVAC technology: Malatestiana Library in Cesena, Italy, UNESCO Memory of the World. *Energy and Building*, *76*, 15–31. DOI: 10.1016/j.enbuild.2014.02.051

Gao, Y., Zhang, Q., Wang, X. R., Huang, Y. M., Meng, F. S., & Tao, W. (2024). Multidimensional knowledge discovery of cultural relics resources in the Tang tomb mural category. *The Electronic Library*, *42*(1), 1–22. DOI: 10.1108/EL-04-2023-0091

Harbin Municipal People's Government. (2024, January 5). *The world's largest ice and snow themed amusement park! Harbin Ice and Snow World has been awarded the title of Guinness World Records.* https://www.harbin.gov.cn/haerbin/c111753/202401/c01_958584.shtml

Kastner, L., Umbach, N., Jusyte, A., Cervera-Torres, S., Fernández, S. R., Nommensen, S., & Gerjets, P. (2021). Designing Visual-Arts Education Programs for Transfer Effects: Development and Experimental Evaluation of (Digital) Drawing Courses in the Art Museum Designed to Promote Adolescents' Socio-Emotional Skills. *Frontiers in Psychology, 11*, 603984. Advance online publication. DOI: 10.3389/fpsyg.2020.603984 PMID: 33536974

Kim, J., & Jin, H. Y. (2024). Interpreting Tactical Urbanism through Innovation-Diffusion Theory: Insights from a Collaborative Design Studio Experience. *Land (Basel), 13*(1), 14. Advance online publication. DOI: 10.3390/land13010014

Kim, M. J., Lee, C. K., & Contractor, N. S. (2019). Seniors' usage of mobile social network sites: Applying theories of innovation diffusion and uses and gratifications. *Computers in Human Behavior, 90*, 60–73. DOI: 10.1016/j.chb.2018.08.046

Kotoula, E., Robinson, D. W., & Bedford, C. (2018). Interactive relighting, digital image enhancement and inclusive diagrammatic representations for the analysis of rock art superimposition: The main Pleito cave (CA, USA). *Journal of Archaeological Science, 93*, 26–41. DOI: 10.1016/j.jas.2018.02.012

Le Fillâtre, V., Robert, E., Petrognani, S., Lesvignes, E., Cretin, C., & Muth, X. (2021). Mapping the walls: High-resolution cartography applied to the analysis of prehistoric cave art in the Grotte du Mammouth (Domme, Dordogne, France). *Journal of Archaeological Science, 127*, 105332. Advance online publication. DOI: 10.1016/j.jas.2021.105332

Lischer-Katz, Z. (2022). The emergence of digital reformatting in the history of preservation knowledge: 1823-2015. *The Journal of Documentation, 78*(6), 1249–1277. DOI: 10.1108/JD-04-2021-0080

Meinecke, C., Hall, C., & Janicke, S. (2022). Towards Enhancing Virtual Museums by Contextualizing Art through Interactive Visualizations. *Acm Journal on Computing and Cultural Heritage, 15*(4), 62. Advance online publication. DOI: 10.1145/3527619

Motta, F. M. D., Barbosa, C. R., & Barbosa, R. R. (2019). BIG DATA AS INNOVATION SOURCE IN MUSEUMS: the case study of The British Museum. *Informacao & Sociedade-Estudos, 29*(1), 83-100. <Go to ISI>://WOS:000462627300006

Noble, K. (2021a). Challenges and Opportunities: Creative Approaches to Museum and Gallery Learning during the Pandemic. *International Journal of Art & Design Education, 40*(4), 676–689. DOI: 10.1111/jade.12380

Noble, K. (2021b). 'Getting Hands On with Other Creative Minds': Establishing a Community of Practice around Primary Art and Design at the Art Museum. *International Journal of Art & Design Education, 40*(3), 615–629. DOI: 10.1111/jade.12371

Rogers, E. M. (2016). *Diffusion of Innovations* (5th ed.). Publishing House of Electronics Industry.

Valente, T. W., Dyal, S. R., Chu, K. H., Wipfli, H., & Fujimoto, K. (2015). Diffusion of innovations theory applied to global tobacco control treaty ratification. *Social Science & Medicine, 145*, 89–97. DOI: 10.1016/j.socscimed.2015.10.001 PMID: 26460508

Xing, Z., Feng, Q. J., Chen, H. R., Dai, Q., Hu, H., Xu, H., Wu, Z. X., & Jiang, Y. G. (2025). A Survey on Video Diffusion Models. *ACM Computing Surveys, 57*(2), 41. Advance online publication. DOI: 10.1145/3696415

Zhao, L. T., Zhang, J. W., Jing, H. R., Wu, J. P., & Huang, Y. J. (2023). A Blockchain-Based cryptographic interaction method of digital museum collections. *Journal of Cultural Heritage, 59*, 69–82. DOI: 10.1016/j.culher.2022.11.001

Zhou, Y. Z., Feng, T., Shuai, S. H., Li, X. D., Sun, L. Y., & Duh, H. B. L. (2022). EDVAM: A 3D eye-tracking dataset for visual attention modeling in a virtual museum. *Frontiers of Information Technology & Electronic Engineering, 23*(1), 101–112. DOI: 10.1631/FITEE.2000318

Chapter 9
From Research to Reality: AI, the HBCU Classroom, and Community Engagement

Cassandra Sligh Conway
http://orcid.org/0000-0001-9357-676X
South Carolina State University, USA

Michelle L. Priester
http://orcid.org/0009-0000-9806-8967
South Carolina State University, USA

Tammara Thomas
http://orcid.org/0009-0009-2532-2210
South Carolina State University, USA

ABSTRACT

Historically Black Colleges and Universities (HBCUs) have been vital in producing many successful graduates (Logan, 2024). HBCUs have historically faced numerous challenges, particularly funding and access to certain resources. AI resources are being advanced increasingly, and some HBCUS may not benefit from these resources because of a lack of funding. The objectives of this chapter are as follows: 1) outline historical and contemporary research and its impact on the HBCU experience (research to realization); 2) explore the impact of AI innovations on HBCUs, focusing on how these technologies can improve educational outcomes in the classroom to adopt a more inclusive and supportive learning environment; 3) provide strategic recommendations for HBCUs to enhance their AI resources, drawing on best practices; and 4) stress the importance of including AI in the classroom and community to support technological advancements at HBCUs, outlining ways to build partnerships and secure funding to sustain these initiatives.

DOI: 10.4018/979-8-2600-2378-5.ch009

INTRODUCTION

Historically, Black Colleges and Universities have lived through many storms in life such as civil rights, inequality, lack of funding, COVID 19, and slavery. Some HBCUS have struggled but kept the doors open and managed to keep the mission and vision alive. These challenges have led to a disparity in educational opportunities compared to predominantly white institutions. HBCUs are resilient and have survived many years. The HBCU is and continues to be a great structure for the education of many in science, medicine and the arts. The ultimate reality is that there is a movement from the research about the HBCUs that takes one into a journey of understanding the background and the future of it. To understand the background and the future, one must review the backbone of the HBCU. The backbone of the HBCU came from the support of so many.

The HBCU Coalition (2024) noted that the first HBCU was established in 1837 and that these institutions expanded significantly following the Civil War. Throughout their history, HBCUs have faced numerous challenges, including educational inequities, systemic oppression, violence, and limited resources. Despite these obstacles, HBCUs have demonstrated remarkable resilience and have continued to thrive while providing educational opportunities and promoting social and economic advancement for African Americans (HBCU Coalition, 2024).

Based on many facts about the HBCU experience, there are some highlights about the greatness of the HBCU experience. Guy-Sheftall and Jackson (2021) noted that, "Historically, HBCUs have produced 50% of Black teachers and doctors and 80% of Black judges. They continue to graduate the largest percentage of African American students with degrees in science, technology, engineering, and mathematics. According to the National Science Foundation, Spelman College has produced the highest number of Black women doctoral recipients in science and engineering. Xavier University of Louisiana and Howard University have been the top producers of African American medical school graduates. HBCUs deliberately focus on brokering access to STEM through a variety of policies, practices, and activities. Howard University now has the distinction of having graduated the first woman vice president in the nation's history. Other outstanding alumni and alumnae include Toni Morrison, Thurgood Marshall, Martin Luther King Jr., Oprah Winfrey, Common, Stacey Abrams, Rosalind Brewer, Andrew Young, Levi Watkins Jr., Jesse Jackson, Samuel L. Jackson, and Chadwick Boseman" (para. 3).

For some students, the HBCU is the only experience that allows them to attend college due to financial constraints or transportation barriers. HBCUs continue to serve many roles such as a sense of support for first generation students who need a university that is near and affordable. Other roles of the HBCU are to remember the historical moments that significantly trace the various stages of a cultural presence and change. HBCUs are commonly universities full of culture, and it is still a sense of pride for a person to attend an HBCU. For instance, most HBCUS are in the South, and few are in the North. Because of the HBCU traditionally being represented in the South, more HBCUs are in rural areas where students are from diverse socioeconomic backgrounds. Therefore, the HBCU is a source of many jobs in some areas.

HBCUs do not receive as much funding as certain universities, so the funding is always a challenge. Currently, the "executive order to pause federal grants and financial aid programs" has affected the HBCU (Gasman, 2025, para 3). According to Gasman (2025), "Despite a legacy of supporting Black students and their surrounding communities, HBCUs have often been vulnerable to shifting political priorities. Under the current Trump administration, executive orders — which have paused or canceled federal grants and equity-focused programs — have disrupted essential funding streams… While an injunction temporarily restored some funding, the risk of future cuts remains high, jeopardizing the economic stability of HBCUs and other colleges and universities. History has shown how interruptions in federal funding disproportionately affect HBCUs, leading to declining enrollment, staff layoffs, and stunted institutional growth. Without immediate action, these funding threats will exacerbate inequities in higher education, weaken career preparation for HBCU graduates, and undermine the economic contributions of these institutions" (para.4). Despite these occurrences, HBCUs have survived over many years without stopping, without being stopped, and is resilient. In competing with other universities, the HBCU must provide a holistic experience that is comparable to what students will find at other universities primarily because the technology must continue to be competitive and useful.

Research does point out that PWIs receive more funding opportunities overall and that there is a significant gap. Research continues to outline how the federal government should support HBCUs (Alexander, 2024). Based on the research of Alexader (2023), "officials found that 16 states have underfunded HBCUs by $13 billion over the course of three decades. That neglect is further underscored by the endowment gap between PWIs and HBUCS which currently stands at 100:1. In essence, every $100 a PWI gets in endowment money, an HBCU gets a $1. When comparing PWI land grant institutions to land grant HBCUs, the disparity is equally a staggering:$45 billion in endowment assets with PWIs compared to $1 billion for HBCUs" (para. 4/5). This is a startling reality, but it is a fact that deserves more attention in public forums, further examination and more ways to resolve such a gap.

HBCU communities tend to thrive by connecting with the community and key stakeholders. The alumni tend to support HBCUS. This in turn, assists in fundraising activities and capital gains in the community, state, and federal government.

As we advance into the 21st century, integrating digital and artificial intelligence (AI) technologies into the classroom and community has become essential for the sustainability and growth of educational institutions worldwide. Further, artificial intelligence tools are transforming educational settings, including health care (Mei-Chin, Li-En, Li-Hwa, Yu-Chun, 2024). For HBCUs, embracing these technologies is not merely about keeping pace with educational trends but ensuring their students receive an education that equips them with the skills necessary to thrive in a technology-driven economy.

AI technology operates by generating content by matching patterns observed, utilizing machine learning algorithms to analyze large amounts of data that are assisted by human intelligence, such as thought, deep learning, adaptation, engagement and sensory understanding (Secinaro, Calendra, Secinaro, 2021). AI technologies offer transformative potential for HBCUs by enhancing teaching and learning processes, streamlining administrative tasks, and opening new avenues for research and collaboration. These technologies can provide personalized learning experiences, expand access to educational resources, and help bridge the digital divide that has historically affected African American communities.

PURPOSE

The objectives of this chapter are as follows: 1) outline historical and contemporary research and its impact on the HBCU experience (research to realization); 2) explore the impact of AI innovations on HBCUs, focusing on how these technologies can improve educational outcomes in the classroom to adopt a more inclusive and supportive learning environment; 3) provide strategic recommendations for HBCUs to enhance their AI resources, drawing on best practices; and 4) stress the importance of including AI in the classroom and community to support technological advancements at HBCUs, outlining ways to build partnerships and secure funding to sustain these initiatives.

By embracing AI technologies, HBCUs can continue to fulfill their mission of providing high-quality education to students and preparing them to be leaders in a rapidly changing world. The integration of these innovations is not just a necessity but an opportunity for HBCUs to reaffirm their role as vital institutions in the American higher education landscape and to ensure that their students are well-equipped to meet future challenges and opportunities. One leader, Dr. Robbie Melton (2025) stated why the HBCU must embrace AI as follows: "The core of our success lies in our heritage of open sharing and collaboration. HBCUs have long championed a culture of communal support, extending beyond the confines of our campuses to influence generations. This tradition is our guiding light as we navigate the complexities of integrating emerging technologies into our curricula. It is through the shared exploration of practices and methods that we can truly harness the potential of AI to serve our students effectively. We stand at the intersection of innovation and legacy, tasked with the responsibility of equipping our learners not just with knowledge but with the wisdom to apply it ethically and inclusively. As educators, our role extends beyond imparting technical skills; we are cultivators of critical thinkers, problem solvers, and, most importantly, compassionate leaders who will shape the future. Our collective commitment to fostering an environment of inclusive excellence and mutual growth is paramount. As we delve into this new era, let us carry forward the ethos of HBCUs— unity, resilience, and a steadfast dedication to empowering our communities. Together, we can create a future where the advancements of digital learning and AI not only reflect the diversity of our society but actively contribute to its enrichment. In this journey towards innovation and beyond, I invite you to join me in embracing our role as pioneers. Let us share, learn, and grow together, ensuring that our students are not only prepared for the challenges ahead but are also imbued with the values that have always set HBCUs apart. Together, we can lead the way in creating a world where technology uplifts, empowers, and unites" (p. 18).

IMPORTANCE OF AI AT HBCUS

In contemporary education, digital technologies are reshaping how learning takes place, providing more flexibility, accessibility, and engagement for students. HBCUs must embrace these technologies to remain competitive in higher education and meet the diverse needs of their student populations. Digital platforms allow for remote learning, which has become increasingly vital in the wake of the COVID-19 pandemic. Moreover, technologies (virtual classrooms, learning management systems, and AI-based educational tools) enable more personalized learning experiences tailored to individual student needs.

AI is a way to take action and achieve goals and outcomes (Boucher, 2020). Several authors noted the benefits of AI. The global workforce is characterized by rapid advancements in the field of artificial intelligence. HBCUs are responsible for ensuring that their graduates possess the digital skills and technological fluency required to succeed in a competitive, global economy. Universities must select and provide courses that focus on AI, machine learning, and digital literacy. These courses can be incorporated into the curriculum across disciplines to prepare students to use skills in different disciplines. By familiarizing students with these technologies early in their academic careers, HBCUs ensure that their graduates are capable of ethical and innovative use of AI technology, thus contributing to the increase in positive outcomes in the classroom and community.

BACKGROUND

AI Innovations at HBCUs

Historically Black Colleges and Universities (HBCUs) have been pillars of higher education for African American students, providing access to quality education, cultural enrichment, and upward mobility for generations. These institutions were established to offer equitable educational opportunities to students of African descent, who were historically excluded from many predominantly white institutions (PWIs). Today, HBCUs are essential to African American students' academic and professional success, arguably outperforming PWIs in student experience, affordability, and post-graduation preparedness areas.

HBCUs are unique in their mission to adopt a supportive, inclusive environment that focuses on academic excellence and nurtures students' social and cultural development. This mission is a testament to the resilience and strength of these institutions. The contributions of HBCUs to African American educational achievement are significant. Although these institutions account for only a small number of U.S. colleges and universities, these universities produce a high number of African American graduates, including many who earn degrees in STEM fields (UNCF, 2024).

Integrating Artificial Intelligence (AI) technologies at HBCUs presents an opportunity to enhance educational outcomes and efficiency. AI innovations offer a more personalized educational experience. For example, AI-driven platforms can adapt to individual learning styles and this, in turn, can maximize the overall learning process. These technologies also enable HBCUs to expand their reach by offering virtual classrooms and online courses catering to non-traditional students such as working adults. This increased accessibility is especially important as HBCUs have historically served as gateways to higher education for underrepresented and underserved communities.

Furthermore, the use of technology in the classroom also prepares students for careers in a digital world. In terms of administrative processes which occur daily in educational settings, digital and AI technologies can be beneficial in significantly improving efficiency at HBCUs in the areas of student admissions, enrollment management, and grading, Utilizing AI technologies in these areas can help to manage and reduce faculty and staff workloads. Additionally, data analytics tools provide key information into student performance and institutional operations which can be critically important to improving retention and graduation rates.

Adopting these technologies also plays a crucial role in advancing HBCUs and bridging the digital divide that historically affected African American communities. By ensuring that students at HBCUs have access to the latest resources, these institutions are leveling the playing field and providing students with opportunities comparable to those available at PWIs. This commitment to technological equity is essential for preparing students for the modern workforce, where digital literacy and technological proficiency are key skills. Furthermore, using AI technologies can enhance the research capabilities of HBCUs, enabling faculty and students to engage in cutting-edge research and innovation. This advancement benefits the academic community and contributes to the broader societal goal of achieving technological equity and inclusion.

According to Davis (2024), "HBCUs have a long legacy of excellence and inclusivity to bring to bear on our exploration and adoption of artificial intelligence. HBCUs excel at embedding cultural competence into their curriculum and preparing graduates for impacting their communities and our greater world. Additionally, HBCUs have long played an outsized role in producing STEM graduates. The McKinsey Institute for Black Economic Mobility reported that while in 2018 Black students only earned seven percent of all STEM baccalaureate degrees, HBCUs produced almost a quarter of those graduates. The need to create racial parity in artificial intelligence is especially critical. By 2045, the McKinsey Institute for Black Economic Mobility reports, generative AI has the potential to widen the Black/White household wealth gap by $43 billion annually – but, if implemented carefully, the technology may have beneficial impacts instead. Additionally, it is imperative that ethical AI systems be developed in consultation with as diverse a group of stakeholders as possible" (para 5). Another author stated that the HBCU must promote AI literacy. Gibbons (2024) emphasized that colleges and universities should incorporate artificial intelligence (AI) courses into their curricula. Including these AI courses so can equip students, faculty, and staff with valuable knowledge and skills that enhance abilities and positively influence the communities they serve.

Students can interact and interface with AI tools in hands-on ways, such as using them for research, developing academic or professional projects, and solving real-world problems. Universities can also offer workshops that encourage the use of AI tools. Universities can also promote collaboration across disciplines through group projects that integrate AI applications. Instruction in AI is also important for preparing students to use technology in a responsible and constructive manner. Finally, inviting companies to partner on AI projects can provide students with access to current technologies, as well as opportunities for internships and mentorship that strengthen real-world learning experiences.

Overall, integrating AI innovations at HBCUs holds great potential for transforming education, improving administrative efficiency, and bridging the digital divide.

By embracing these technologies, HBCUs can continue to fulfill their mission of providing high-quality education to African American students while also preparing them to thrive in an increasingly digital world. Because HBCUs are instrumental in the lives of students, they can easily remain at the forefront of educational excellence, fostering the next generation of leaders, innovators, and changemakers.

STRATEGIC RECOMMENDATIONS/AI

This section of the paper recommends key strategies that can be useful at HBCUs to enhance AI capabilities. Some recommendations are noted as personal learning, comprehensive educational experiences, and accessibility for non-traditional students.

Personalized Learning in the Classroom

Integrating Artificial Intelligence (AI) into educational settings can revolutionize personalized learning, especially at HBCUs. AI-driven platforms can adapt to individual learning styles and provide customized learning experiences that cater to each student's unique needs. An example is measuring data concerning the use of materials for a course and identifying areas where students are unsure, then providing support resources to improve learning and engagement with the course. Technology plays a vital role in enhancing academic performance and creating engaging learning environments.

A solid personalized learning pathway might be through AI-based tutoring systems that provide instant feedback and modify teaching modes in response to a learner's performance. For instance, if a student shows difficulty in a specific mathematical concept, the system can offer additional exercises or alternative explanations until the student demonstrates mastery. Such systems ensure that each student progresses at their own pace, reducing frustration and increasing the chance of academic success.

In addition, AI can assist in creating personalized study plans by assessing a student's strengths and weaknesses. For example, platforms like Coursera or Khan Academy use AI to recommend courses based on past performance and learning preferences. Implementing similar strategies at HBCUs could help guide students through their academic journey more effectively, ensuring they take courses that align with their career goals and academic needs. This approach enhances learning outcomes and helps students develop a stronger sense of ownership over their educational paths.

Accessibility for Non-Traditional Students

HBCUs have long served as gateways to higher education for underrepresented communities, including non-traditional students such as working adults and those in remote areas. Expanding virtual classrooms and online courses has further broadened access to education for these groups. Online learning platforms offer flexibility in scheduling and location, enabling students to balance their studies with work and family responsibilities. This is particularly beneficial for those who may not have the means or ability to attend on-campus classes.

The impact of AI is evident in the increased enrollment and retention rates among non-traditional students at HBCUs. Many institutions report that offering online degree programs and flexible learning options has attracted a more diverse student body, including military personnel, full-time workers, and parents. These programs provide educational opportunities and contribute to students' economic mobility, especially those who may not have pursued higher education under conventional circumstances.

The Office of Educational Technology (2024) stated certain benefits of AI to students with disabilities. By using AI, "...teaching requires significant time in planning as well to account for the breadth of needs across their rosters—especially for inclusive learning environments and students with IEPs and 504 plans. AI could help teachers with recommendations that are tuned to their situation and their ways of practicing teaching and support with adapting found materials to fit their exact classroom needs. For students with an IEP, AI could help with finding components to add to lesson plans to fully address standards and expectations and to meet each student's unique requirements. Even beyond finding components, AI might help adapt standardized resources to better fit specific needs—for example, providing a voice assistant that allows a student with a visual difficulty to hear material and respond to it or permitting a group of students to present their project using American Sign Language (ASL) which could be audibly voiced for other students using an AI ASL-to-Spoken-English translation capability. Indeed, coordinating IEPs is time-consuming work that might benefit from supportive automation and customized interactivity that can be provided by AI" (p. 71).

Tech Partnerships

Collaborations with technology giants like Microsoft and Google have been instrumental in enhancing the digital capabilities of HBCUs. These partnerships provide access to the latest software, training programs, and cloud computing services, enabling HBCUs to offer state-of-the-art educational experiences. For example, Microsoft's partnership with several HBCUs includes access to its Azure cloud platform, enabling students to gain hands-on experience with the same tools industry professionals use. These collaborations have enhanced academic experience and equipped students with valuable skills that increase their employability in a competitive job market.

Automation

Automation transforms administrative processes at HBCUs by streamlining operations such as admissions, enrollment management, and grading. AI-powered systems can manage large volumes of data, reducing the manual workload on administrative staff and enabling them to focus on more strategic tasks. Some authors note that managing data systems can change the trajectory of a company or organization (Nadal et al., 2022). For example, automated admissions systems can sort and evaluate thousands of applications based on predefined criteria, ensuring a fair and efficient selection process. This reduces the potential for human error and bias, making the process more transparent and equitable. Automated grading systems are another example of how AI can enhance administrative efficiency.

Data Analytics

Data analytics is a powerful tool for improving institutional decision-making at HBCUs. By analyzing trends in student performance, enrollment, and resource allocation, administrators can gain valuable insights that inform strategic planning and policy development. Ensuring access to the latest AI resources is essential for leveling the playing field. Partnerships with technology companies like Microsoft and Google have been instrumental in enhancing technological access and equity. Microsoft's partnership with several HBCUs includes access to cloud platforms. These partnerships often include professional development opportunities for faculty, ensuring that educators are well-informed and using cutting-edge technology.

Impact on the Classroom: Teaching/Research Capabilities

Collaborations with technology companies also play a crucial role in supporting the classroom and research initiatives. These partnerships often provide access to specialized software and training in artificial intelligence. For example, Google's partnership with several HBCUs includes workshops and certification programs that equip students and faculty with the skills to conduct cutting-edge research. Such initiatives enhance the research output of HBCUs and contribute to the broader academic community by addressing pressing societal challenges.

Foster Connections

AI allows multiple people to engage in communication within teams, facilitating collaboration and helping to achieve shared goals and outcomes. Focusing on project goals in committee work can streamline productivity, allowing members to concentrate on outcomes rather than administrative tasks. An example of this would be the AI feature in platforms like Zoom that offer tools to support engagement. This platform is user-friendly, and with minimal training and experience a user can take advantage of AI's ability to take notes and effectively capture all the ideas and comments of those involved in a project.

Students and faculty can benefit from the AI minutes that is a feature in Zoom. For example, during a particular class, the faculty member can record a lecture on a specific topic, and then include the minutes shared. This is an effective tool because the faculty member does not waste a lot of time, and it allows the exchange of information to occur at a faster rate. What took 2 days now can be accomplished in minutes.

Community Engagement

Community engagement "refers to the process of building relationships and partnerships between individuals, organizations, and communities to achieve a common goal. It involves collaborating with local residents, stakeholders, and other organizations to identify and address community needs, improve communication, and foster a sense of ownership and responsibility" (CLRN Team, para. 2). As part of the process of supporting technological advancement at HBCUs, community engagement must be included. This engagement will allow collaborative partners to be active, and share their opinions, concerns, and hopes (Talo, 2018) for how AI will be utilized at HBCUs. Benneworth, Ćulum, Farnell, et al. (2018) contend that the promotion of community engagement depends on the level of commitment by HBCU administrators and the community at large. Benneworth, et al (2018) also states that the "framework of the engagement should be developed in such a way as to reinforce attempts to promote university-community engagement that have internal and external benefits and discourage more instrumental and bad-faith approaches" (pg.146). Hazelkorn (2016) expounds on this concept further by reiterating how community engagement models are important and suggests that HBCUs should work towards community engagement in a manner that is conducive to its character and ethical standards. The three models include the social justice model, the economic development model, and the public good model.

The social justice models' central point of attention is on the social disadvantage of students and emphasizes community empowerment and engagement. Given the lack of AI resources that students at HBCUs generally have, additional assistance from the community can aid in the technological development and growth of the students. Barrett et al. (2023) report that although "mainstreaming digital technologies, the digital divide continues to widen, leaving underserved communities behind, especially as the most basic modes of health communication and community engagement shift online" (pg. 425).

The economic development model focuses on the innovativeness of technology and the ability to make marginal profits. This model is implemented by the development of entrepreneurs and local businesses with an emphasis on technology. The public good model is concerned with the well-being of the community and focuses on fortifying the community through education and awareness of the utilization of AI and other technologies (Hazelkorn, 2016). The more knowledgeable of AI future employers and partners are, the more willing they may be to provide funding and support for technological initiatives. Collaborative efforts between HBCUs and the community empower students to be equipped for careers in a technologically advancing environment.

For community engagement to be effective for HBCUs and their partners, all parties must be willing to share information and resources (Barrett et al., 2023). The sharing of this information will help to assess the needs of HBCUs and determine possible solutions. Furthermore, the implementation of technology in these efforts will help community engagement in various ways such as improved communication; increased collaboration, wider audience reach, and prevention of participation barriers (Caperon et al., 2022). This can be accomplished via various digital community engagements such as online community panels, participatory forums, and digital storytelling (Barrett et al., 2023).

For AI innovations to have a long-term effect on the classroom and community, key faculty must be allowed to attend conferences on integrating AI in the classroom and ways to foster connections in the community. Faculty can attend these conferences with the expectations that the faculty share the information with other faculty upon return to the university.

Faculty members can include key AI industry contacts as speakers in certain classes. Certain modules can be developed in the class schedule that allows speakers to provide opportunities to share the latest and groundbreaking AI technology. Faculty can partner with these industry trailblazers to make a direct impact on curriculum development. Additionally, invite an AI staff from an external agency to serve on the program's advisory board. This person can review the curriculum and make suggestions that can impact the program. The program can then include some program modifications that can impact student enrollment, the classroom environment, and encourage more financial support from industries and community agencies.

Moving from the Classroom to Community Engagement

The traditional classroom has significant impacts in today's society; however, if AI is not used in in the classroom at some level the population that will suffer is the student. The students will be in a classroom that is impactful, but the students will not be able to serve in the cutting-edge technological advancements. The most significant impact of not using AI in the classroom is graduating a student that is not ready for the real world of work. Chace (2020) explained that artificial intelligence can enhance learning by supporting technologies such as virtual reality, making education more immersive and engaging. These tools can allow students to experience historical events in a more interactive way. AI may also make education more enjoyable and effective by supporting more kinesthetic and active learning experiences, such as interactive forums, dialogue-based learning, and movement-based engagement. Nandhini (2025) stated that AI should be used to complement traditional teaching, but not as a stand-alone. Both traditional and AI can be integrated to provide the ultimate classroom experience. For example, Nandhini noted the positive reason to integrate AI into the classroom as follows: "...Beyond teachers, AI can act as a 'mentor' that offers feedback and guidance outside class hours, assisting students during their homework or late-night study sessions when teachers are unavailable. Certain AI systems integrate game mechanics into learning, encouraging students to persist after failures by rewarding effort and perseverance, not just correct answers. Unlike traditional teaching, AI-based systems can be available to students at any time, offering support beyond classroom hours, which is particularly helpful for self-paced learning and distance education. AI tools can help students with different learning challenges, like using speech-to-text for those who find writing difficult or providing visual help for those with poor eyesight. This makes learning easier for everyone...(para. 3)...By working together, AI and human teachers can create a more efficient and effective educational system, using the strengths of both to improve learning. Schools and educators will need to adjust to this new approach, using AI tools while keeping the human aspects of teaching intact (para. 4)."

Moving from the AI classroom to community engagement requires collaboration among faculty, staff, and the administrators with key agency connections. Noyes and Girdharry (2024) revealed how Ai can be an opportunity for faculty to connect with other faculty, staff, and community. These authors reported that "Because AI will have far-reaching effects on education, faculty must be the ones to lead this transformation. Specifically, faculty urgently need to build their own AI communities of practice and engage with other educators to understand both the opportunities and risks presented by AI. Faculty must come together to probe how AI can accelerate both student learning and research innovation" (para. 3). Moreover, Melo (2025) noted that including AI in the classroom can promote student learning by providing a personalized experience. AI systems are quick and can provide immediate instruction and learning patterns that are used in feedback and supplementary information. This approach to AI in the classroom can provide a deeper dive into understanding the benefits of technology. This, in turn, allows for more opportunities to understand and be a part of the digital world.

Personal Journey

In this section, a personal journey of a faculty member is shared to understand how impactful AI is in sustaining the classroom and it is key to establishing community engagement. Further, the faculty member states how the HBCU must be open to providing AI opportunities to faculty to provide opportunities to enhance the classroom, need for more funding opportunities and resources, and the connection with the community. The student summarized what AI tools were beneficial to student learning and benefits of AI.

"In my years of teaching in academia, I learned many essential things about the classroom and engaging and interacting with others at the community level. There were times when the resources were short but my care for the students was never short, and I took great care in making sure every student had what was needed in the classroom to be successful on many levels. I do think that there are times that lack of funding did not make things easier, but I relied on other areas to get through it. When I first entered the profession, of course AI was not the buzz word. However, as time moved forward the integration of AI became the buzz word. At my university, there are resources available like Zoom, Teams, but the integration of AI is still lacking. There are not enough staff available to provide training related to AI, and the burden is placed on the faculty member to use AI via free resources or face peril or pass. Free resources allow me to gain more experience with AI; however, if the university does not keep up with technology, it limits what faculty can produce and collaborate with students. With this type of experience, the faculty member and the student may become disillusioned or become motivated to use AI despite the circumstances. To gain valuable AI experience, I go outside the HBCU and attend conferences, workshops, and trainings on AI so I will not be left behind. When I return from these conferences, I share what I learned and email different faculty so that they can read the information. If the university would provide some funding for faculty to attend conferences, a few faculty could attend different trainings and be given the charge to share the information in a Brown Bag luncheon or virtually.

Some of the students who are coming from undergraduate experiences use AI, but it can be a different experience if CHATGPT is not used appropriately. I think CHATGPT is a resource that can help students, but students should not use it word for word because it can be viewed as plagiarism by some professors. On the syllabus, if CHATGPT is used, it should state what it is acceptable and what is not acceptable. I do recommend that professors check the students' work against such software like Grammarly or other available sources.

In some circumstances, faculty have the skills but no way to gain the required certifications or further requirements to compete with other faculty members who receive funding. This experience may be the experience of some faculty at HBCUs. Because of this, the students may come into the classroom knowing more AI technology than the professors. Therefore, professors may seek assistance from external resources at the university and community.

In several years of teaching, I have noticed a lack of professional development funding. This lack of funding can cause faculty to try to find more affordable ways to obtain professional development and trainings. To assist in the lack of funding across many levels, universities can determine how the faculty AI experiences can be affordable while working at the HBCU. For example, faculty may feel that these experiences not available, and the faculty may choose to go outside the university for basic AI training. If the training is not at the university level, then the faculty will need to move to another source of obtaining critical AI skills. For example, PESI is an affordable way to gain trainings in many areas. Additionally, some organizations provide free webinars that are available.

I reach out to the community and invite them to speak in my classes. However, the connection must go beyond inviting speakers to classes to have a long-lasting effect that impacts AI technology at HBCUs. The shift must move to inviting panels where speakers can join from diverse universities to bring attention to major topics for discussions regarding HBCUs, AI technology, and the classroom. Why not host an annual conference or panel discussion on the topic of the impact of AI at HBCUs and break the topics down into different subtopics. Instead of having a one-time conference choose to save funds and raise the funds for this event annually. If the funding does not come to us, then we should solicit the community to help raise the funds and use key congress figures and senators to be a part of these discussions. The AI industry lists key personnel at some companies that can provide some lectures and training to faculty and students. Once I reached the organizations, some agreed to contact specific offices to donate some helpful resources and offered to speak with others at the university to fund some AI opportunities for students and faculty.

As a faculty member, I feel that my past years at an HBCU were more mission focused than work. I do believe the saying, "let them know you by your work." I have worked diligently at an HBCU, and as time goes by, I now understand my mission. I do wonder how the significant cuts of 2025 in federal funding will affect HBCUs. I never believed these cuts would close the door of HBCUs. I often wonder what we can do to gain more endowment money to support the mission and purpose of HBCUs. It is a great institution of learning, and I hope that before I retire, I see the world changing in a positive way towards the HBCU, the students who attend HCBUS, and I hope that a greater amount of AI technology will be placed in the classroom. I am now closer to the end of my career where I can retire on years but not age. I will continue a few more years to see the HBCU continue to blossom and as the legacy of South Carolina State states, "Hail, hail, dear SCSU," we will continue the glorious fight to keep the dream of the HBCU alive and thriving."

AI Source in the Classroom

To incorporate the best AI sources in the classroom, it is essential to integrate AI software. There are some essential tips to consider. Ross (2025) emphasized several key considerations for the use of generative AI in education. The author noted that AI tools can be used to support feedback, develop ideas, and enhance critical thinking skills. Ross also highlighted that AI should be viewed as a collaborative tool and not a replacement for independent thinking and personal effort. The teaching staff can share these tips so that a mutual learning environment can be created with students and staff learning together. This sharing of AI information is an inclusive part of the learning process that is shared and viewed as a positive experience that yields a sense of community.

Recommendations to HBCUs

Sustainability

While there have been reports of increased enrollment in many Historically Black Colleges and Universities post-pandemic, statistics reflect that the total number of Black students enrolled in all degree-granting postsecondary institutions, including HBCUs and non-HBCUS, was higher in 2022 (Census.gov, 2024; Postsecondary National Policy Institute, 2024; National Center for Education Statistics, 2023). The enrollment rates at HBCUs have indeed fluctuated over time, which are the result of social, economic, and policy factors. Further, HBCUs have historically been underfunded compared to other colleges and universities (Department of Education and Agriculture, 2023; The White House, 2024). Taking into consideration the challenges that HBCUs face to remain competitive and viable it is critical that HBCUs take advantage of innovative strategies to increase student enrollment and sustainability. According to NJRA (2024), digital marketing "drives traffic, generates leads, and fosters…growth" (par. 3).Further, digital marketing can support the planning of intentional events that can highlight the benefit of AI and digital technologies such as tech fairs and workshops as an innovative way to showcase digital and artificial intelligence advances. It is imperative to provide students with the tools they need to thrive and succeed in today's educational environment and the rapidly growing digital landscape.

As HBCUs navigate future technological advancements, embracing AI will be essential for improving educational outcomes, fostering research opportunities, and preparing students for a technology-driven job market. The university must be receptive to the establishment of direct collaborations with key corporations regionally and internationally. To effectively leverage AI at HBCUs, establishing strong connections with key players in the AI industry is crucial. This can lead to collaborative research opportunities, internships for students, and enhanced funding for programs. In order for AI to be included in the classroom there must be a buy-in. To effectively integrate AI into the HBCU infrastructure, it is essential to have a strategic plan that encompasses its presence in the classrooms, administrative processes, and marketing efforts. This integrative approach ensures that AI becomes a fundamental aspect of the university's mission and vision. It is important to attend key conferences, and webinars, and submit grant proposals that acknowledge the needs of the HBCU. It is equally important to hire university presidents, faculty, and staff who have a background in higher education so there is no disconnect in the vision and responsibilities of the HBCU.

Supporting AI in the classrooms at HBCUs must be a vision that starts at the top that proceeds to the other levels. The university must have a shared vision about including AI in the classroom and the community. Community engagement is essential to make more connections with key players in the AI industry. Faculty and administrators at universities can connect with other disciplines, also. For example, departments like business, computer science, and mathematics, can strengthen efforts and forge collaborations.

Davis (2024) identified several priorities for Historically Black Colleges and Universities (HBCUs) in relation to artificial intelligence. Additional priorities include identifying key offices to carry out the vision of AI, advancing teaching practices, integrating AI in curricula, and promoting global AI literacy. Davis also emphasized the importance of forming partnerships with corporations, expanding learning opportunities outside of the traditional classroom, and supporting AI research and projects. By identifying and using the strategies noted by Davis, HBCUs can continue to be powerful institutions of higher education.

Ongoing Communication with Key Corporations

Nixon-Saintil (2025) stated that corporations are forging milestones to help students and universities. IBM stated that "faculty training, new free generative AI courses for beginners, and Hands on Labs," can serve to recreate technologies at HBCUs (Nixon-Saintil, 2025, para.5/6). "Together with universities and our network of IBM SkillsBuild partners, we are investing in the pipeline of technological talent for our future, and preparing them for the marketplace… We will continue to: enhance our resources for university faculty, including opportunities to connect across the faculty community, expand our catalog of new AI coursework for students, and connect with teams across universities to build new innovative AI and open source…" (Nixon-Saintil, 2025, para5/6).

Other countries use AI with students and technological advancements appear to advance faster and faster. Faculty and administrators must be key attendees at international conferences related to AI and discussions about integrating AI into the classroom and community.

Other Considerations

In exploring the integration of AI at HBCUs inside the classroom and the community, it is essential to consider numerous factors that can influence success. These may include funding opportunities, faculty training, community engagement, and the ethical implications of AI technology. Addressing these considerations can enhance the overall effectiveness of AI initiatives within the university. Here are some questions that may be of interest when considering the integration of AI at HBCUs. These could also include inquiries related to funding opportunities, partnerships with tech companies, and the development of relevant curricula to enhance student learning outcomes. What AI conference should the university attend? How can participation in conferences benefit students and faculty? What strategies can the university use to enhance its visibility at these events? What key leadership staff should attend these events? What faculty might attend these events to provide trainings? How can staff and faculty train others through a train the trainer model? Are there interdisciplinary grants that can be applied for by encouraging participation from faculty from different programs? How can the HBCU partner with PWIS in applying for grants? What key faculty can receive release time dedicated to producing more classes that use AI and how can these same faculty produce more grants through partnerships with key organizations?

It is a known fact that there are major AI corporations around the world. Having a seat at the table means that the HBCU is actively participating in discussions and decision-making processes regarding AI initiatives and collaboration. Ensuring that HBCUs maintain a strong presence could lead to opportunities for representation in industry panels, partnerships with tech companies, and involvement in AI research projects. Moreover, maintaining connections can give access to regular updates, newsletters, and community engagement. The HBCU must be open to developing international connections where key speakers and lead people can come into the different disciplines represented on campus. For example, 1890 programs might be able to open more doors to develop more lasting connections, especially since there are so many different disciplines working together in 1890 programs.

CONCLUSION

In conclusion, the literature available on AI at HBCUs and community engagement takes one from research to realization. AI is a great way to communicate, and it can provide so many opportunities to HBCUs in the classroom and the community. Historically Black Colleges and Universities will serve students well by providing opportunities for exposure to AI innovations and technologies. Access to innovative resources may also increase positive perceptions associated with the use and minimize abandonment of helpful digital technologies that are supported by AI interfaces. Additional areas to consider are, "What is the reality of AI technology moving into the HCBU with the same level of funding as other universities?" "When does the research move into reality regarding AI in the classroom and communities?" "When will the blinders be taken off at many levels?" "When will HBCUs receive a fair chance at being a powerhouse for AI technology? What will it take to get to this point?" "Are university administrators at the table when there are AI discussions that impact the HBCU?" For the HBCU to be included, it is essential to be at the table. Not being at the table when these discussions occur leaves the HBCU in the background and not in the forefront of advancement. Last, it stands to reason that the benefits of these technologies always outweigh the cons.

REFERENCES

Alexander, R. (2024, September 3). Beyond affirmative action: HBCUs and the time for equitable funding. Columbia Political Review. https://www.cpreview.org/articles/2024/9/beyond-affirmative-action-hbcus-and-the-time-for-equitable-funding

Barrett, S. E., Ochapa, M. O., & Warren, T. (2023). Digital technology in community engagement: Impacts and implications. *Journal of Health Care for the Poor and Underserved*, *34*(1), 425–430. DOI: 10.1353/hpu.2023.0027 PMID: 37464503

Benneworth, P., Ćulum, B., Farnell, T., Kaiser, F., Seeber, M., Šćukanec, N., Vossensteyn, H., & Westerheijden, D. (2018). Mapping and critical synthesis of current state-of-the-art on community engagement in higher education. European Commission. https://research.utwente.nl/en/publications/mapping-and-critical-synthesis-of-current-state-of-the-art-on-com

Boucher, P. (2020). Artificial intelligence: How does it work, why does it matter, and what can we do about it? European Parliament. https://www.europarl.europa.eu/thinktank/en/document/EPRS_STU(2020)641547

Caperon, L., Saville, F., Ahern, S., & Ali, G. (2022). Developing a socio-ecological model for community engagement in a health programme in an underserved urban area. *PLoS One*, *17*(9), e0275092. Advance online publication. DOI: 10.1371/journal.pone.0275092 PMID: 36155664

Chace, C. (2020, October 29). The impact of artificial intelligence on education. Forbes. https://www.forbes.com/sites/calumchace/2020/10/29/the-impact-of-artificial-intelligence-on-education/

CLRN Team. (2024). What does community engagement mean? Community Learning and Research Network. https://clrn.org/what-does-community-engagement-mean/

Coalition, H. B. C. U. (2021). How HBCUs have always led the way for social change. https://www.hbcucoalition.org/post/how-hbcus-have-always-led-the-way-for-social-change

Coulson, A. (2024). Community engagement trends for 2025. SocialPoint. https://socialpoint.com

Davis, V. (2024). Leading the revolution: The crucial role of HBCUs in the AI leadership. WCET Frontiers. https://wcet.wiche.edu/frontiers/2024/07/25/leading-the-revolution-the-crucial-role-of-hbcus-in-steering-ai-leadership/

First, H. B. C. U. (2024). Why choose an HBCU? https://hbcufirst.com/resources/why-choose-an-hbcu

Gibson, G. (2024). How generative AI is shaking up education: The role of HBCUs in leading the charge. HBCU Lifestyle. https://hbculifestyle.com/generative-ai-hbcus-education-leadership/

Hazelkorn, E. (2016). Contemporary debates part 2: Initiatives, and governance and organizational structures. In Goddard, J., Hazelkorn, E., & Vallance, P. (Eds.), *The civic university: The policy and leadership challenges* (pp. 34–64). Edward Elgar Publishing., DOI: 10.4337/9781784717728.00011

Mei-Chin, S., Le-En, L., Li-Hwa, L., & Yu-Chu, C. (2024). Assessing question characteristic influences on ChatGPT's performance and response-explanation consistency: Insights from Taiwan's nursing licensing exam. *International Journal of Nursing Studies, 153*, 104717. Advance online publication. DOI: 10.1016/j.ijnurstu.2024.104717 PMID: 38401366

National Center for Education Statistics. (2023). Fast facts: Historically Black colleges and universities. https://nces.ed.gov/fastfacts/display.asp?id=667

Nixon-Saintil. [Initial]. (2025). IBM is equipping the next generation of student developers across universities with AI skills. IBM Newsroom. https://newsroom.ibm.com/blog-IBM-is-equipping-the-next-generation-of-student-developers-across-universities-with-AI-skills

Noyes, E., & Girdharry, K. (2024, December 3). Why faculty should lead the AI revolution. AACSB Insights. https://www.aacsb.edu/insights/articles/2024/12/why-faculty-should-lead-the-ai-revolution

Postsecondary National Policy Institute. (2024). Black students in higher education. https://pnpi.org/factsheets/black-students/

Ross, M. (2025). Tips for using AI, from grad students and professors. Usable Knowledge. https://www.gse.harvard.edu/ideas/usable-knowledge/25/01/tips-for-using-ai-from-grad-students-and-professors

Secinaro, S., Calandra, D., Secinaro, A., Muthurangu, V., & Biancone, P. (2021). The role of artificial intelligence in healthcare: A structured literature review. *BMC Medical Informatics and Decision Making, 21*(1), 125. Advance online publication. DOI: 10.1186/s12911-021-01488-9 PMID: 33836752

Talo, C. (2018). Community-based determinants of community engagement: A meta-analysis research. *Social Indicators Research, 140*(2), 571–596. DOI: 10.1007/s11205-017-1778-y

The White House. (2024, May 16). The economics of HBCUs. Council of Economic Advisers. https://www.whitehouse.gov/cea/written-materials/

United Negro College Fund. (2024). The numbers don't lie: HBCUs are changing the college landscape. https://uncf.org/the-latest/the-numbers-dont-lie-hbcus-are-changing-the-college-landscape

United Negro College Fund. (2024). The impact of HBCUs on diversity in STEM fields. https://uncf.org/the-latest/the-impact-of-hbcus-on-diversity

United Negro College Fund. (2024). The 2024 consumer economic impact report. https://uncf.org/hbcuimpactreport/2024

ADDITIONAL READING

eJournal of Public Affairs. (2024, September 29). *Considering AI for the classroom to boost civic learning and democratic engagement.* Retrieved from https://www.ejournalofpublicaffairs.org/considering-ai-for-the-classroom-to-boost-civic- learning-and-democratic-engagement/

McCauley, C., Webb, J., Abdelrahim, S., & Mahmoud-Tabana, S. (2023). A community engaged framing: Building successful community engagement for schools and families of bilingual students through inquiry. *School Community Journal, 33*(1), 11–36.

The Chicago School Library. (2025, June 27). *Using Ai for searching the literature - conducting a literature review.* Retrieved from https://library.thechicagoschool.edu/c.php?g=1296999&p=10017149

U.S. Department of Education. (.gov). (n.d.). *Artificial intelligence and the future of teaching and learning.* Retrieved from https://www.ed.gov/sites/ed/files/documents/ai-report/ai- report.pdf

Xu, G., Xue, M., & Zhao, J. (2023). The association between artificial intelligence awareness and employee depression: The mediating role of emotional exhaustion and the moderating role of perceived organizational support. *International Journal of Environmental Research and Public Health, 20*(6), 5147. DOI: 10.3390/ijerph20065147 PMID: 36982055

KEY TERMS AND DEFINITIONS

Artificial Intelligence: AI is defined as "systems that display intelligent behavior by analyzing their environment and taking action – with some degree of autonomy – to achieve specific goals" (Boucherm, 2020).

College Students: College students are defined as students enrolled and attending a college or university.

Community Engagement: Community engagement is "a sense of shared responsibility and creates a stronger united community for building long term relationships" (Coulson, 2024).

Graduate students: Graduate students are students enrolled in a master's or higher level program.

Historically Black College/University: HBCUs were primarily created to meet the needs of Black students. There were not many schools available to provide education. "By the late 1800's there were over 86 black colleges. However, whites and especially former plantation owners that relied on cheap black labor saw a danger in teaching blacks anything that could make them question their place in society. As a result, they started to resort to extremely violent means to drive teachers away and to destroy many black colleges. Between 1866 and 1972, over 20,000 people were killed because of the educational equity quest. Yet, despite the ongoing violence and the limited resources, black colleges survived" (HBCU Coalition, 2021).

Innovative resources: Innovative resources are sources that enhance AI and positively impact advancements.

Personalized Learning: Personalized learning is learning that involves an individual approach that best meets the needs of the learner.

Predominantly White Institutions: PWIs are institutions that are predominantly comprised of white faculty, students, and staff.

Undergraduate students: Undergraduate students are those students enrolled in a bachelor's program.

Compilation of References

Addanki, M., Patra, P., & Kandra, P. (2022). Recent advances and applications of artificial intelligence and related technologies in the food industry. *Applied Food Research*, *2*(2), 100126. DOI: 10.1016/j.afres.2022.100126

Adirim, T. (2025). Digital health technologies: Virtual health and remote patient monitoring. In Digital health, AI and generative AI in healthcare (pp. 11–26). DOI: 10.1007/978-3-031-83526-1_2

Ahmad, N., Fatima, I., Alam, M., Khan, I. R., & Mahmoud, A. (2025). Transforming healthcare through generative AI. In Generative intelligence in healthcare (pp. 259–279). DOI: 10.1201/9781003539483-11

Ahmadnia, H., Gholizadeh, M., Bavafa, M., & Rahbarianyazd, R. (2012). *Art of Feng Shui and its Relationship with Modern Interior Design*. academia.edu

Aijaz, N., Lan, H., Raza, T., Yaqub, M., Iqbal, R., & Pathan, M. S. (2025). Artificial intelligence in agriculture: Advancing crop productivity and sustainability. *Journal of Agriculture and Food Research*, *20*, 101762. Advance online publication. DOI: 10.1016/j.jafr.2025.101762

Akinnagbe, O. B. (2024). Human-AI Collaboration: Enhancing Productivity and Decision-Making. *International Journal of Education, Management, and Technology*, *2*(3), 387–417. DOI: 10.58578/ijemt.v2i3.4209

Alabdali, S. A., Pileggi, S. F., & Cetindamar, D. (2023). Influential factors, enablers, and barriers to adopting smart technology in rural regions: A literature review. *Sustainability (Basel)*, *15*(10), 7908. Advance online publication. DOI: 10.3390/su15107908

Aldoseri, A., Al-Khalifa, K., & Hamouda, A. M. (2023). Re-thinking data strategy and integration for artificial intelligence: Concepts, opportunities, and challenges. *Applied Sciences (Basel, Switzerland)*, *13*(12), 7082. https://www.mdpi.com/2076-3417/13/12/7082. DOI: 10.3390/app13127082

Alexander, R. (2024, September 3). Beyond affirmative action: HBCUs and the time for equitable funding. Columbia Political Review. https://www.cpreview.org/articles/2024/9/beyond-affirmative-action-hbcus-and-the-time-for-equitable-funding

Alexander, C. S., Yarborough, M., & Smith, A. (2024). Who is responsible for 'responsible AI'?: Navigating challenges to build trust in AI agriculture and food system technology. *Precision Agriculture*, *25*(1), 146–185. DOI: 10.1007/s11119-023-10063-3

Al-Medfa, M. (2022). [Title not provided]. DOI: 10.2196/preprints.38137

Almeida, D., & Barr, N. (2025). Innovations in health data protection: Ethical, legal, and technological perspectives in a global context. In Navigating privacy, innovation, and patient empowerment through ethical healthcare technology (pp. 171–196). DOI: 10.4018/979-8-3693-7630-0.ch007

Alowais, S. A., Alghamdi, S. S., Alsuhebany, N., Alqahtani, T., Alshaya, A. I., Almohareb, S. N., Aldairem, A., Alrashed, M., Bin Saleh, K., Badreldin, H. A., Al Yami, M. S., Al Harbi, S., & Albekairy, A. M. (2023). Revolutionizing healthcare: The role of artificial intelligence in clinical practice. *BMC Medical Education*, *23*(1), 239. Advance online publication. DOI: 10.1186/s12909-023-04698-z PMID: 37740191

Anjorin, K. F., Raji, M. A., Olodo, H. B., & Oyeyemi, O. P.Kikelomo Fadilat AnjorinMustafa Ayobami RajiHameedat Bukola OlodoOluwaseun Peter Oyeyemi. (2024). Harnessing artificial intelligence to develop strategic marketing goals. *International Journal of Management & Entrepreneurship Research*, *6*(5), 1625–1650. DOI: 10.51594/ijmer.v6i5.1127

Anozie, U. C., Onyenahazi, O. B., Ekeocha, P. C., Adekola, A. D., Ukadike, C. A., & Oloko, O. A.Uchechukwu Christopher AnozieObinna Barnabas OnyenahaziPrince Chukwuemeka EkeochaAdeleke Damilola AdekolaChifum Ann UkadikeOmodunni Adejoke Oloko. (2024). Advancements in artificial intelligence for omnichannel marketing and customer service: Enhancing predictive analytics, automation, and operational efficiency. *International Journal of Science and Research Archive*, *12*(2), 1621–1629. DOI: 10.30574/ijsra.2024.12.2.1436

Aquilani, C., Confessore, A., Bozzi, R., Sirtori, F., & Pugliese, C. (2022). Review: Precision livestock farming technologies in pasture-based livestock systems. *Animal*, *16*(1), 100429. Advance online publication. DOI: 10.1016/j.animal.2021.100429 PMID: 34953277

Aronsson, I.-L., & Price, S. (2024). Culture, heritage, memory: Toward a resonant cultural solution for resettlement. *Human Organization*, *83*(3), 216–227. DOI: 10.1080/00187259.2024.2307845

Ashley, C., & Tuten, T. (2015). Creative Strategies in Social Media Marketing: An Exploratory Study of Branded Social Content and Consumer Engagement. *Psychology and Marketing*, *32*(1), 15–27. DOI: 10.1002/mar.20761

Asif, F., & Qureshi, M. F. (2024). AI and quality and patient safety. In Advances in healthcare information systems and administration (pp. 217–232). DOI: 10.4018/979-8-3693-7051-3.ch009

Asuncion, J. A. (2024). Revolutionizing Marketing: Harnessing Artificial Intelligence for Enhanced Customer Engagement and Insights. *2024 International Conference on IoT, Communication and Automation Technology (ICICAT)*, 767–772. DOI: 10.1109/ICICAT62666.2024.10922912

Atapattu, A. J., Perera, L., Nuwarapaksha, T., Udumann, S. S., & Dissanayaka, N. (2024). Challenges in achieving artificial intelligence in agriculture. In Artificial intelligence in agriculture (pp. 7–34). DOI: 10.1007/978-981-97-5878-4_2

Avaji, G. M., & N, D. G. (2024). IoT innovations in healthcare: Enhancing patient care and operational efficiency. *International Journal of Research Publication and Reviews*, *5*(5), 11479–11485. DOI: 10.55248/gengpi.5.0524.1414

Azith Teja Ganti, V. K. (2023). Exploring the intersection of bioethics and AI-driven clinical decision-making: Navigating the ethical challenges of deep learning applications in personalized medicine and experimental treatments. *Journal of Material Sciences & Manufacturing Research*, *4*(6), 1–10. DOI: 10.47363/JMSMR/2023(4)192

Babatunde, S. O., Odejide, O. A., Edunjobi, T. E., & Ogundipe, D. O.Sodiq Odetunde BabatundeOpeyemi Abayomi OdejideTolulope Esther EdunjobiDamilola Oluwaseun Ogundipe. (2024). THE ROLE OF AI IN MARKETING PERSONALIZATION: A THEORETICAL EXPLORATION OF CONSUMER ENGAGEMENT STRATEGIES. *International Journal of Management & Entrepreneurship Research*, *6*(3), 936–949. DOI: 10.51594/ijmer.v6i3.964

Bačiulienė, V., Bilan, Y., Navickas, V., & Civín, L. (2023). The aspects of artificial intelligence in different phases of the food value and supply chain. *Foods*, *12*(8), 1654. DOI: 10.3390/foods12081654 PMID: 37107449

Bairwa, H., & Jangid, R. (2024). Pneumonia detection from chest X-rays using the CheXNet deep learning algorithm. *Preprints*. DOI: 10.20944/preprints202407.0104.v1

Balamurugan, , M. (2024). AI-Driven Adaptive Content Marketing: Automating Strategy Adjustments for Enhanced Consumer Engagement. *International Journal For Multidisciplinary Research*, *6*(5), 27940. Advance online publication. DOI: 10.36948/ijfmr.2024.v06i05.27940

Barceló, J., Del Castillo, F., Kayikci, D., & Urbistondo, B. (2022). Neural networks for archaeological classification and typology: An overview. *PCA European Journal of Post-Classical Archaeologies*, *12*, 7–32.

Barney, J. (1991). Firm resources and sustained competitive advantage. *Journal of Management*, *17*(1), 99–120. DOI: 10.1177/014920639101700108

Barredo Arrieta, A., Díaz-Rodríguez, N., Del Ser, J., Bennetot, A., Tabik, S., Barbado, A., Garcia, S., Gil-Lopez, S., Molina, D., Benjamins, R., Chatila, R., & Herrera, F. (2020). Explainable artificial intelligence (XAI): Concepts, taxonomies, opportunities and challenges toward responsible AI. *Information Fusion*, *58*, 82–115. DOI: 10.1016/j.inffus.2019.12.012

Barrett, S. E., Ochapa, M. O., & Warren, T. (2023). Digital technology in community engagement: Impacts and implications. *Journal of Health Care for the Poor and Underserved*, *34*(1), 425–430. DOI: 10.1353/hpu.2023.0027 PMID: 37464503

Barua, U., Sabir Hossain, M., & Shamsul Arefin, M. (2019). Real Estate Recommendation Using Historical Data and Surrounding Environments. *International Journal of Information Engineering and Electronic Business*, *11*(5), 33–39. DOI: 10.5815/ijieeb.2019.05.05

Basri, W. (2020). Examining the impact of artificial intelligence (AI)-assisted social media marketing on the performance of small and medium enterprises: Toward effective business management in the Saudi Arabian context. *International Journal of Computational Intelligence Systems*, *13*(1), 142. DOI: 10.2991/ijcis.d.200127.002

Batra Deep Manishkumar, P. (2024). Revolutionizing healthcare platforms: The impact of AI on patient engagement and treatment efficacy. *International Journal of Science and Research (Raipur, India)*, *13*(2), 273–280. DOI: 10.21275/SR24201070211

Baur, K., Rosenfelder, M., & Lutz, B. (2023). Automated real estate valuation with machine learning models using property descriptions. *Expert Systems with Applications*, *213*, 119147. DOI: 10.1016/j.eswa.2022.119147

Benneworth, P., Ćulum, B., Farnell, T., Kaiser, F., Seeber, M., Šćukanec, N., Vossensteyn, H., & Westerheijden, D. (2018). Mapping and critical synthesis of current state-of-the-art on community engagement in higher education. European Commission. https://research.utwente.nl/en/publications/mapping-and-critical-synthesis-of-current-state-of-the-art-on-com

Benti, N. E., Chaka, M. D., Semie, A. G., Warkineh, B., & Soromessa, T. (2024). Transforming agriculture with machine learning, deep learning, and IoT: Perspectives from Ethiopia—challenges and opportunities. *Discover Agriculture*, *2*(1), 63. Advance online publication. DOI: 10.1007/s44279-024-00066-7

Beyene, E., Bedemo, A., & Gebremeskel, A. (2024). Determinants of digital technology development in sub-Saharan African countries: Evidence from panel data analysis. *Energy Informatics*, *7*(1), 21. Advance online publication. DOI: 10.1186/s42162-024-00324-4

Bilal, M., Zhang, Y., Cai, S., Akram, U., & Halibas, A. (2024). Artificial intelligence is the magic wand making customer-centric a reality! An investigation into the relationship between consumer purchase intention and consumer engagement through affective attachment. *Journal of Retailing and Consumer Services*, *77*, 103674. DOI: 10.1016/j.jretconser.2023.103674

Bjorklund, D. (2022). Children's evolved learning abilities and their implications for education. *Educational Psychology Review*, *34*(4), 2243–2273. Advance online publication. DOI: 10.1007/s10648-022-09688-z PMID: 35730061

Bo, Z. H. F. (2024). Digital technology empowers public cultural services: value presupposition, blocking factors and method adjustment. Study&Exploration(09), 56-62. https://kns.cnki.net/kcms2/article/abstract?v=qKY3Y0De9e7i23-aJAdLH9qxTsP_RFV_uaiLLcGh8UCzXLyS0v8lQLdyp7krbVKjW2r2mlDpR_52vDls6frdGJyGAZIsW0e_IJzG4p-KZzKEd6k5FgzEyiL7XdG4EfRypPjX1AH6ewyNbkv4xYINPHdhuEx4-wHj-8FAsU_26lSlm1nWN5Ri-pCqlZ3f4ZTnfjfLMgiwUpQ=&uniplatform=NZKPT&language=CHS

Bolton, R. N. (2011). Comment: Customer Engagement: Opportunities and Challenges for Organizations. *Journal of Service Research*, *14*(3), 272–274. DOI: 10.1177/1094670511414582

Borissova, V. (2018). Cultural heritage digitization and related intellectual property issues. *Journal of Cultural Heritage*, *34*, 145–150. DOI: 10.1016/j.culher.2018.04.023

Boucher, P. (2020). Artificial intelligence: How does it work, why does it matter, and what can we do about it? European Parliament. https://www.europarl.europa.eu/thinktank/en/document/EPRS_STU(2020)641547

Bouderbane, A. (2024). Using AI tools to prompt knowledge appropriately and ethically in higher education. In General aspects of applying generative AI in higher education (pp. 327–345). DOI: 10.1007/978-3-031-65691-0_17

Bourdieu, P. (1986). The Forms of Capital. In Richardson, J. G. (Ed.), *Handbook of Theory and Research for the Sociology of Education* (pp. 241–258). Greenwood Press.

Braden, L. E. A., & Teekens, T. (2020). Historic networks and commemoration: Connections created through museum exhibitions. *Poetics*, *81*, 101446. Advance online publication. DOI: 10.1016/j.poetic.2020.101446

Braun, V., & Clarke, V. (2012). Thematic Analysis. In H. Cooper, P. M. Camic, D. L. Long, A. T. Panter, D. Rindskopf, & K. J. Sher (Eds), APA handbook of research methods in psychology, Vol. 2: Research designs: Quantitative, qualitative, neuropsychological, and biological. *Washington, DC: American Psychological Association.*, 57–71. DOI: 10.1037/13620-004

Braun, V., & Clarke, V. (2006). Using thematic analysis in psychology. *Qualitative Research in Psychology*, *3*(2), 77–101. DOI: 10.1191/1478088706qp063oa

Braun, V., & Clarke, V. (2021). Thematic analysis: A practical guide. *Sage (Atlanta, Ga.)*.

Brodie, R. J., Ilic, A., Juric, B., & Hollebeek, L. (2013). Consumer engagement in a virtual brand community: An exploratory analysis. *Journal of Business Research*, *66*(1), 105–114. DOI: 10.1016/j.jbusres.2011.07.029

Buhalis, D., & Sinarta, Y. (2019). Real-time co-creation and nowness service: Lessons from tourism and hospitality. *Journal of Travel & Tourism Marketing*, *36*(5), 563–582. DOI: 10.1080/10548408.2019.1592059

Buragohain, D., Meng, Y., Deng, C., Li, Q., & Chaudhary, S. (2024). Digitalizing cultural heritage through metaverse applications: Challenges, opportunities, and strategies. *Heritage Science*, *12*(1), 295. Advance online publication. DOI: 10.1186/s40494-024-01403-1

Burton, R., Kuczera, C., & Schwarz, G. (2008). Exploring farmers' cultural resistance to voluntary agri-environmental schemes. *Sociologia Ruralis*, *48*(1), 16–37. DOI: 10.1111/j.1467-9523.2008.00452.x

Çakır, A. M. (2024). Human AI Collaboration in Decision-Making Auto Systems With AI. *Human-Computer Interaction*, *8*(1), 123. DOI: 10.62802/b4z5p105

Campbell, C., Sands, S., Ferraro, C., Tsao, H.-Y., & Mavrommatis, A. (2020). From data to action: How marketers can leverage AI. *Business Horizons*, *63*(2), 227–243. DOI: 10.1016/j.bushor.2019.12.002

Canady, V. A. (2019). Study explores use of AI to treat patients with depression. *Mental Health Weekly*, *29*(42), 3–4. DOI: 10.1002/mhw.32116

Cantor, P., Osher, D., Berg, J., Steyer, L., & Rose, T. (2019). Malleability, plasticity, and individuality: How children learn and develop in context. *Applied Developmental Science*, *23*(4), 307–330. DOI: 10.1080/10888691.2017.1398649

Cao, T. D., & Nguyen, Q. H. (2021). An Approach for Building Effective Real Estate Chatbots in Vietnamese. In *Soft Computing for Biomedical Applications and Related Topics* (Vol. 899, pp. 221–229). Springer., DOI: 10.1007/978-3-030-49536-7_19

Caperon, L., Saville, F., Ahern, S., & Ali, G. (2022). Developing a socio-ecological model for community engagement in a health programme in an underserved urban area. *PLoS One*, *17*(9), e0275092. Advance online publication. DOI: 10.1371/journal.pone.0275092 PMID: 36155664

Carpendale, J., & Lewis, C. (2004). Constructing an understanding of mind: The development of children's social understanding within social interaction. *The Behavioral and Brain Sciences*, *27*(1), 79–96. DOI: 10.1017/S0140525X04000032 PMID: 15481944

Carrasco Ramírez, J. G. (2024). AI in healthcare: Revolutionizing patient care with predictive analytics and decision support systems. *Journal of Artificial Intelligence General Science*, *1*(1), 31–37. DOI: 10.60087/jaigs.v1i1.p37

Casillo, M., Colace, F., Gupta, B. B., Lorusso, A., Santaniello, D., & Valentino, C. (2024). The role of AI in improving interaction with cultural heritage: An overview. Handbook of Research on AI and ML for Intelligent Machines and Systems, 107-136.

Castellotti, S., D'Agostino, O., & Del Viva, M. M. (2024). Effectiveness of labels in digital art experience: Psychophysiological and behavioral evidence. *Frontiers in Psychology*, *15*, 1342667. Advance online publication. DOI: 10.3389/fpsyg.2024.1342667 PMID: 39011289

Castelo, N., Bos, M. W., & Lehmann, D. R. (2019). Task-Dependent Algorithm Aversion. *JMR, Journal of Marketing Research*, *56*(5), 809–825. DOI: 10.1177/0022243719851788

Ch'ng, E., Cai, S. D., Leow, F. T., & Zhang, T. E. (2019). Adoption and use of emerging cultural technologies in China's museums. *Journal of Cultural Heritage*, *37*, 170–180. DOI: 10.1016/j.culher.2018.11.016

Chace, C. (2020, October 29). The impact of artificial intelligence on education. Forbes. https://www.forbes.com/sites/calumchace/2020/10/29/the-impact-of-artificial-intelligence-on-education/

Chaffey, D., & Smith, P. (2022). *Digital marketing excellence: Planning, optimizing and integrating online marketing* (6th ed.). Routledge., DOI: 10.4324/9781003009498

Chandra Kaushik, A., & Raj, U. (2020). AI-driven drug discovery: A boon against COVID-19? *AI Open*, *1*, 1–4. DOI: 10.1016/j.aiopen.2020.07.001

Chen, C. (2005). CiteSpace II: Detecting and visualizing emerging trends and transient patterns in scientific literature. *Journal of the American Society for Information Science and Technology*, *57*(3), 359–377. DOI: 10.1002/asi.20317

Chen, Y., Prentice, C., Weaven, S., & Hisao, A. (2022). The influence of customer trust and artificial intelligence on customer engagement and loyalty – The case of the home-sharing industry. *Frontiers in Psychology*, *13*, 912339. DOI: 10.3389/fpsyg.2022.912339 PMID: 35992434

Chen, Z. (2023). Artificial Intelligence-Virtual Trainer: Innovative Didactics Aimed at Personalized Training Needs. *Journal of the Knowledge Economy*, *14*(2), 2007–2025. DOI: 10.1007/s13132-022-00985-0 PMID: 40479149

CHINA CENTRAL TELEVISION. (2024, January 11). *Harbin ice sculptors: Store ice one year in advance and build the Harbin Ice and Snow World within 15 days.* https://news.cctv.com/2024/01/11/ARTIRhIt4TDjrhYWtXSy7qAJ240111.shtml

Chinanews.com. (2024, December 20). *The immersive AI digital art museum named "One Ice and Snow, One World" has made its debut at the Harbin Ice and Snow World.* https://www.toutiao.com/article/7450392175508062783/?upstream_biz=doubao&source=m_redirect

Chinchanachokchai, S., Thontirawong, P., & Chinchanachokchai, P. (2021). A tale of two recommender systems: The moderating role of consumer expertise on artificial intelligence based product recommendations. *Journal of Retailing and Consumer Services*, *61*, 102528. Advance online publication. DOI: 10.1016/j.jretconser.2021.102528

Cho, A. (2020a). *How to Recognize Good Feng Shui in a Home*. Thespruce.Com. https://www.thespruce.com/signs-of-a-good-feng-shui-house-1274498

Cho, A. (2020b). *How to Use Feng Shui for Wealth and Prosperity in Your Home*. Http://Www.Thespruce.Com. https://www.thespruce.com/feng-shui-money-tips-and-wealth-1275336

Choi, M., Choi, Y., Nosrati, S., Hailu, T. B., & Kim, S. (2023). Psychological dynamics in the metaverse: Evaluating perceived values, attitude, and behavioral intention in metaverse events. *Journal of Travel & Tourism Marketing*, *40*(7), 602–618. DOI: 10.1080/10548408.2023.2276435

Ciesielska, M., Wolanik Boström, K., & Öhlander, M. (2018). Observation methods. In Qualitative methodologies in organization studies (pp. 33–52). DOI: 10.1007/978-3-319-65442-3_2

CLRN Team. (2024). What does community engagement mean? Community Learning and Research Network. https://clrn.org/what-does-community-engagement-mean/

Coalition, H. B. C. U. (2021). How HBCUs have always led the way for social change. https://www.hbcucoalition.org/post/how-hbcus-have-always-led-the-way-for-social-change

Colavizza, G., Blanke, T., Jeurgens, C., & Noordegraaf, J. (2021). Archives and AI: An overview of current debates and future perspectives. [JOCCH]. *ACM Journal on Computing and Cultural Heritage*, *15*(1), 1–15.

ContentSquare. (2024). *11 Best Qualitative Data Analysis Tools and Software*. Contentsquare.Com. https://contentsquare.com/guides/qualitative-data-analysis/tools/

Conway, J., & Architecture, B. A. (2018). *Artificial Intelligence and Machine Learning: Current Applications in Real Estate*.

Coulson, A. (2024). Community engagement trends for 2025. SocialPoint. https://socialpoint.com

Croce, V., Caroti, G., De Luca, L., Piemonte, A., & Véron, P. (2023). Neural radiance fields (nerf): Review and potential applications to digital cultural heritage. *The International Archives of the Photogrammetry, Remote Sensing and Spatial Information Sciences*, *48*, 453–460. DOI: 10.5194/isprs-archives-XLVIII-M-2-2023-453-2023

Dacin, T., Goodstein, J., & Scott, R. (2002). Institutional theory and institutional change: Introduction to the special research forum. *Academy of Management Journal*, *45*(1), 45–56. https://www.jstor.org/stable/3069284. DOI: 10.2307/3069284

Dahake, N. S., Gajghate, A., Dahake, P. S., & Bansod, S. (2024). A transformative impact of IoT and AI by predictive analytics on modern marketing strategies: A study of its significance & accuracy. In 2024 2nd International Conference on Advances in Computation, Communication and Information Technology (ICAICCIT) (pp. 1150–1158). https://doi.org/DOI: 10.1109/ICAICCIT64383.2024.10912388

Dahake, N. S., Suresh Dahake, P., & Gajghate, A. (2024). Beyond the hype: A critical evaluation of predictive analytics accuracy in high-stakes business decisions. In 2024 2nd DMIHER International Conference on Artificial Intelligence in Healthcare, Education and Industry (IDICAIEI) (pp. 1–6). DOI: 10.1109/IDICAIEI61867.2024.10842857

Dahake, P. S., & Somani, N. (2024). Harnessing Predictive Analytics for Accurate Consumer Behaviour Forecasting: A Comprehensive Review. *2024 2nd DMIHER International Conference on Artificial Intelligence in Healthcare, Education and Industry (IDICAIEI)*, 1–6. DOI: 10.1109/IDICAIEI61867.2024.10842743

Dahake, P. S., Chandak, S., & Dahake, N. S. (2024). Predictive Analytics in Performance Appraisal Systems: Enhancing Objectivity and Fairness. *2024 2nd DMIHER International Conference on Artificial Intelligence in Healthcare, Education and Industry (IDICAIEI)*, 1–5. DOI: 10.1109/IDICAIEI61867.2024.10842916

Dahake, P. S., Jain, M., Anusha, B., Poojitha, S., Singh, N., & Sharma, A. (2024). Integrating Organisational Theory, Resource-Based View, and Big Data Culture to Improve Manufacturing Performance. In D. Jelonek, N. Kumar, M. Chahar, R. Kinkladze, & L. Knop, *Recent Trends In Engineering and Science for Resource Optimization and Sustainable Development* (1st ed., pp. 156–159). CRC Press. DOI: 10.1201/9781003596721-36

Dahake, P. S., Mohare, R., Chandak, S., & Bhadade, P. (2024). Predictive Analytics in Assessing Employee Retention Rates: An Analysis of Effectiveness and Accuracy of Different Predictive Model. *2024 2nd International Conference on Advances in Computation, Communication and Information Technology (ICAICCIT)*, 1159–1164. DOI: 10.1109/ICAICCIT64383.2024.10912426

Dahake, P. S., Parchake, S., Somani, N., & Ingale, A. R. (2024). Predictive analytics in automotive consumer behaviour: An insight into personal use 4-wheeler purchase decisions. In 2024 2nd International Conference on Advances in Computation, Communication and Information Technology (ICAICCIT) (pp. 1143–1149). https://doi.org/DOI: 10.1109/ICAICCIT64383.2024.10912242

Dahake, P. S., Bagaregari, P., & Dahake, N. S. (2024). Shaping the Future of Retail: A Comprehensive Review of Predictive Analytics Models for Consumer Behavior. In Inder, S., Min, B., & Sharma, S. (Eds.), (pp. 143–160). Advances in Business Strategy and Competitive Advantage. IGI Global., DOI: 10.4018/979-8-3693-1734-1.ch011

Dahake, P. S., Chandak, S., Mohare, R. V., Wadhwani, K., & Bhadade, P. (2023). The crystal ball of marketing: How predictive analytics is reshaping the industry? In *2023 Second International Conference On Smart Technologies For Smart Nation (SmartTechCon)* (pp. 304–311). https://doi.org/DOI: 10.1109/SmartTechCon57526.2023.10391334

Dahake, P. S., & Dahake, N. S. (2024). The Role of Big Data in Predictive Analytics for E-Commerce: Examine How Big Data Contributes to Better Predictions in Online Shopping. In Tarnanidis, T., Papathanasiou, J., Ismyrlis, V., & Kittu Manda, V. (Eds.), (pp. 403–426). Advances in Marketing, Customer Relationship Management, and E-Services. IGI Global., DOI: 10.4018/979-8-3693-9122-8.ch018

Dahake, P. S., Mohare, R. V., & Somani, N. (2023). Exploring Machine Learning's Potential in Predicting Consumer Behavior for Captivating Marketing. *2023 Second International Conference On Smart Technologies For Smart Nation (SmartTechCon)*, 1347–1352. DOI: 10.1109/SmartTechCon57526.2023.10391558

Darling-Hammond, L., Flook, L., Cook-Harvey, C., Barron, B., & Osher, D. (2020). Implications for educational practice of the science of learning and development. *Applied Developmental Science*, *24*(2), 97–140. DOI: 10.1080/10888691.2018.1537791

Das, U. (2025). Generative AI for drug discovery and protein design: The next frontier in AI-driven molecular science. *Preprints*. DOI: 10.20944/preprints202504.0512.v1

Davenport, T., Guha, A., Grewal, D., & Bressgott, T. (2020). How artificial intelligence will change the future of marketing. *Journal of the Academy of Marketing Science*, *48*(1), 24–42. DOI: 10.1007/s11747-019-00696-0

Davis, V. (2024). Leading the revolution: The crucial role of HBCUs in the AI leadership. WCET Frontiers. https://wcet.wiche.edu/frontiers/2024/07/25/leading-the-revolution-the-crucial-role-of-hbcus-in-steering-ai-leadership/

De Bari, M. (2024). A study for innovation of methodologies and processes in digitisation and digital creation for improving the interaction with the Cultural Heritage. *Procedia Computer Science*, *237*, 187–194. DOI: 10.1016/j.procs.2024.05.095

De Luca, L. M., & Atuahene-Gima, K. (2007). Market Knowledge Dimensions and Cross-Functional Collaboration: Examining the Different Routes to Product Innovation Performance. *Journal of Marketing*, *71*(1), 95–112. DOI: 10.1509/jmkg.71.1.095

De Mauro, A., Sestino, A., & Bacconi, A. (2022). Machine learning and artificial intelligence use in marketing: A general taxonomy. *Italian Journal of Marketing*, *2022*(4), 439–457. DOI: 10.1007/s43039-022-00057-w

Departure of Culture and Tourism of Heilongjiang Province. (2024, December 26). *Top - tier Harbin has something new! The "One Ice - Snow, One World" immersive AI digital art museum has made a stunning debut at Harbin Ice - Snow World.* https://wlt.hlj.gov.cn/wlt/c114172/202412/c00_31796073.shtml

Dhanta, R., Mwale, M., & Dhanta, R. (2024). Transforming agriculture with modern AI: Harnessing artificial intelligence to revolutionize farming. In Handbook of research on AI and machine learning applications in smart agriculture (pp. 350–360). DOI: 10.4018/979-8-3693-1902-4.ch021

Dibbern, T., Romani, L. A. S., & Massruhá, S. M. F. S. (2024). Main drivers and barriers to the adoption of digital agriculture technologies. *Smart Agricultural Technology*, *8*, 100459. Advance online publication. DOI: 10.1016/j.atech.2024.100459

Duan, Y., Edwards, J. S., & Dwivedi, Y. K. (2019). Artificial intelligence for decision making in the era of Big Data – Evolution, challenges and research agenda. *International Journal of Information Management*, *48*, 63–71. DOI: 10.1016/j.ijinfomgt.2019.01.021

Durai, S., Manoharan, G., Priya, T. S., Jayanthi, R., Razak, A., & Ashtikar, S. P. (2024). Quantifying the impacts of artificial intelligence implementations in marketing. In Irani, H. R., & Nozari, H. (Eds.), (pp. 120–144). Advances in marketing, customer relationship management, and e-services. IGI Global., DOI: 10.4018/979-8-3693-1339-8.ch008

Durmus Senyapar, H. N. (2024). Artificial Intelligence in Marketing Communication: A Comprehensive Exploration of the Integration and Impact of AI. *Technium Social Sciences Journal*, *55*, 64–81. DOI: 10.47577/tssj.v55i1.10651

Dursun, B. (2023). A Qualitative Research Technique: Interview. *Journal Of Interdisciplinary Educational Research*, *7*(14), 100–113. DOI: 10.57135/jier

Dwivedi, R., Nerur, S., & Balijepally, V. (2023). Exploring artificial intelligence and big data scholarship in information systems: A citation, bibliographic coupling, and co-word analysis. *International Journal of Information Management Data Insights*, *3*(2), 100185. Advance online publication. DOI: 10.1016/j.jjimei.2023.100185

Dwivedi, Y. K., Hughes, L., Ismagilova, E., Aarts, G., Coombs, C., Crick, T., Duan, Y., Dwivedi, R., Edwards, J., Eirug, A., Galanos, V., Ilavarasan, P. V., Janssen, M., Jones, P., Kar, A. K., Kizgin, H., Kronemann, B., Lal, B., Lucini, B., & Williams, M. D. (2021). Artificial intelligence (AI): Multidisciplinary perspectives on emerging challenges, opportunities, and agenda for research, practice and policy. *International Journal of Information Management*, *57*, 101994. Advance online publication. DOI: 10.1016/j.ijinfomgt.2019.08.002

Dwivedi, Y. K., Sharma, A., Rana, N. P., Giannakis, M., Goel, P., & Dutot, V. (2023). Evolution of artificial intelligence research in Technological Forecasting and Social Change: Research topics, trends, and future directions. *Technological Forecasting and Social Change*, *192*, 122579. Advance online publication. DOI: 10.1016/j.techfore.2023.122579

Dyer, J. H., & Singh, H. (1998). The relational view: Cooperative strategy and sources of interorganizational competitive advantage. *Academy of Management Review*, *23*(4), 660–679. DOI: 10.2307/259056

Eisenmann, T., Parker, G., & Van Alstyne, M. (2006). Strategies for two-sided markets. *Harvard Business Review*. https://www.binhphuongnguyen.com/wp-content/uploads/2021/07/HBR-Strategy-for-2-side-market-Egg-and-Chicken-Dilemma-Startup.pdf

Ejjami,, R.. (2024). Leveraging AI to Enhance Marketing and Customer Engagement Strategies in the French Market. *International Journal For Multidisciplinary Research*, *6*(3), 23147. Advance online publication. DOI: 10.36948/ijfmr.2024.v06i03.23147

Ekperi, E., Owigho, O., & Ovwigho, B. (2024). Socio-cultural effects of artificial intelligence in agriculture by farmers, extension agents and lecturers in Delta State, Nigeria. *FUDMA Journal of Agriculture and Agricultural Technology*, *10*(1), 10–18.

Erevelles, S., Fukawa, N., & Swayne, L. (2016). Big Data consumer analytics and the transformation of marketing. *Journal of Business Research*, *69*(2), 897–904. DOI: 10.1016/j.jbusres.2015.07.001

Esteva, A., & Topol, E. (2019). Can skin cancer diagnosis be transformed by AI? *Lancet*, *394*(10211), 1795. DOI: 10.1016/S0140-6736(19)32726-6

Evrard, Y., & Krebs, A. (2018). The authenticity of the museum experience in the digital age: The case of the Louvre. *Journal of Cultural Economics*, *42*(3), 353–363. DOI: 10.1007/s10824-017-9309-x

Fabbri, K., & Pretelli, M. (2014). Heritage buildings and historic microclimate without HVAC technology: Malatestiana Library in Cesena, Italy, UNESCO Memory of the World. *Energy and Building*, *76*, 15–31. DOI: 10.1016/j.enbuild.2014.02.051

FAO. (2024). *Mobile-based AI solutions for food security: Case studies from Africa and Asia*. Food and Agriculture Organization of the United Nations. https://www.fao.org/publications/ai-mobile-tools

Farinu, U. (2025). Fairness, accountability, and transparency in AI: Ethical challenges in data-driven decision-making. *SSRN*. DOI: 10.2139/ssrn.5128174

First, H. B. C. U. (2024). Why choose an HBCU? https://hbcufirst.com/resources/why-choose-an-hbcu

Foster, L., Szilagyi, K., Wairegi, A., Oguamanam, C., & de Beer, J. (2023). Smart farming and artificial intelligence in East Africa: Addressing indigeneity, plants, and gender. *Smart Agricultural Technology*, *3*, 100132. Advance online publication. DOI: 10.1016/j.atech.2022.100132

Freeman, R. E. (1984). *Strategic management: A stakeholder approach.* Pitman.

Froschauer, U., & Lueger, M. (2020). Das qualitative Interview zur Analyse sozialer Systeme. *Utb.*, S. 18. https://books.google.de/books?hl=de&lr=&id=NqfNDwAAQBAJ&oi=fnd&pg=PP1&dq=qualitative+interviews&ots=2MYVGTJn2z&sig=QGSvpuxu5MMHE38DzyKXx1nRLME#v=onepage&q=Trans&f=false

Füller, J. (2010). Refining Virtual Co-Creation from a Consumer Perspective. *California Management Review*, *52*(2), 98–122. DOI: 10.1525/cmr.2010.52.2.98

Füller, J., Mühlbacher, H., Matzler, K., & Jawecki, G. (2009). Consumer Empowerment Through Internet-Based Co-creation. *Journal of Management Information Systems*, *26*(3), 71–102. DOI: 10.2753/MIS0742-1222260303

Gaikwad, S., & Rajvardhan, . (2025). Overview on biased news reporting of Indian television with legal aspect. *The Scientific Temper*, *16*(04), 4017–4028. DOI: 10.58414/SCIENTIFICTEMPER.2025.16.4.03

Gajjar, N., Rathod, K., & Jani, K. (2022). A systematic literature review on Robotic Process Automation security (Version 1). arXiv. https://doi.org//arXiv.2212.05544DOI: 10.48550

Ganeshkumar, M., Karunakaran, S., Muzhumathi, R., & Suguna, R. (2024). AI-Driven Marketing Insights: Harnessing Machine Learning for Enhanced Consumer Engagement and Personalization. *2024 First International Conference on Software, Systems and Information Technology (SSITCON)*, 1–6. DOI: 10.1109/SSITCON62437.2024.10796558

Gao, Y., Zhang, Q., Wang, X. R., Huang, Y. M., Meng, F. S., & Tao, W. (2024). Multidimensional knowledge discovery of cultural relics resources in the Tang tomb mural category. *The Electronic Library*, *42*(1), 1–22. DOI: 10.1108/EL-04-2023-0091

Gardezi, M., Joshi, B., Rizzo, D. M., Ryan, M., Prutzer, E., Brugler, S., & Dadkhah, A. (2024a). Artificial intelligence in farming: Challenges and opportunities for building trust. *Agronomy Journal*, *116*(3), 1217–1228. DOI: 10.1002/agj2.21353

Gautam, R., Kaur, P., & Sharma, M. (2025). Predictive analytics for early cancer detection using machine learning and generative AI. In Generative intelligence in healthcare (pp. 150–175). DOI: 10.1201/9781003539483-7

Georgescu, S. D., & Anastaiu, I. E. (2022, February 18). The Interview as a Qualitative Research Instrument. *Proceedings of the 15th International Management Conference "Managing People and Organizations in a Global Crisis."* DOI: 10.24818/IMC/2021/05.14

Ghimire, A., Thapa, S., Jha, A. K., Adhikari, S., & Kumar, A. (2020). Accelerating business growth with Big Data and artificial intelligence. In 2020 Fourth International Conference on I-SMAC (IoT in Social, Mobile, Analytics and Cloud) (I-SMAC) (pp. 441–448). https://doi.org/DOI: 10.1109/I-SMAC49090.2020.9243318

Gibler, K., & Nelson, S. (1998). Consumer Behavior Applications to Real Estate. *Academia.Edu.*

Gibson, G. (2024). How generative AI is shaking up education: The role of HBCUs in leading the charge. HBCU Lifestyle. https://hbculifestyle.com/generative-ai-hbcus-education-leadership/

Gîrbacia, F. (2024). An Analysis of Research Trends for Using Artificial Intelligence in Cultural Heritage. *Electronics (Basel)*, *13*(18), 3738. DOI: 10.3390/electronics13183738

Guadalupe, A. I. S. (2018). User-Centered Design to a Digital Cultural Heritage Installation Based on Afro-Peruvian Musical Legacy. 2018 IEEE Games, Entertainment, *Media Conference (GEM)*

Gugenishvili, I. (2019). Framework for the National Culture and User Engagement in Social Media. *Association of Marketing Theory and Practice Proceedings 2019.* https://digitalcommons.georgiasouthern.edu/amtp-proceedings_2019/48

Gungunawat, A., Khandelwal, N., & Gupta, N. (2024). AI-Powered Personalization in Digital Marketing: Transforming Consumer Engagement and Strategy. *RESEARCH REVIEW International Journal of Multidisciplinary*, *9*(11), 183–191. DOI: 10.31305/rrijm.2024.v09.n11.026

Guo Wei, H. R. (2022). Research Characteristics and Trends of Cultural and Creative Product Design Based on Bibliometrics. *PACKAGING ENGINEERING*, *43*(14), 326–334. DOI: 10.19554/j.cnki.1001-3563.2022.14.040

Gupta, K. (2024). Transforming healthcare: The role of artificial intelligence in revolutionising patient care. *MMJ*, *1*(4). Advance online publication. DOI: 10.62830/MMJ1-04-6a

Gursoy, D., Chi, O. H., Lu, L., & Nunkoo, R. (2019). Consumers acceptance of artificially intelligent (AI) device use in service delivery. *International Journal of Information Management*, *49*, 157–169. DOI: 10.1016/j.ijinfomgt.2019.03.008

Haag, F., Hopf, K., Vasconcelos, P. M., & Staake, T. (2022). *Augmented cross-selling through explainable AI — a case from energy retailing* (Version 1). arXiv. DOI: 10.48550/ARXIV.2208.11404

Haase, J., & Pokutta, S. (2024). *Human-AI Co-Creativity: Exploring Synergies Across Levels of Creative Collaboration* (Version 2). arXiv. DOI: 10.48550/ARXIV.2411.12527

Haenlein, M., & Kaplan, A. (2019). A brief history of artificial intelligence: On the past, present, and future of artificial intelligence. *California Management Review*, *61*(4), 5–14. DOI: 10.1177/0008125619864925

Halabhavi, B. M. (2024). AI-driven predictive analytics in healthcare: Improving patient outcomes and resource allocation. *African Journal of Biomedical Research*, *27*(4S), 9288–9297. DOI: 10.53555/AJBR.v27i4S.5427

Haleem, A., Javaid, M., Asim Qadri, M., Pratap Singh, R., & Suman, R. (2022). Artificial intelligence (AI) applications for marketing: A literature-based study. *International Journal of Intelligent Networks*, *3*, 119–132. DOI: 10.1016/j.ijin.2022.08.005

Harbin Municipal People's Government. (2024, January 5). *The world's largest ice and snow themed amusement park! Harbin Ice and Snow World has been awarded the title of Guinness World Records.* https://www.harbin.gov.cn/haerbin/c111753/202401/c01_958584.shtml

Harmeling, C. M., Moffett, J. W., Arnold, M. J., & Carlson, B. D. (2017). Toward a theory of customer engagement marketing. *Journal of the Academy of Marketing Science*, *45*(3), 312–335. DOI: 10.1007/s11747-016-0509-2

Hatture, S. M., & Kadakol, N. (2021). Clinical diagnostic systems based on machine learning and deep learning. In Demystifying big data, machine learning, and deep learning for healthcare analytics (pp. 159–183). DOI: 10.1016/B978-0-12-821633-0.00011-8

Haupt, M., Freidank, J., & Haas, A. (2025). Consumer responses to human-AI collaboration at organizational frontlines: Strategies to escape algorithm aversion in content creation. *Review of Managerial Science*, *19*(2), 377–413. DOI: 10.1007/s11846-024-00748-y

Hazelkorn, E. (2016). Contemporary debates part 2: Initiatives, and governance and organizational structures. In Goddard, J., Hazelkorn, E., & Vallance, P. (Eds.), *The civic university: The policy and leadership challenges* (pp. 34–64). Edward Elgar Publishing., DOI: 10.4337/9781784717728.00011

Hazzan-Bishara, A., Kol, O., & Levy, S. (2025). The factors affecting teachers' adoption of AI technologies: A unified model of external and internal determinants. *Education and Information Technologies*, *30*(11), 15043–15069. Advance online publication. DOI: 10.1007/s10639-025-13393-z

Henderlong, J., & Lepper, M. R. (2002). The effects of praise on children's intrinsic motivation: A review and synthesis. *Psychological Bulletin*, *128*(5), 774–795. DOI: 10.1037/0033-2909.128.5.774 PMID: 12206194

Hengstler, M., Enkel, E., & Duelli, S. (2016). Applied artificial intelligence and trust—The case of autonomous vehicles and medical assistance devices. *Technological Forecasting and Social Change*, *105*, 105–120. DOI: 10.1016/j.techfore.2015.12.014

Hermann, E. (2022). Leveraging artificial intelligence in marketing for social good—An ethical perspective. *Journal of Business Ethics*, *179*(1), 43–61. DOI: 10.1007/s10551-021-04843-y PMID: 34054170

Hilditch, D. J. (1995). *At the heart of the world: Merleau-Ponty and the existential phenomenology of embodied and embedded intelligence in everyday coping*. Washington University in St. Louis.

Hollebeek, L. D. (2011). Demystifying customer brand engagement: Exploring the loyalty nexus. *Journal of Marketing Management*, *27*(7–8), 785–807. DOI: 10.1080/0267257X.2010.500132

Hollebeek, L. D., Glynn, M. S., & Brodie, R. J. (2014). Consumer Brand Engagement in Social Media: Conceptualization, Scale Development and Validation. *Journal of Interactive Marketing*, *28*(2), 149–165. DOI: 10.1016/j.intmar.2013.12.002

Hollebeek, L. D., & Macky, K. (2019). Digital Content Marketing's Role in Fostering Consumer Engagement, Trust, and Value: Framework, Fundamental Propositions, and Implications. *Journal of Interactive Marketing*, *45*(1), 27–41. DOI: 10.1016/j.intmar.2018.07.003

Hollebeek, L. D., Menidjel, C., Sarstedt, M., Jansson, J., & Urbonavicius, S. (2024). Engaging consumers through artificially intelligent technologies: Systematic review, conceptual model, and further research. *Psychology and Marketing*, *41*(4), 880–898. DOI: 10.1002/mar.21957

Holtorf, C. (2018). Embracing change: How cultural resilience is increased through cultural heritage. *World Archaeology, 50*(4), 639–650. DOI: 10.1080/00438243.2018.1510340

Hoyer, W. D., Kroschke, M., Schmitt, B., Kraume, K., & Shankar, V. (2020). Transforming the customer experience through new technologies. *Journal of Interactive Marketing, 51*(1), 57–71. DOI: 10.1016/j.intmar.2020.04.001

Hs, M., & Minhat, H. S. (2015). An Overview on the methods of interviews in qualitative research. In *International Journal of Public Health and Clinical Sciences* (Vol. 2, Issue 1).

Huang, M.-H., & Rust, R. T. (2021a). A strategic framework for artificial intelligence in marketing. *Journal of the Academy of Marketing Science, 49*(1), 30–50. DOI: 10.1007/s11747-020-00749-9

Jain, S., & Nigam, H. (2025). AI-powered diagnostics and personalized treatment: Enhancing patient outcomes in modern healthcare. *Journal of Neonatal Surgery, 14*(6S), 349–359. DOI: 10.52783/jns.v14.2241

Janiesch, C., Zschech, P., & Heinrich, K. (2021). Machine learning and deep learning. *Electronic Markets, 31*(3), 685–695. DOI: 10.1007/s12525-021-00475-2

Jantanukul, W. (2024). AI and Human Synergy: Utilizing AI to Enhance Teaching and Learning. *Journal of Education and Learning Reviews, 1*(4), 1–12. DOI: 10.60027/jelr.2024.749

Jarrahi, M. H. (2018). Artificial intelligence and the future of work: Human-AI symbiosis in organizational decision making. *Business Horizons, 61*(4), 577–586. DOI: 10.1016/j.bushor.2018.03.007

Jaswal, S., Sharma, S., & Parihar, N. (2024). Artificial intelligence-driven predictive analytics for early detection of chronic diseases in primary healthcare. In *2024 International Conference on Innovative Computing, Intelligent Communication and Smart Electrical Systems (ICSES)* (pp. 1–6). DOI: 10.1109/ICSES63760.2024.10910314

Javaid, M., Haleem, A., Khan, I. H., & Suman, R. (2023). Understanding the potential applications of artificial intelligence in agriculture sector. *Advanced Agrochem, 2*(1), 15–30. DOI: 10.1016/j.aac.2022.10.001

Jones-Esan, L., Somasiri, N., & Lorne, K. (2024). Enhancing healthcare delivery through digital health interventions: A systematic review on telemedicine and mobile health applications in low and middle-income countries (LMICs). *Research Square*. DOI: 10.21203/rs.3.rs-5189203/v1

Joseph, O. O., Akintola, A. S., Offiong, E., Olajuwon, O. O., Offia, U. I., & Faniyan, A. A. (2025). AI-Enhanced Consumer Insights: Leveraging Behavioural Analytics for Hyper-Personalised Marketing Strategies. *Journal of Artificial Intelligence. Machine Learning and Data Science, 3*(1), 2361–2368. DOI: 10.51219/JAIMLD/Omotoso-Oluwayomi-Joseph/511

Joshi, H., Bansal, J., & Shukla, P. (2025). Revolutionizing patient access through AI in healthcare. In Advances in artificial intelligence for healthcare applications (pp. 208–227). DOI: 10.1201/9781003503385-15

Julien, C., Leide, J. E., & Bouthillier, F. (2008). Controlled user evaluations of information visualization interfaces for text retrieval: Literature review and meta-analysis. *Journal of the American Society for Information Science and Technology, 59*(6), 1012–1024. DOI: 10.1002/asi.20786

Kaiser, A., & Burger, P. (2022). Understanding diversity in farmers' routinized crop protection practices. *Journal of Rural Studies, 89*, 149–160. DOI: 10.1016/j.jrurstud.2021.12.002

Kambala, M. (2024). AI-powered healthcare: Transforming patient outcomes with machine learning. *Journal of Medical Science and Clinical Research, 12*(8), 34–47. DOI: 10.18535/jmscr/v12i08.07

Kastner, L., Umbach, N., Jusyte, A., Cervera-Torres, S., Fernández, S. R., Nommensen, S., & Gerjets, P. (2021). Designing Visual-Arts Education Programs for Transfer Effects: Development and Experimental Evaluation of (Digital) Drawing Courses in the Art Museum Designed to Promote Adolescents' Socio-Emotional Skills. *Frontiers in Psychology, 11*, 603984. Advance online publication. DOI: 10.3389/fpsyg.2020.603984 PMID: 33536974

Ke, X. L. J. (2020). Digital Communication of Chinese Intangible Cultural Heritage: New Challenges and New Solutions. Cultural Heritage(01), 140-146. https://kns.cnki.net/kcms2/article/abstract?v=qKY3Y0De9e4T346Jsn9f2RN88NfWfhsdlGTfxWnYDbDYsz_DxIGNPB9d6lyriH7bSgTbOcOCUwaY5VrtKOAKZC7VECmpfbYCm7TJSCQrI7qwOsXWGdYJIP2G1q8TwZRhaIUZsNWGUjBa8XUsJGeSmL14aXKsr8cu6-tAH0j7B3cPvqcyecnR5G73kdL_gmSc__GZ3fMXH48=&uniplatform=NZKPT&language=CHS

Keegan, B. J., Dennehy, D., & Naudé, P. (2024a). Implementing artificial intelligence in traditional B2B marketing practices: An activity theory perspective. *Information Systems Frontiers, 26*(3), 1025–1039. DOI: 10.1007/s10796-022-10294-1 PMID: 35637917

Kim, J., & Jin, H. Y. (2024). Interpreting Tactical Urbanism through Innovation-Diffusion Theory: Insights from a Collaborative Design Studio Experience. *Land (Basel)*, *13*(1), 14. Advance online publication. DOI: 10.3390/land13010014

Kim, M. J., Lee, C. K., & Contractor, N. S. (2019). Seniors' usage of mobile social network sites: Applying theories of innovation diffusion and uses and gratifications. *Computers in Human Behavior*, *90*, 60–73. DOI: 10.1016/j.chb.2018.08.046

Knowles, M. S. (1978). Andragogy: Adult learning theory in perspective. *Community College Review*, *5*(3), 9–20. DOI: 10.1177/009155217800500302

Kolb, D. A. (1984). *Experiential learning: Experience as the source of learning and development*. Prentice-Hall.

Kollia, I., Stevenson, J., & Kollias, S. (2021). AI-enabled efficient and safe food supply chain [Preprint]. *arXiv*.https://arxiv.org/abs/2105.00333?utm_source=chatgpt.com

Kolomaznik, M., Petrik, V., Slama, M., & Jurik, V. (2024). The role of socio-emotional attributes in enhancing human-AI collaboration. *Frontiers in Psychology*, *15*, 1369957. Advance online publication. DOI: 10.3389/fpsyg.2024.1369957 PMID: 39474095

Kosaraju, D. (2024). Predictive analytics in healthcare: Leveraging AI to anticipate disease outbreaks and enhance patient outcomes. *Galore International Journal of Health Sciences and Research*, *8*(3), 73–79. DOI: 10.52403/gijhsr.20230312

Kotha, , S. (2024). The Convergence of Artificial Intelligence and Human Marketing: A Framework for Enhanced Customer Insights and Personalization. *International Journal For Multidisciplinary Research*, *6*(6), 30266. Advance online publication. DOI: 10.36948/ijfmr.2024.v06i06.30266

Kotoula, E., Robinson, D. W., & Bedford, C. (2018). Interactive relighting, digital image enhancement and inclusive diagrammatic representations for the analysis of rock art superimposition: The main Pleito cave (CA, USA). *Journal of Archaeological Science*, *93*, 26–41. DOI: 10.1016/j.jas.2018.02.012

Krishna Pasupuleti, M. (2024). Quantum AI and precision genomics: Redefining cancer care and personalized medicine. In Precision genomics and quantum AI: Revolutionizing cancer treatment and personalized health (pp. 40–55). DOI: 10.62311/nesx/46080

Kshetri, N., Dwivedi, Y. K., Davenport, T. H., & Panteli, N. (2024). Generative artificial intelligence in marketing: Applications, opportunities, challenges, and research agenda. *International Journal of Information Management*, *75*, 102716. DOI: 10.1016/j.ijinfomgt.2023.102716

Kühl, N., Mühlthaler, M., & Goutier, M. (2020). Supporting customer-oriented marketing with artificial intelligence: Automatically quantifying customer needs from social media. *Electronic Markets*, *30*(2), 351–367. DOI: 10.1007/s12525-019-00351-0

Kulkarni, A., Shivananda, A., & Kulkarni, A. (2021). Natural language processing and artificial intelligence overview. In Natural language processing projects (pp. 1–37). DOI: 10.1007/978-1-4842-7386-9_1

Kulkarni, Y., Mahamuni, A., Sane, S., Kalshetti, P., Patil, K., & Jarad, R. S. (2024). Analysing How Marketing Management and Artificial Intelligence are used to Change Customer Engagement. *2024 International Conference on Trends in Quantum Computing and Emerging Business Technologies*, 1–6. DOI: 10.1109/TQCEBT59414.2024.10545064

Kumar, V., Iqbal, M. I., & Rathore, R. (2025). Natural language processing (NLP) in disease detection. In AI in disease detection (pp. 53–75). DOI: 10.1002/9781394278695.ch3

Kumar, A., Bapat, G., Kumar, A., Hota, S. L., Abishek, G. D., & Vaz, S. (2024). Unlocking Brand Excellence: Harnessing AI Tools for Enhanced Customer Engagement and Innovation. *RAiSE*, *2023*, 204. DOI: 10.3390/engproc2023059204

Kumar, A., & Joshi, S. (2022). Applications of AI in healthcare sector for enhancement of medical decision making and quality of service. In *2022 International Conference on Decision Aid Sciences and Applications (DASA)* (pp. 37–41). DOI: 10.1109/DASA54658.2022.9765041

Kumar, V., Dixit, A., Javalgi, R. G., & Dass, M. (2016). Research framework, strategies, and applications of intelligent agent technologies (IATs) in marketing. *Journal of the Academy of Marketing Science*, *44*(1), 24–45. DOI: 10.1007/s11747-015-0426-9

Kumar, V., Ramachandran, D., & Kumar, B. (2021). Influence of new-age technologies on marketing: A research agenda. *Journal of Business Research*, *125*, 864–877. DOI: 10.1016/j.jbusres.2020.01.007

Kushwaha, A. K., & Kar, A. K. (2020). Language model-driven chatbot for business to address marketing and selection of products. In Sharma, S. K., Dwivedi, Y. K., Metri, B., & Rana, N. P. (Eds.), *Re-imagining diffusion and adoption of information technology and systems: A continuing conversation* (Vol. 617, pp. 16–28). Springer., DOI: 10.1007/978-3-030-64849-7_3

Kushwaha, A. K., & Kar, A. K. (2024). MarkBot – A language model-driven chatbot for interactive marketing in post-modern world. *Information Systems Frontiers*, *26*(3), 857–874. DOI: 10.1007/s10796-021-10184-y

Lago-Olveira, S., El-Areed, S. R. M., Moreira, M. T., & González-García, S. (2023). Improving environmental sustainability of agriculture in Egypt through a life-cycle perspective. *The Science of the Total Environment*, *890*, 164335. Advance online publication. DOI: 10.1016/j.scitotenv.2023.164335 PMID: 37209738

Lamanna, M., Bovo, M., & Cavallini, D. (2025). Wearable collar technologies for dairy cows: A systematized review of the current applications and future innovations in precision livestock farming. *Animals (Basel)*, *15*(3), 458. Advance online publication. DOI: 10.3390/ani15030458 PMID: 39943229

Laszkiewicz, A., & Kalinska-Kula, M. (2023). Virtual influencers as an emerging marketing theory: A systematic literature review. *International Journal of Consumer Studies*, *47*(6), 2479–2494. DOI: 10.1111/ijcs.12956

Le Fillâtre, V., Robert, E., Petrognani, S., Lesvignes, E., Cretin, C., & Muth, X. (2021). Mapping the walls: High-resolution cartography applied to the analysis of prehistoric cave art in the Grotte du Mammouth (Domme, Dordogne, France). *Journal of Archaeological Science*, *127*, 105332. Advance online publication. DOI: 10.1016/j.jas.2021.105332

Le. (2024). *Ra m t ng d ng Xhero, bư c ti n trong ng d ng công ngh hi n đ i | Nh p s ng s | Báo Văn Hóa Online*. Baovanhoa.Vn. https://baovanhoa.vn/nhip-song-so/ra-mat-ung-dung-xhero-buoc-tien-trong-ung-dung-cong-nghe-hien-dai-106004.html

Le, B. (2022). *The behavioral intention to adopt proptech services in Vietnam real estate market*. Swiss School of Management.

Lee, D., Hosanagar, K., & Nair, H. (2017). Advertising Content and Consumer Engagement on Social Media: Evidence from Facebook. SSRN *Electronic Journal*. DOI: 10.2139/ssrn.2290802

Lee, J., Davari, H., Singh, J., & Pandhare, V. (2018). Industrial artificial intelligence for industry 4.0-based manufacturing systems. *Manufacturing Letters*, *18*, 20–23. DOI: 10.1016/j.mfglet.2018.09.002

Lee, M., Kwon, W., & Back, K.-J. (2021). Artificial intelligence for hospitality big data analytics: Developing a prediction model of restaurant review helpfulness for customer decision-making. *International Journal of Contemporary Hospitality Management*, *33*(6), 2117–2136. DOI: 10.1108/IJCHM-06-2020-0587

Lee, S. (1986). *FS definitions and concepts.* Cornell University ProQuest Dissertations Publishing.

Lee, Y. S. (2024). Qualitative and mixed methods. *Translational Orthopedics, 229–232*, 229–232. Advance online publication. DOI: 10.1016/B978-0-323-85663-8.00010-6

Leung, C. H., & Chan, W. T. Y. (2024). Bridging ancient wisdom and modern management: Enhancing I Ching learning with ChatGPT. *Asian Journal of Contemporary Education, 8*(1), 19–32. DOI: 10.55493/5052.v8i1.5028

Lin, C.-C., & Chen, C.-L. (2012). *An Estimation of the Impact of Feng-Shui on Housing Prices in Taiwan: A Quantile Regression Application* (Vol. 15, Issue 3).

Liquin, E. G., & Gopnik, A. (2022). Children are more exploratory and learn more than adults in an approach-avoid task. *Cognition, 218*, 104940. Advance online publication. DOI: 10.1016/j.cognition.2021.104940 PMID: 34715584

Lischer-Katz, Z. (2022). The emergence of digital reformatting in the history of preservation knowledge: 1823-2015. *The Journal of Documentation, 78*(6), 1249–1277. DOI: 10.1108/JD-04-2021-0080

Liu, J., & Babar, M. (2022). Digitally Protecting and Disseminating the Intangible Cultural Heritage in Information Technology Era. *Mobile Information Systems, 2022*, 1–10. DOI: 10.1155/2022/1115655

Liu, Z. (2022). Literature review on image restoration. *Journal of Physics: Conference Series.*

Li, Y., Li, Y., Chen, Q., & Chang, Y. (2024). Humans as teammates: The signal of human–AI teaming enhances consumer acceptance of chatbots. *International Journal of Information Management, 76*, 102771. DOI: 10.1016/j.ijinfomgt.2024.102771

Ljepava, N. (2022). AI-enabled marketing solutions in marketing decision making: AI application in different stages of marketing process. TEM Journal, 11(3), 1308–1315. https://doi.org/DOI: 10.18421/TEM113-40

Lorenz, F. (2021). Dissertation_Felix_Lorenz_Pflichtexemplar. *Library of University of Regensburg.*

Mak, M. Y. (1998). *ANZASCA-1998-Mak_FengShui.* Science and Design.

Malthouse, E., & Copulsky, J. (2023). Artificial intelligence ecosystems for marketing communications. *International Journal of Advertising, 42*(1), 128–140. DOI: 10.1080/02650487.2022.2122249

Manning, L., Brewer, S., Craigon, P. J., Frey, J., Gutierrez, A., Jacobs, N., Kanza, S., Munday, S., Sacks, J., & Pearson, S. (2022). Artificial intelligence and ethics within the food sector: Developing a common language for technology adoption across the supply chain. *Trends in Food Science & Technology*, *125*, 33–42. DOI: 10.1016/j.tifs.2022.04.025

Manser Payne, E. H., Dahl, A. J., & Peltier, J. (2021). Digital servitization value co-creation framework for AI services: A research agenda for digital transformation in financial service ecosystems. *Journal of Research in Interactive Marketing*, *15*(2), 200–222. DOI: 10.1108/JRIM-12-2020-0252

Mariani, M. M., Perez-Vega, R., & Wirtz, J. (2022). AI in marketing, consumer research and psychology: A systematic literature review and research agenda. *Psychology and Marketing*, *39*(4), 755–776. DOI: 10.1002/mar.21619

Mark, B., & Bizz, P. (2025). Ethical implications of AI-driven decision making in healthcare information systems. *Preprints*. DOI: 10.20944/preprints202503.0078.v1

Martínez, A., Schmuck, C., Pereverzyev, S.Jr, Pirker, C., & Haltmeier, M. (2020). A machine learning framework for customer purchase prediction in the non-contractual setting. *European Journal of Operational Research*, *281*(3), 588–596. DOI: 10.1016/j.ejor.2018.04.034

May, K., Taylor, J. S., & Binding, C. (2023). Stratigraphic Analysis and The Matrix: Connecting and reusing digital records and archives of archaeological investigations. *Internet Archaeology*, *61*. Advance online publication. DOI: 10.11141/ia.61.2

McCall, C. J., DeCaprio, D., & Gartner, J. (2022). The measurement and mitigation of algorithmic bias and unfairness in healthcare AI models developed for the CMS AI health outcomes challenge. *medRxiv*. DOI: 10.1101/2022.09.29.22280537

Mehrad, A., Nguyen, K., & Dihii, J. B. (2024). Social Cognition Approach: Using Observation and Interview Methods via Qualitative Research. *Journal of Education for Sustainable Innovation*, *2*(1), 11–17. DOI: 10.56916/jesi.v2i1.718

Mei-Chin, S., Le-En, L., Li-Hwa, L., & Yu-Chu, C. (2024). Assessing question characteristic influences on ChatGPT's performance and response-explanation consistency: Insights from Taiwan's nursing licensing exam. *International Journal of Nursing Studies*, *153*, 104717. Advance online publication. DOI: 10.1016/j.ijnurstu.2024.104717 PMID: 38401366

Meinecke, C., Hall, C., & Janicke, S. (2022). Towards Enhancing Virtual Museums by Contextualizing Art through Interactive Visualizations. *Acm Journal on Computing and Cultural Heritage*, *15*(4), 62. Advance online publication. DOI: 10.1145/3527619

Melash, A. A., Bogale, A. A., Migbaru, A. T., Chakilu, G. G., Percze, A., Ábrahám, É. B., & Mengistu, D. K. (2023). Indigenous agricultural knowledge: A neglected human based resource for sustainable crop protection and production. *Heliyon, 9*(1), e12978. Advance online publication. DOI: 10.1016/j.heliyon.2023.e12978 PMID: 36711305

Mhlanga, D. (2025). AI and predictive analytics and risk assessment in healthcare. In AI in hospital administration (pp. 89–104). DOI: 10.1201/9781003475804-6

Mikalef, P., & Gupta, M. (2021). Artificial intelligence capability: Conceptualization, measurement calibration, and empirical study on its impact on organizational creativity and firm performance. *Information & Management*, *58*(3), 103434. Advance online publication. DOI: 10.1016/j.im.2021.103434

Miller, P. H. (2022). Developmental theories: Past, present, and future. *Developmental Review*, *66*, 101049. Advance online publication. DOI: 10.1016/j.dr.2022.101049

Miller, T. (2019). Explanation in artificial intelligence: Insights from the social sciences. *Artificial Intelligence*, *267*, 1–38. DOI: 10.1016/j.artint.2018.07.007

Mishra, M. (2021). Machine learning techniques for structural health monitoring of heritage buildings: A state-of-the-art review and case studies. *Journal of Cultural Heritage*, *47*, 227–245. DOI: 10.1016/j.culher.2020.09.005

Mostafa, R. B., & Kasamani, T. (2022). Antecedents and consequences of chatbot initial trust. *European Journal of Marketing*, *56*(6), 1748–1771. DOI: 10.1108/EJM-02-2020-0084

Motta, F. M. D., Barbosa, C. R., & Barbosa, R. R. (2019). BIG DATA AS INNOVATION SOURCE IN MUSEUMS: the case study of The British Museum. *Informacao & Sociedade-Estudos, 29*(1), 83-100. <Go to ISI>://WOS:000462627300006

Munro, K. (2022). Natural language processing (NLP). In The handbook of data science and AI (pp. 273–316). DOI: 10.3139/9781569908877.010

Mustak, M., Salminen, J., Plé, L., & Wirtz, J. (2021). Artificial intelligence in marketing: Topic modeling, scientometric analysis, and research agenda. *Journal of Business Research, 124*, 389–404. DOI: 10.1016/j.jbusres.2020.10.044

Naber, A. (2015). Qualitative experiment as a participating method in innovation research. In *Historical Social Research* (Vol. 40, Issue 3, pp. 233–257). GESIS - Leibniz Institute for the Social Sciences. DOI: 10.12759/hsr.40.2015.3.233-257

Naidoo, G. (2024). The potential of artificial intelligence in South African rural development. *African Journal of Science, Technology, Innovation and Development, 17*(2), 207–218.

Nair, K., & Gupta, R. (2021). Application of AI technology in modern digital marketing environment. *World Journal of Entrepreneurship, Management and Sustainable Development, 17*(4), 383–396. DOI: 10.1108/WJEMSD-08-2020-0099

Nandish, M. (2023). How has the rapid growth of artificial intelligence impacted business management? A case study of prominent organizations that have experienced significant incline in growth due to adaptation to AI advancements. *International Journal of Research Publication and Reviews, 4*(6), 4124–4130. DOI: 10.55248/gengpi.4.623.47437

National Center for Education Statistics. (2023). Fast facts: Historically Black colleges and universities. https://nces.ed.gov/fastfacts/display.asp?id=667

Nghiem-Phu, B. (2022). Consumer behaviour towards purchasing Feng Shui goods: An empirical study from Vietnam. *Journal of Asian Finance, 9*(1), 83–0092. DOI: 10.13106/jafeb.2022.vol9.no1.0083

Nguyen, T., & Vu, N. (2023). Spiritual factors in Vietnamese consumers' purchasing decisions. *International Journal of Business and Management, 06*(10), 15–27. DOI: 10.5281/zenodo.8407732

Nikolola-Alexieva, V., Valeva, K., & Pashev, S. (2024). Artificial intelligence in the food industry. *BIO Web of Conferences, 102,* 04002. https://doi.org/DOI: 10.1051/bioconf/202410204002

Nissar, H., & M, M. S. (2025). AI-enhanced robotic surgery: A machine learning approach to improving surgical precision and patient safety. TechRxiv. DOI: 10.36227/techrxiv.174430796.67882981/v1

Nixon-Saintil. [Initial]. (2025). IBM is equipping the next generation of student developers across universities with AI skills. IBM Newsroom. https://newsroom.ibm.com/blog-IBM-is-equipping-the-next-generation-of-student-developers-across-universities-with-AI-skills

Noble, C. H., & Mokwa, M. P. (1999). Implementing marketing strategies: Developing and testing a managerial theory. *Journal of Marketing, 63*(4), 57–73. DOI: 10.1177/002224299906300406

Noble, K. (2021a). Challenges and Opportunities: Creative Approaches to Museum and Gallery Learning during the Pandemic. *International Journal of Art & Design Education, 40*(4), 676–689. DOI: 10.1111/jade.12380

Noble, K. (2021b). 'Getting Hands On with Other Creative Minds': Establishing a Community of Practice around Primary Art and Design at the Art Museum. *International Journal of Art & Design Education*, *40*(3), 615–629. DOI: 10.1111/jade.12371

Noyes, E., & Girdharry, K. (2024, December 3). Why faculty should lead the AI revolution. AACSB Insights. https://www.aacsb.edu/insights/articles/2024/12/why-faculty-should-lead-the-ai-revolution

Ntjamba, F. C., & Ashipala, D. O. (2024). Impact on and ethical considerations of artificial intelligence on human healthcare. In Advances in computational intelligence and robotics (pp. 1–36). DOI: 10.4018/979-8-3693-9158-7.ch001

Ntjamba, F. C., & Ashipala, D. O. (2025). Beyond reality: Integrating AI and Metaverse technologies for enhanced scientific collaboration. In Advances in computational intelligence and robotics (pp. 99–118). DOI: 10.4018/979-8-3373-0340-6.ch005

Ntjamba, F. C., & Ashipala, D. O. (2025). Ethical dimensions of AI in the Metaverse. In Advances in computational intelligence and robotics (pp. 195–212). DOI: 10.4018/979-8-3373-0340-6.ch010

Ntjamba, F. C., & Ashipala, D. O. (2025). Innovative pathways: AI and the Metaverse in interdisciplinary research and knowledge discovery. In Advances in computational intelligence and robotics (pp. 119–138). DOI: 10.4018/979-8-3373-0340-6.ch006

Nurchim, N., & Nurmalitasari, N. (2025). Empowering Asian students through artificial intelligence: A workshop on predicting plant growth to support smart farming. *International Journal of Community Service*, *5*(1), 30–36. DOI: 10.51601/ijcs.v5i1.839

O'Connor, J. (2007). *The cultural and creative industries: a review of the literature*. Arts Council England.

O'Reilly, C. A.III, & Tushman, M. L. (2008). Ambidexterity as a dynamic capability: Resolving the innovator's dilemma. *Research in Organizational Behavior*, *28*, 185–206. DOI: 10.1016/j.riob.2008.06.002

Olagbende, R. (2025). Natural language processing (NLP) techniques for Afan Oromo text analysis. *Preprints*. DOI: 10.20944/preprints202503.0841.v1

Ouariach, S., & Raghav, A. (2025). Legal accountability in AI-enhanced healthcare. In Computational intelligence in healthcare law (pp. 72–85). DOI: 10.1201/9781003538462-5

Padmavathy, A., & Alamelu, R.A. (2025). HARNESSING PERSONALIZED EMOTIONAL AI FOR ENHANCED CONSUMER ENGAGEMENT IN INTELLIGENT MARKETING. *Donglixue Yu Kongzhi Xuebao*, 13–22. DOI: 10.71058/jodac.v9i1002

Pan, S., Luo, L., Wang, Y., Chen, C., Wang, J., & Wu, X. (2024). Unifying large language models and knowledge graphs: A roadmap. *IEEE Transactions on Knowledge and Data Engineering*, *36*(7), 3580–3599. DOI: 10.1109/TKDE.2024.3352100

Pan, Y., & Zhang, L. (2021). Roles of artificial intelligence in construction engineering and management: A critical review and future trends. *Automation in Construction*, *122*, 103517. Advance online publication. DOI: 10.1016/j.autcon.2020.103517

Paschen, J. (2019). Investigating the emotional appeal of fake news using artificial intelligence and human contributions. *Journal of Product and Brand Management*, *29*(2), 223–233. DOI: 10.1108/JPBM-12-2018-2179

Paul, J., Ueno, A., & Dennis, C. (2023). ChatGPT and consumers: Benefits, Pitfalls and Future Research Agenda. *International Journal of Consumer Studies*, *47*(4), 1213–1225. DOI: 10.1111/ijcs.12928

Philp, M., Jacobson, J., & Pancer, E. (2022). Predicting social media engagement with computer vision: An examination of food marketing on Instagram. *Journal of Business Research*, *149*, 736–747. DOI: 10.1016/j.jbusres.2022.05.078

Phong-Thuy-Dai-Nam. (2024). *Hư ng d n s d ng App Phong Th y Xhero Đ i Nam*. Phongthuydainam.Vn. https://phongthuydainam.vn/news/675134b439cf021bcf2fe4b7

Ping, M. (2025). *Digitalization of cultural heritage and construction of cultural identity*. Jiangsu Social Sciences., DOI: 10.13858/j.cnki.cn32-1312/c.20250319.004

Portalés, C., Rodrigues, J. M. F., Rodrigues Gonçalves, A., Alba, E., & Sebastián, J. (2018). Digital Cultural Heritage. *Multimodal Technologies and Interaction*, *2*(3), 58. Advance online publication. DOI: 10.3390/mti2030058

Postsecondary National Policy Institute. (2024). Black students in higher education. https://pnpi.org/factsheets/black-students/

Prentice, C., & Nguyen, M. (2020). Engaging and retaining customers with AI and employee service. *Journal of Retailing and Consumer Services*, *56*, 102186. DOI: 10.1016/j.jretconser.2020.102186

Prentice, C., Weaven, S., & Wong, I. A. (2020). Linking AI quality performance and customer engagement: The moderating effect of AI preference. *International Journal of Hospitality Management*, *90*, 102629. DOI: 10.1016/j.ijhm.2020.102629

Pujianto, S. (2024). Effectiveness of employee training and development in increasing productivity and innovation in multinational companies. *Global International Journal of Innovative Research, 2*(9), 2245–2256. DOI: 10.59613/global.v2i9.325

Punia, A., Bhakuni, P., & Sharma, N. (2025). Development of AI- and NLP-driven chatbots and virtual assistants for mental health support. In Advances in psychology, mental health, and behavioral studies (pp. 229–262). DOI: 10.4018/979-8-3693-9341-3.ch010

Quach, S., Thaichon, P., Martin, K. D., Weaven, S., & Palmatier, R. W. (2022). Digital technologies: Tensions in privacy and data. *Journal of the Academy of Marketing Science, 50*(6), 1299–1323. DOI: 10.1007/s11747-022-00845-y PMID: 35281634

Rachh, R., Allagi, S., & Shravan, B. (2021). Machine learning algorithms for prediction of heart disease. In Demystifying big data, machine learning, and deep learning for healthcare analytics (pp. 247–275). DOI: 10.1016/B978-0-12-821633-0.00010-6

Raffik, R. (2023). Role of robotics and automation in food industries: An overview. In *2023 2nd International Conference on Advancements in Electrical, Electronics, Communication, Computing and Automation (ICAECA)* (pp. 1–5). IEEE. https://doi.org/DOI: 10.1109/ICAECA56562.2023.10199301

Rallis, I., Voulodimos, A., Bakalos, N., Protopapadakis, E., Doulamis, N., & Doulamis, A. (2020). Machine learning for intangible cultural heritage: a review of techniques on dance analysis. Visual Computing for Cultural Heritage, 103-119.

Ramadan, A., Dewedar, O., El-Shafie, A., & Abdelbaset, M. (2023). Improvement of irrigation water management using simulation models and artificial intelligence under dry environment conditions in Egypt: A review. *Egyptian Journal of Chemistry, 0*(0), 0. Advance online publication. DOI: 10.21608/ejchem.2023.151289.6552

Ramakrishnan, N. (2002). The traits of the personable (Version 1). arXiv. https://doi.org/DOI: 10.48550/arXiv.cs/0205022

Rana, J., Gaur, L., Singh, G., Awan, U., & Rasheed, M. I. (2022). Reinforcing customer journey through artificial intelligence: A review and research agenda. *International Journal of Emerging Markets, 17*(7), 1738–1758. DOI: 10.1108/IJOEM-08-2021-1214

Rane, J., Kaya, Ö., Mallick, S., & Rane, N. (2024). Smart farming using artificial intelligence, machine learning, deep learning, and ChatGPT: Applications, opportunities, challenges, and future directions. In AI and machine learning for smart farming (pp. 218–272). DOI: 10.70593/978-81-981271-7-4_6

Rashid, M., & Sharma, M. (2025). AI-assisted diagnosis and treatment planning. In AI in disease detection (pp. 313–336). DOI: 10.1002/9781394278695.ch14

Rini, A. S., Wandrial, S., Lutfi, L., Jaya, I., & Satrionugroho, B. (2024). Data-Driven Marketing: Harnessing Artificial Intelligence to Personalize Customer Experience and Enhance Engagement. *Join: Journal of Social Science, 1*(6), 282–295. DOI: 10.59613/akx6j040

Rivas, P., & Zhao, L. (2023). Marketing with ChatGPT: Navigating the ethical terrain of GPT-based chatbot technology. *AI, 4*(2), 375–384. DOI: 10.3390/ai4020019

Roberts, T. G. (2006). A philosophical examination of experiential learning theory for agricultural educators. *Journal of Agricultural Education, 47*(1), 17–29. DOI: 10.5032/jae.2006.01017

Rogers, E. M. (2016). *Diffusion of Innovations* (5th ed.). Publishing House of Electronics Industry.

Ross, M. (2025). Tips for using AI, from grad students and professors. Usable Knowledge. https://www.gse.harvard.edu/ideas/usable-knowledge/25/01/tips-for-using-ai-from-grad-students-and-professors

Russell, S., & Norvig, P. (1995). Artificial intelligence: A modern approach. *Choice (Chicago, Ill.), 33*(3), 33–1577. DOI: 10.5860/CHOICE.33-1577

Ryan, G. W., & Bernard, H. R. (2003). Techniques to identify themes. *Field Methods, 15*(1), 85–109. DOI: 10.1177/1525822X02239569

Ryoo, Y., Bakpayev, M., Jeon, Y. A., Kim, K., & Yoon, S. (2025). High hopes, hard falls: Consumer expectations and reactions to AI-human collaboration in advertising. *International Journal of Advertising*, •••, 1–33. DOI: 10.1080/02650487.2025.2458996

Sachdev, R. (2020). Towards security and privacy for edge AI in IoT/IoE based digital marketing environments. In *2020 Fifth International Conference on Fog and Mobile Edge Computing (FMEC)* (pp. 341–346). https://doi.org/DOI: 10.1109/FMEC49853.2020.9144755

Salehi, H., & Burgueño, R. (2018). Emerging artificial intelligence methods in structural engineering. *Engineering Structures, 171*, 170–189. DOI: 10.1016/j.engstruct.2018.05.084

Samala, N., Katkam, B. S., Bellamkonda, R. S., & Rodriguez, R. V. (2022). Impact of AI and robotics in the tourism sector: A critical insight. *Journal of Tourism Futures, 8*(1), 73–87. DOI: 10.1108/JTF-07-2019-0065

Samek, W., Montavon, G., Lapuschkin, S., Anders, C. J., & Muller, K.-R. (2021). Explaining deep neural networks and beyond: A review of methods and applications. *Proceedings of the IEEE, 109*(3), 247–278. DOI: 10.1109/JPROC.2021.3060483

Sandhiya, V., & Bhuvaneswari, M. (2024). Qualitative Research Analysis. In *In book* (pp. 289–310). Design and Validation of Research Tools and Methodologies., DOI: 10.4018/979-8-3693-1135-6.ch014

Sang, L. (2013). *The Principles of Feng Shui - Larry Sang - Google Books.* Books. Google.Com. https://books.google.com.vn/books?hl=vi&lr=&id=bCiLW7_DdI8C&oi=fnd&pg=PR12&dq=the+principles+of+feng+shui&ots=fUu-PgWIut&sig=c_ec_Jwt4rmLV3ynBFoZycrCwWQ&redir_esc=y#v=onepage&q=the%20principles%20of%20feng%20shui&f=false

Sankranthi, K. (2025). AI and Human Collaboration: Creating Personalized Shopping Experiences. *Journal of Computer Science and Technology Studies*, *7*(4), 156–164. DOI: 10.32996/jcsts.2025.7.4.18

Sarath Kumar Boddu, R., Santoki, A. A., Khurana, S., Vitthal Koli, P., Rai, R., & Agrawal, A. (2022). An analysis to understand the role of machine learning, robotics and artificial intelligence in digital marketing. *Materials Today: Proceedings*, *56*, 2288–2292. DOI: 10.1016/j.matpr.2021.11.637

Sashi, C. M. (2012). Customer engagement, buyer-seller relationships, and social media. *Management Decision*, *50*(2), 253–272. DOI: 10.1108/00251741211203551

Saunders, M. N. (2023). Research Methods for Business Students. *Pearson, 9*. https://www.researchgate.net/publication/367780349

Sawhney, M., Verona, G., & Prandelli, E. (2005). Collaborating to create: The Internet as a platform for customer engagement in product innovation. *Journal of Interactive Marketing*, *19*(4), 4–17. DOI: 10.1002/dir.20046

Schauer, A., & Schnurr, D. (2025). *Fostering trustworthy human-AI collaboration through explainable AI and uncertainty quantification.* AEA Randomized Controlled Trials., DOI: 10.1257/rct.15730-1.0

Secinaro, S., Calandra, D., Secinaro, A., Muthurangu, V., & Biancone, P. (2021). The role of artificial intelligence in healthcare: A structured literature review. *BMC Medical Informatics and Decision Making*, *21*(1), 125. Advance online publication. DOI: 10.1186/s12911-021-01488-9 PMID: 33836752

Selwyn, E. J., Tiwari, A., Pandurang, G. G., Das, S. S., & Vardhan, R. (2025). Healthsyncc: An AI-powered mobile application for health management. In Modern technologies in healthcare (pp. 241–266). DOI: 10.1201/9781003481959-11

Sharpe, S. L., & Elwood, S. A. (2024). E-learning design for older adults in the United States. *Social Sciences (Basel, Switzerland)*, *13*(10), 522. Advance online publication. DOI: 10.3390/socsci13100522

Singh, A. (2025). LLM-assisted drug discovery: Examining the potential of LLMs in identifying new therapeutic targets, predicting drug efficacy, and streamlining the drug development process. *SSRN*. DOI: 10.2139/ssrn.5213574

Singla, L., Nandrajog, A. B., Singh, N., Ahuja, K., & Mehta, S. (2024). AI and Consumer Behavior: Innovations in Marketing Strategy and Consumer Engagement. *2024 15th International Conference on Computing Communication and Networking Technologies (ICCCNT)*, 1–5. DOI: 10.1109/ICCCNT61001.2024.10725771

Sivasankari, M. (2025). Artificial Intelligence in Retail Marketing: Optimizing Product Recommendations and Customer Engagement. *Journal of Informatics Education and Research*, *5*(1). Advance online publication. DOI: 10.52783/jier.v5i1.2105

Slater, S. F., & Narver, J. C. (2000). Intelligence Generation and Superior Customer Value. *Journal of the Academy of Marketing Science*, *28*(1), 120–127. DOI: 10.1177/0092070300281011

Slijper, T., Urquhart, J., Poortvliet, P. M., Soriano, B., & Meuwissen, M. P. M. (2022). Exploring how social capital and learning are related to the resilience of Dutch arable farmers. *Agricultural Systems*, *198*, 103385. Advance online publication. DOI: 10.1016/j.agsy.2022.103385

Soleymani, A., Jahangir, H., & Nehdi, M. L. (2023). Damage detection and monitoring in heritage masonry structures: Systematic review. *Construction & Building Materials*, *397*, 132402. DOI: 10.1016/j.conbuildmat.2023.132402

Sparrow, R., Howard, M., & Degeling, C. (2021). Managing the risks of artificial intelligence in agriculture. *NJAS: Impact in Agricultural and Life Sciences*, *93*(1), 172–196. DOI: 10.1080/27685241.2021.2008777

Srivastava, G., & Bag, S. (2024). Modern-day marketing concepts based on face recognition and neuro-marketing: A review and future research directions. *Benchmarking*, *31*(2), 410–438. DOI: 10.1108/BIJ-09-2022-0588

Steils, N. (2021). Qualitative Experiments for Social Sciences. *New Trends in Qualitative Research*, *6*, 24–31. DOI: 10.36367/ntqr.6.2021.24-31

Su, H., Li, G., & Fischer, G. S. (2018). Sensors, actuators, and robots for MRI-guided surgery and interventions. In The encyclopedia of medical robotics (pp. 201–231). DOI: 10.1142/9789813232303_0008

Suchman, M. C. (1995). Managing legitimacy: Strategic and institutional approaches. *Academy of Management Review*, *20*(3), 571. DOI: 10.2307/258788

Sun, H., Zafar, M. Z., & Hasan, N. (2022). Employing natural language processing as artificial intelligence for analyzing consumer opinion toward advertisement. *Frontiers in Psychology*, *13*, 856663. Advance online publication. DOI: 10.3389/fpsyg.2022.856663 PMID: 35769737

Sun, X., & Song, Y. (2025). Unlocking the Synergy: Increasing productivity through Human-AI collaboration in the industry 5.0 Era. *Computers & Industrial Engineering*, *200*, 110657. DOI: 10.1016/j.cie.2024.110657

Syaharuddin, H., Hidayanti, N. F., Iswanto, D., & Ningsih, N. H. I. (2025). Artificial Intelligence in Personalized Marketing: Strategies for Enhancing Consumer Engagement. *Advance Sustainable Science Engineering and Technology*, *7*(2), 02502014. DOI: 10.26877/df5dwp69

T.R.I. (2024). *Hi u bi t phong thu - L i th l n c a môi gi i b t đ ng s n*. Vi n Nghiên C u B t Đ ng S n (Real Estate Research and Training Institute). https://tri.edu.vn/hieu-biet-phong-thuy/

Talib, M. A., Majzoub, S., Nasir, Q., & Jamal, D. (2021). A systematic literature review on hardware implementation of artificial intelligence algorithms. *The Journal of Supercomputing*, *77*(2), 1897–1938. DOI: 10.1007/s11227-020-03325-8

Talo, C. (2018). Community-based determinants of community engagement: A meta-analysis research. *Social Indicators Research*, *140*(2), 571–596. DOI: 10.1007/s11205-017-1778-y

Tambe, P., Cappelli, P., & Yakubovich, V. (2019). Artificial Intelligence in Human Resources Management: Challenges and a Path Forward. *California Management Review*, *61*(4), 15–42. DOI: 10.1177/0008125619867910

Taneja, A., Nair, G., Joshi, M., Sharma, S., Sharma, S., Jambrak, A. R., Roselló-Soto, E., Barba, F. J., Castagnini, J. M., Leksawasdi, N., & Phimolsiripol, Y. (2023). Artificial intelligence: Implications for the agri-food sector. *Agronomy (Basel)*, *13*(5), 1397. DOI: 10.3390/agronomy13051397

Taylor, D. C. M., & Hamdy, H. (2013). Adult learning theories: Implications for learning and teaching in medical education: AMEE Guide No. 83. *Medical Teacher*, *35*(11), e1561–e1572. DOI: 10.3109/0142159X.2013.828153 PMID: 24004029

Technology Review Insights, M. I. T. (2025). *Powering the food industry with AI*. MIT Technology Review. https://www.technologyreview.com/2025/03/19/1112920/powering-the-food-industry-with-ai/

Teece, D., Pisano, G., & Shuen, A. (1997). Dynamic capabilities and strategic management. *Strategic Management Journal, 18*(7), 509–533. https://www.jstor.org/stable/3088148. DOI: 10.1002/(SICI)1097-0266(199708)18:7<509::AID-SMJ882>3.0.CO;2-Z

Thapa, A., Nishad, S., Biswas, D., & Roy, S. (2023). A comprehensive review on artificial intelligence assisted technologies in food industry. *Food Bioscience, 56*, 103231. DOI: 10.1016/j.fbio.2023.103231

The White House. (2024, May 16). The economics of HBCUs. Council of Economic Advisers. https://www.whitehouse.gov/cea/written-materials/

Thevißen, P., Santiago Wolf, L., & Stommel, D. (2023). *AI or human? Applicants decisions in discrimination settings*. AEA Randomized Controlled Trials., DOI: 10.1257/rct.12308

Throsby, D. (2018). Art, politics, money: Revisiting Australia's cultural policy. Platform papers(55).

Tiago, M. T. P. M. B., & Veríssimo, J. M. C. (2014). Digital marketing and social media: Why bother? *Business Horizons*, *57*(6), 703–708. DOI: 10.1016/j.bushor.2014.07.002

Tilden, F. (1967). *Interpreting our Heritage. Chapel Hill.* North Carolina Press.

Tomaškinová, J., West, L., Chircop, T., Judge, J., Vries, S., Marinescu-Muster, R., Hiddink, F., Delicata, F., Pierucci, M., Montalto, M., Camilleri, S., Hermanova, S., Středová, H., Mrkvicova, E., & Stastna, M. (2024). Shaping the future: Navigating the AI era with an innovative training model. In EDULEARN24 proceedings (p. 6303). DOI: 10.21125/edulearn.2024.1486

Trist, E. L. (1981). *The evolution of socio-technical systems: A conceptual framework and an action research program*(Occasional Paper No. 2). Ontario Quality of Working Life Centre.

Trist, E. (1981). The evolution of sociotechnical systems: A conceptual framework and an action research program. In Van De Ven, A. H., & Joyce, W. F. (Eds.), *Perspectives on Organization Design and Behavior* (pp. 19–75).

Trivyzadakis, N. (2023). A tool for quantitative research on the social awareness for the digitization of cultural heritage: A pilot application. *Journal of Archaeological Science, Reports*, *51*, 104202. Advance online publication. DOI: 10.1016/j.jasrep.2023.104202

Truong, T. (2020). Phú M Gold City mang vư ng khí đón Phú quý | Báo Dân trí. *Dan Tri Newspaper.* https://dantri.com.vn/kinh-doanh/phu-my-gold-city-mang-vuong-khi-don-phu-quy-20201118084003622.htm

Ullal, M. S., Hawaldar, I. T., Soni, R., & Nadeem, M. (2021). The role of machine learning in digital marketing. *SAGE Open*, *11*(4), 21582440211050394. Advance online publication. DOI: 10.1177/21582440211050394

United Negro College Fund. (2024). The 2024 consumer economic impact report. https://uncf.org/hbcuimpactreport/2024

United Negro College Fund. (2024). The impact of HBCUs on diversity in STEM fields. https://uncf.org/the-latest/the-impact-of-hbcus-on-diversity

United Negro College Fund. (2024). The numbers don't lie: HBCUs are changing the college landscape. https://uncf.org/the-latest/the-numbers-dont-lie-hbcus-are-changing-the-college-landscape

Valente, T. W., Dyal, S. R., Chu, K. H., Wipfli, H., & Fujimoto, K. (2015). Diffusion of innovations theory applied to global tobacco control treaty ratification. *Social Science & Medicine*, *145*, 89–97. DOI: 10.1016/j.socscimed.2015.10.001 PMID: 26460508

Varghese, M., Raj, S., & Venkatesh, V. (2022). *Influence of AI in human lives* (Version 1). arXiv. DOI: 10.48550/ARXIV.2212.12305

Vladimirovna, G., & Chu, H. (2019). The gardens and architecture of the imperial burial complexes of Vietnam as a result of the influence of Chinese culture on the countries of Southeast Asia. *Amazonia Investiga*, *8*(21), 491–499.

Wæraas, A. (2022). Thematic analysis: Making values emerge from texts. In *Researching Values: Methodological Approaches for Understanding Values Work in Organisations and Leadership* (pp. 153–170). Springer International Publishing. DOI: 10.1007/978-3-030-90769-3_9

Wagner, C., & Majchrzak, A. (2006). Enabling Customer-Centricity Using Wikis and the Wiki Way. *Journal of Management Information Systems*, *23*(3), 17–43. DOI: 10.2753/MIS0742-1222230302

Wang, J., & Li, A. (2023). Revolutionizing healthcare with AI: Predicting future medication models and improving patient outcomes. In Machine learning & applications (pp. 25–35). DOI: 10.5121/csit.2023.131003

Wang, C. L. (Ed.). (2023). *The Palgrave handbook of interactive marketing*. Springer., DOI: 10.1007/978-3-031-14961-0

Wang, Y., Huang, X., & Ouyang, S. (2014). Automatic Fengshui evaluation method and evaluation system used for residential house type. In *Worldwide Patent Application #CN103679618A*. Current Assignee Hangzhou Group's Nuclear Information Technology Co Ltd.

Wang, Z., Xue, L., & Wei, L. (2024). *Overcoming medical overuse with AI assistance: An experimental investigation*. AEA Randomized Controlled Trials., DOI: 10.1257/rct.13376-1.0

WFP. (2025). *AI-powered food distribution models: Optimising last-mile delivery in crisis zones*. World Food Programme. https://www.wfp.org/publications/ai-last-mile-distribution

WFP. (2025). Does artificial intelligence hold key to ending hunger? https://www.wfp.org/stories/does-artificial-intelligence-hold-key-ending-hunger#:~:text=AI

Whig, P., Dutta, P. K., Yathiraju, N., Bhattacharya, P., & Batra, I. (2024). Enhancing healthcare operations: Harnessing generative AI for efficiency and innovation. In Studies in systems, decision and control (pp. 273–287). DOI: 10.1007/978-3-031-75771-6_18

Whig, P., Dutta, P. K., Yathiraju, N., Bhattacharya, P., & Sharma, S. (2024). Transforming healthcare operations with generative AI: Optimizing resource allocation, workflow management, and predictive analytics. In Studies in systems, decision and control (pp. 289–308). DOI: 10.1007/978-3-031-75771-6_19

Wirtz, J., Den Ambtman, A., Bloemer, J., Horváth, C., Ramaseshan, B., Van De Klundert, J., Gurhan Canli, Z., & Kandampully, J. (2013). Managing brands and customer engagement in online brand communities. *Journal of Service Management*, *24*(3), 223–244. DOI: 10.1108/09564231311326978

Wongkitrungrueng, A., & Assarut, N. (2020). The role of live streaming in building consumer trust and engagement with social commerce sellers. *Journal of Business Research*, *117*, 543–556. DOI: 10.1016/j.jbusres.2018.08.032

World Health Organization. (2025). *Ethics and governance of artificial intelligence for health: Guidance on large multi-modal models*. https://www.who.int/publications/i/item/9789240084759

World Health Organization. (2025). *Global standards for AI in food safety: Risk management, transparency, and accountability protocols*. World Health Organization. https://www.who.int/publications/ai-food-safety-standards

Wu, S.-J. (2019). *Feng Shui: A comparison of the original concept and its current Westernized version*. proquest.com

www.suzhou.gov.cn. (2025, January 7). The Immersive Digital Experience Exhibition of the Palace Museum Opens. https://www.suzhou.gov.cn/szsrmzf/szyw/202501/78b4e73690c241e6b1abd216905052e8.shtml

Xing, Z., Feng, Q. J., Chen, H. R., Dai, Q., Hu, H., Xu, H., Wu, Z. X., & Jiang, Y. G. (2025). A Survey on Video Diffusion Models. *ACM Computing Surveys*, *57*(2), 41. Advance online publication. DOI: 10.1145/3696415

Xu, Y., Liu, X., Cao, X., Huang, C., Liu, E., Qian, S., Liu, X., Wu, Y., Dong, F., Qiu, C.-W., Qiu, J., Hua, K., Su, W., Wu, J., Xu, H., Han, Y., Fu, C., Yin, Z., Liu, M., & Zhang, J. (2021). Artificial intelligence: A powerful paradigm for scientific research. *Innovation (Cambridge (Mass.))*, *2*(4), 100179. Advance online publication. DOI: 10.1016/j.xinn.2021.100179 PMID: 34877560

Yang, H., Jiao, W., Zouyi, L., Diao, H., & Xia, S. (2025). Artificial intelligence in the food industry: Innovations and applications. *Discover Artificial Intelligence*, *5*(1), 60. Advance online publication. DOI: 10.1007/s44163-025-00296-8

Yang, S., Hou, M., & Li, S. (2023). Three-dimensional point cloud semantic segmentation for cultural heritage: A comprehensive review. *Remote Sensing (Basel)*, *15*(3), 548. DOI: 10.3390/rs15030548

Yau, K.-L. A., Saad, N. M., & Chong, Y.-W. (2021). Artificial Intelligence Marketing (AIM) for Enhancing Customer Relationships. *Applied Sciences (Basel, Switzerland)*, *11*(18), 8562. DOI: 10.3390/app11188562

Yisu, S. (2024). Evolution iteration and development direction of audio visual communication in the Palace Museum in the "digital intelligence era". (05), 52-55. DOI: 10.13994/j.cnki.stj.2024.05.028

Yuan, X., Hu, J., & Zhang, Q. (2024). *A Comparative Analysis of Cultural Alignment in Large Language Models in Bilingual Contexts*. https://doi.org/https://doi.org/10.31219/osf.io/6hpcf DOI: 10.31219/osf.io/6hpcf

Yu, Y., & Wang, H. (2022). The Application of Artificial Intelligence in the Dissemination of Local Culture in Hainan Province: Taking the Dissemination of Bai Yuchan's Thoughts as an Example. *Advances in Multimedia*, *2022*, 1–10. Advance online publication. DOI: 10.1155/2022/7255853

Zahra Ouariach, F., & Ahmed, S. (2025). Ethical and legal dimensions of AI in healthcare. In Computational intelligence in healthcare law (pp. 86–98). DOI: 10.1201/9781003538462-6

Zartashea, F. (2023). Bias and fairness in AI-driven healthcare: Addressing disparities in machine learning models. *International Journal of Science and Research Archive*, *9*(1), 835–846. DOI: 10.30574/ijsra.2023.9.1.0359

Zatsu, V., Shine, A. E., Tharakan, J. M., Peter, D., Ranganathan, T. V., Alotaibi, S. S., Mugabi, R., Muhsinah, A. B., Waseem, M., & Nayik, G. A. (2024). Revolutionizing the food industry: The transformative power of artificial intelligence—A review. *Food Chemistry: X*, *24*, 101867. DOI: 10.1016/j.fochx.2024.101867 PMID: 39431210

Zawacki-Richter, O., Marín, V. I., Bond, M., & Gouverneur, F. (2019). Systematic review of research on artificial intelligence applications in higher education – where are the educators? *International Journal of Educational Technology in Higher Education*, *16*(1), 39. DOI: 10.1186/s41239-019-0171-0

Zhang Chaozhi, Y. J. (2022). Construction and negotiation: Rethinking the relationship between cultural heritage and Tourism. Tourism Tribune/Lvyou Xuekan, 37(11).

Zhang, J., Schomer, P. D., Marc, Zhang, Lei, Wu, D., & Boyle, J. (2014). A Metric Matrix Establishment for Cases Studies on the Effectiveness of the Key Environmental Protection Policies for Transportation Pollution Control. *Inter-Noise 2014*.

Zhang, C., & Lu, Y. (2021). Study on artificial intelligence: The state of the art and future prospects. *Journal of Industrial Information Integration*, *23*, 100224. Advance online publication. DOI: 10.1016/j.jii.2021.100224

Zhang, Q., & Xiong, Y. (2024). Harnessing AI potential in E-Commerce: Improving user engagement and sales through deep learning-based product recommendations. *Current Psychology*, *43*(38), 30379–30401. DOI: 10.1007/s12144-024-06649-3

Zhang, S. (2018). The heritage practices in a Chinese historic neighbourhood: The manifestation of traditional Feng Shui in Langzhong, China. *International Journal of Heritage Studies*, *24*(5), 531–546. DOI: 10.1080/13527258.2017.1390778

Zhao, L. T., Zhang, J. W., Jing, H. R., Wu, J. P., & Huang, Y. J. (2023). A Blockchain-Based cryptographic interaction method of digital museum collections. *Journal of Cultural Heritage*, *59*, 69–82. DOI: 10.1016/j.culher.2022.11.001

Zhao, M., Wu, X., Liao, H.-T., & Liu, Y. (2020). Exploring research fronts and topics of Big Data and Artificial Intelligence application for cultural heritage and museum research. *IOP Conference Series. Materials Science and Engineering*, *806*(1), 012036. DOI: 10.1088/1757-899X/806/1/012036

Zhou, Y. Z., Feng, T., Shuai, S. H., Li, X. D., Sun, L. Y., & Duh, H. B. L. (2022). EDVAM: A 3D eye-tracking dataset for visual attention modeling in a virtual museum. *Frontiers of Information Technology & Electronic Engineering*, *23*(1), 101–112. DOI: 10.1631/FITEE.2000318

Zhou, Z., Chen, X., Li, E., Zeng, L., Luo, K., & Zhang, J. (2019). Edge intelligence: Paving the last mile of artificial intelligence with edge computing. *Proceedings of the IEEE*, *107*(8), 1738–1762. DOI: 10.1109/JPROC.2019.2918951

Ziakis, C., & Vlachopoulou, M. (2023). Artificial intelligence in digital marketing: Insights from a comprehensive review. *Information (Basel)*, *14*(12), 664. Advance online publication. DOI: 10.3390/info14120664

Zwitter, A. (2014). Big data ethics. *Big Data & Society*, *1*(2), 2053951714559253. Advance online publication. DOI: 10.1177/2053951714559253

About the Contributors

Poshan (Sam) Yu is a Chinese-Australian professional with an astonishingly diverse background and an extensive collection of remarkable accomplishments. In the academic sphere, he currently teaches at the University of Victoria Program at Soochow University. Previously, he held the position of an external professor of financial technology and finance at SKEMA Business School. He also serves as a visiting researcher at the Australian Studies Centre of Shanghai University, a guest professor at emlyon business school China, and an adjunct faculty member at George Washington University's master (finance) program at Renmin University of China. As a visiting professor at Krirk University in Thailand and an EU-Asia research coordinator at the European Business Institute in Luxembourg, he brings a truly global perspective to his teaching and research activities. He is a member of the IP & Innovation Researchers of Asia (IPIRA) Network and the Global South Dialogue on Economic Crime (GSDEC), further enriching his academic engagement. His prior role as a casual academic at the University of South Australia Business School has significantly enhanced and diversified his academic portfolio. From September 2020 to February 2021, he held the position of visiting scholar at The Chinese University of Hong Kong on two occasions. His scholarly pursuits are extensive and multifaceted. As an editor for the European Academic Journal (Special Issue) - European Business Institute in Luxembourg, an associate editor for Humanities & Social Sciences Communications (an SSCI-indexed journal), and an associate editor for the International Journal of Interdisciplinary Research in Social Science, he plays a crucial role in the dissemination of knowledge. He is an ad hoc reviewer for a number of prestigious journals and academic presses, including Applied Economics, Applied Artificial Intelligence, Cogent Economics & Finance, Humanities and Social Sciences Communications, Journal of Chinese Economic and Business Studies, Journal of Digital Learning and Education, The Singapore Economic Review, ILMA Journal of Social Sciences & Economics, Journal of Environmental Issues and Climate Change, Advances in Global Economic and Business Journal, The

Interdisciplinary Social Sciences Journal Collection, Common Ground Research Networks, Imprint - Xi'an Jiaotong-Liverpool University, Innovative Teaching and Learning, IBT Journal of Business Studies, Journal of Infrastructure Policy and Development, and Innovative Informatics and Artificial Intelligence Research. Moreover, he is an international editorial advisory board member for several edited books such as Advancing STEM Education and Innovation in a Time of Distance Learning (Scopus-indexed), Multidisciplinary Perspectives Towards Building a Digitally Competent Society (Scopus-indexed), Handbook of Research on SDGs for Economic Development, Social Development, and Environmental Protection, and Implications for Entrepreneurship and Enterprise Development in the Blue Economy. He has released more than 12 journal articles, among which 6 are listed as ABDC: B/SSCI. He has also contributed over 70 peer-reviewed book chapters imprinted by renowned publishers like IGI Global, Routledge, and Springer. He has edited 22 books, with 14 English-edited books indexed by Scopus and 6 Chinese-edited books. Notably, the book about Innovation & Exploration: The Development of FinTech in Jiangsu China has received the 14th Outstanding Achievement Award of Philosophy and Social Sciences in Suzhou and the 3rd prize of the 2018 Jiangsu Provincial Education Science Research Achievement Award by Jiangsu Provincial Department of Education. His receipt of four Xi'an Jiaotong-Liverpool University (XJTLU) National University Teaching Innovation Awards from 2016 to 2020 and being recognized as a top contender in the same national competition in 2016 led to his appointment as the Institute of Leadership and Education Advanced Development (ILEAD) Research Fellow by XJTLU. In 2021, he received The Outstanding Contribution Award for Industrial Research Development in China, CIE100 issued by Wisdom China Magazine: Beijing. Beyond academia, he is actively involved in various industries. He is an entrepreneur and venture capitalist, collaborating with alumni and senior faculty members from Cambridge University. Additionally, he is a guest financial news commentator on over 30 media platforms in China. In the business world, he holds significant positions. He is a regional partner in China for FasterCapital based in Dubai, United Arab Emirates. He also serves as vice president at Tinbowah Investment in Beijing and as an honorary vice president of the Belt and Road Blockchain Association. Furthermore, he mentors businesses for AIC RAISE in Coimbatore, India, the Griffith Industry Mentoring Program in Australia, and University College London Innovation & Enterprise. In 2023, he was invited to act as a venture capital panel judge and mentor at 'Y&Archer A Stream 2023' conference in Jeju, South Korea. His research interests are broad and encompass financial technology, regulatory technology, blockchain, web 3.0 in financial services, generative AI, public-private partnerships, mergers and acquisitions, private equity, venture capital, start - ups, family offices, IoTs, supply chain finance, intellectual property, art finance, wine, back-door listing, intangible cultural heritage, and Chinese economy & policy. His extensive experience in international fintech, edtech, startup, social media, and cleantech companies across India, the UK, Australia, Hong Kong, and China is truly remarkable. His three-year tenure in significant leadership positions within the Chinese-Australian media marketing and public relations sector, along with his role as an OzAsia Festival Ambassador in Australia, has exerted a profound influence on his current modus operandi as an academic and business consultant. Additionally, it has considerably broadened his perspective within the international multicultural arts community and enlarged his business network. This multifaceted experience has furnished him with a more comprehensive and inclusive outlook on academia, industry, and research. He is fully dedicated to making substantial contributions in these areas and is keen on furthering his development in the future. Incidentally, Sam has an inclination towards wine, arts, and reading.

James Mulli is the Founder, Program Director and Academic Dean of the European Business Institute, Luxembourg. He lectures in Applied Blockchain Technologies and Artificial Intelligence integration, post-graduate Corporate Finance & Business Statistics, and Managerial Accounting. He has held lectureships at in Luxembourg, China, the USA, Kenya and Germany. Dr. Mulli is a former director of two successful New York based multinational conglomerates and has valuable experience in the field as an NASD Series 7 compliance officer for the broker dealer Heyka Capital Management. He worked as a Portfolio Manager for the Soros Fund Management's Quantum Industrial Partners and holds a Doctorate and Masters degree from Saint Johns University, New York, and an MBA from New Hampshire Plymouth State University with specialization in Investment and Finance. Dr. Mulli is an international educator, and educational technology consultant with a keen interest in looking over the horizon at how technology is shaping the world. His search for trends in emerging technologies allows him to be an educator who applies skills to help students thrive. He is the founder of Technology Enabled Learning - Africa, and European Business Institute, Luxembourg, both non-for-profit educational providers whose initiative is to support existing Institutions in developing e-learning . His belief is that education is the silver bullet for development and through existing expertise and support infrastructure, EBI can provide the needed “jump-start” to institutions in Europe, Africa, Asia and beyond. Dr. Mulli is a Rockefeller Foundation Grant Recipient, and a Prince Fellow Teaching Excellence recipient. He is a textbook author and has published numerous newspaper and magazine articles.

Zain Ali Shah Syed is currently affiliated with Soochow University.

***

Laila Umme holds a PhD in Computer Science and Information Technology from NED University of Engineering and Technology and a MS/BS in Computer Engineering from Sir Syed University of Engineering & Technology. Her PhD is in pure Software Engineering, focusing on "open source software". Her research interests include open-source software, IoT, computer programming, software engineering, Artificial Intelligence, and Machine Learning. Currently She is working as an Associate Professor in Computer Science Department at IoBM. She has 20 years of teaching experience and has published over 20 publications in international and national journals and conferences.

Daniel Opotamutale Ashipala (MN Sc) is a senior lecturer and Head of Department for the School of nursing and Public Health (SoNPH), General Nursing Science department in the Faculty of Health Sciences and Veterinary Medicine (FHSVM) at the University of Namibia, Namibia. His research interests include Primary health Care, Health care services and Nursing education, and Mental health. He has published a quite number of peer reviewed articles on the delivery of teaching and learning in higher education studies. He is a doctoral candidate at Stellenbosch University in Cape Town, South Africa. His doctoral work focuses on developing a framework to facilitate the implementation of a task shifting approach for nurses in the Namibia PHC services.

Diletta Cerasuolo holds a BSc in Business Management and an MSc in International Business Management from Royal Holloway, University of London. Her academic and professional experience spans business research, editorial design, visual communication, and luxury brand management. She has contributed to interdisciplinary research in higher education and curriculum development through work with the Quality Assurance Agency and Royal Holloway, co-authoring research on disciplinary affiliation identities in shared modules and presenting findings at academic conferences. She has also undertaken independent interdisciplinary research, with publications exploring consumer behaviour through socio-economic and psychoanalytical perspectives. Alongside her research activities, she works in editorial design and multimedia content production for scientific publications. Her research interests include artificial intelligence, consumer behaviour, branding, business strategy, emerging technologies, and contemporary visual culture.

Cassandra Sligh Conway is a Full Professor at a Historically Black College and University (HBCU) with extensive experience in higher education leadership, program administration, and community engagement. Throughout her career, she has served in a variety of roles, including Chair of the Department of Human Services, Program Coordinator, Director of Field Placement, and Grant Consultant.

Dr. Sligh Conway's research and professional interests focus on providing services to individuals with disabilities, rehabilitation counseling, mentoring students and faculty, disaster preparedness and prevention, and strengthening community resources and engagement. She is committed to advancing educational opportunities, promoting inclusive practices, and fostering partnerships that enhance the well-being of communites with unique needs. An accomplished grant writer and consultant, Dr. Sligh Conway has successfully secured more than $1.5 million in grant funding to support educational initiatives, community programs, research, and workforce development efforts. Her work reflects a dedication to improving outcomes for underserved communities while advancing scholarship, leadership, and service in higher education.

Nihar Suresh Dahake is working as an Assistant Professor at G H Raisoni College of Engineering, Nagpur in the Department of MBA. He received his B.Com, M.Com, MBA, from the Rashtrasant Tukdoji Maharaj Nagpur University and and Ph.D. from Sant gadge baba Amravati University, Amravati. Dr.Nihar Dahake has Qualified SET(Commerce) and SET (Management). He has 14 Years of Teaching Experience and 8 years of Research Experience. His areas of expertise are Financial Accounting, Consumer behavior, Marketing Management and Financial Management.

Parihar Suresh Dahake is working as an Assistant Professor at Ramdeobaba University in the School of Management. He received his BTech, MBA, and Ph.D. from the Rashtrasant Tukdoji Maharaj Nagpur University and PGDIBO, M.Com from Indira Gandhi National Open University (IGNOU). Dr.Parihar dahake has Qualified SET(Commerce) and SET (Management). He has 15 Years of Teaching Experience and 14 years of Research Experience. His areas of expertise are Consumer behavior, Organizational Behaviour, Marketing management, and IMC.

Dang Thi Thanh Tam, Founder and CEO - Rubie Marble Joint Stock Company (Prop-Tech), Ho Chi Minh City, Vietnam. Experienced business leader with 20+ year background in retail marketing and project management across global brands like Intel, Microsoft and Indochina Land. Currently pursuing a Doctorate in Business Administration with a focus on Business and Data Analytics.

Guan Jiawen a is an independent researcher specializing in financial management, Chinese business, and economic analysis.

Rahul Vijay Mohare is currently working as Assistant Professor at Ramdeobaba University in the School of Management, Maharashtra, India, He received his B.Sc, and Ph.D.(Computer Science),PhD(Management) from the Rashtrasant

Tukdoji Maharaj Nagpur. His areas of expertise are Business Analytics,Marketing Analytics,Data Mining & Power BI &SQL. He has 20 years of teaching and research experience.

Bitange Ndemo is a professor of Entrepreneurship with great interest in the role of Emerging technologies especially data and AI, in economic development. He is currently serving as the Ambassador of Kenya to the Kingdom of Belgium and the European Union.

Fransisco Chindongo Ntjamba holds a Bachelor of Nursing Science (Clinical) (Honours) from the University of Namibia, School of Nursing and Public Health, General Nursing Department in the Faculty of Health Sciences and Veterinary Medicine (FHSVM). He is a NIMART-trained & a Provider Initiated Testing and Counseling (PITC) nurse, specializing in sexual reproductive health and focusing on HIV prevention, treatment initiation, and ongoing care. In addition to his qualifications, Fransisco Ntjamba is an academic researcher who has published research studies in nursing education, mental health, and interdisciplinary innovations, including Artificial Intelligence on human healthcare. His career is defined by a commitment to advancing healthcare through both direct patient care and research, with key interests in Primary Health Care, healthcare services, nursing education, and mental health. He is dedicated to improving healthcare through evidence-based practices and community outreach initiatives, particularly in HIV prevention and treatment.

Michelle L Priester is a Full Professor at South Carolina State University in the Rehabilitation Counseling program. Rehabilitation Counseling is a graduate program designed to prepare master level students to counsel, work with individuals with various disabilities, and place them in successful employment. Michelle holds a graduate certificate in Administration and Supervision. She also has a Bachelor of Arts degree in Psychology and a Master of Science degree in Rehabilitation Administration and Supervision. In 2004, she completed her studies and graduated with a RhD in Rehabilitation Education from Southern Illinois University at Carbondale with a focus in rehabilitation counseling and research. Her research interests include: pedagogy; organizational culture; mental health and addictions; and women's issues. She has presented at several national conferences and published articles in these areas of interest. Dr. Maultsby is also an experienced grant writer and has secured funding in the amount of $3.5 million dollars.

Tammara Thomas received her doctorate in Rehabilitation and Counseling Education at the University of Iowa. She also received her Master of Science degree in Rehabilitation and Administration Services from Southern Illinois University at

Carbondale, Illinois. Her areas of interest include rehabilitation counseling, clinical mental health counseling, addictions and co-occurring disorders, gender-specific treatment, program development, program evaluation and management, ethics, training, and consultation. She is currently an Associate Professor and has educated both graduate and undergraduate students in the areas of rehabilitation counseling, clinical mental health counseling, and addiction studies.

Yupian Zhang is an independent researcher specializing in the Chinese economy, FinTech, and AI.

Index

A

B

C

P

S

T

www.ingramcontent.com/pod-product-compliance
Lightning Source LLC
LaVergne TN
LVHW062321300926
849161LV00007B/1527